D0085568

Fieldbus and Networking in Process Automation

Fieldbus and Networking in Process Automation

Sunit Kumar Sen

CRC Press is an imprint of the
Taylor & Francis Group, an **informa** business

CRC Press
Taylor & Francis Group
6000 Broken Sound Parkway NW, Suite 300
Boca Raton, FL 33487-2742

Printed on acid-free paper
Version Date: 20140404

International Standard Book Number-13: 978-1-4665-8676-5 (Hardback)

Visit the Taylor & Francis Web site at
http://www.taylorandfrancis.com

and the CRC Press Web site at
http://www.crcpress.com

Contents

Preface

Fieldbus, particularly the wireless fieldbus, offers a multitude of benefits in the field of process control and automation. Wireless fieldbus is fast emerging and is trying to carve out a niche among the different fieldbus offerings in the market. Fieldbus replaces the point-to-point technology with digital communication networking, offers increased data availability, and is easily configurable and interoperable. It is a modest attempt on the part of the author to discuss the different fieldbuses in the market, their utilities along with their shortcomings, the fieldbus configurations, the installation techniques, the safety aspects in hostile environmental conditions, and other relevant issues pertaining to fieldbuses.

Fieldbus and Networking in Process Automation provides a clear, concise, and comprehensive coverage of fieldbuses as used in the process control and automation industries. Fieldbus and networking is an emerging area and is increasingly being applied in process industries. It will be very helpful for engineering students in the area of instrumentation, process, electrical, electronics, and computer science disciplines, and will give them adequate exposure about the different fieldbus technologies in use today.

The book starts with an introduction about data communication followed by networking, network models, and networks as applied in process automation. The three most used fieldbuses, viz., HART, Foundation Fieldbus, and PROFIBUS, followed by several others are then discussed in detail. Intrinsic safety in fieldbuses is a major area of concern and is discussed comprehensively. Chapter 17, "Wiring, Installation, and Commissioning," gives an overview of cabling, surge protection systems, device connection techniques, and different fieldbus components and configurations. Chapter 18, "Wireless Communication," discusses different wireless standards, their coexistence issues, and wireless sensor networks.

WHART and ISA 100.11a—the latest offerings in the wireless arena for networking in process automation and control—are discussed in a threadbare manner in the last two chapters. Wireless fieldbuses are yet to establish themselves firmly as far as their industrial applications are concerned. Despite their minor shortcomings and drawbacks, the two standards offer reliable and secure wireless communication in the field of industrial automation for noncritical monitoring and control applications. A comparison between these two emergent standards has been made so that readers will become conversant with them about their application potentials in a given industrial environment.

Author

Sunit Kumar Sen is a professor of instrumentation engineering in the Department of Applied Physics, University of Calcutta. He graduated from St. Xavier's College, Kolkata, in 1972 with honors in physics and secured first class. Subsequently, he did his BTech and MTech degrees from the University of Calcutta in 1975 and 1977, respectively. He obtained his PhD (Tech) degree from the same university in 1993.

In 1978, he joined Bokaro Steel Plant (under SAIL) and served for more than five years as assistant manager, instrumentation (operation). In 1984, he joined the Department of Applied Physics as a lecturer. He teaches digital electronics, microprocessors, digital communication, industrial instrumentation, electrical networks, fieldbus, etc. He has around 34 research papers in national and international journals. He has published two books: *Understanding 8085/8086 Microprocessors and Peripheral ICs through Questions and Answers (2006, 1st edition; 2010, 2nd edition),* and *Measurement Techniques in Industrial Instrumentation (2012),* both published by New Age International (P) Limited, New Delhi. He is a life member of IETE, India, and a member of IEEE.

His research interests include new designs for PRBS generators, new designs and development of various types of ADCs such as sigma delta ADCs, pipeline ADCs with improved comparator error correction, designs of novel cyclic architectures in pipeline ADCs, etc.

He was head of the Department of Applied Physics and also USIC, University of Calcutta from 2008 through 2010.

1 Data Communication

1.1 INTRODUCTION

Data communication refers to the transfer of information from one place to another. The term "data" means an information that is digital in nature, i.e., binary ones and zeroes. Analog data includes TVs, radios, and telephone systems. In the vast majority of cases, communication is digital in nature, although at the source and at the final user point it is analog in nature.

A communication system includes a transmitter, a receiver, and a link connecting these two. The link can be copper wire, optical fiber, or microwave. The link is parallel in nature for short distances, while it is serial for long distances. Data communications involving digital data are mostly serial in nature. A simplified data communication model is shown in Figure 1.1. The input information x is the source of data, which may be a computer (data) or a telephone (analog). This gives out a digital data stream $c(t)$ and it is fed to a transmitter block that acts as a modem whose output $m(t)$ is as shown in Figure 1.1. The modem output passes through the transmission medium. $n(t)$ is the signal received by the receiver modem from which digital data $d(t)$ is extracted, and this is then stored at the final destination point. The output information, y, should closely match the input information, x, for a good transmission system.

1.2 COMPARISON BETWEEN DIGITAL AND ANALOG COMMUNICATION

Analog signals are more susceptible to undesired amplitude, frequency, and phase variations. These quantities are of no concern in digital communication where a received signal is analyzed to determine whether a signal is above or below a threshold level. In the presence of noise, analog signals are more susceptible to corruption, particularly if the magnitude of the signal is low. For such cases, if the impact of noise is not lessened, there are chances that the original signal cannot be retrieved, while a degraded digital pulse can be restored to its original form by employing repeaters along the passage to its final destination. Digital pulses can be stored in memory for later use, which is not at all possible in analog communication. The transmission rate of digital pulses can very easily be varied

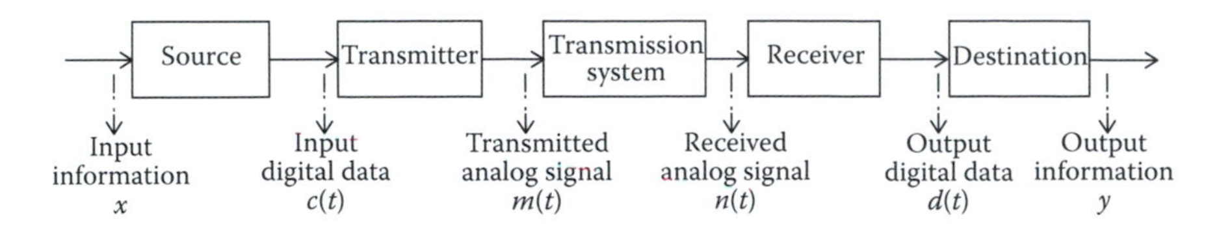

FIGURE 1.1 Simplified data communication model. (From W. Stallings. *Data & Computer Communications*, 6th Edition. Pearson Inc., New Delhi, India, p. 8, 2000.)

to suite different interface conditions. Again, digital pulses can easily be multiplexed and processed. A digital transmission scheme can very easily be evaluated for its performance in terms of bit error rate (BER), which is not so with its analog counterpart. The robustness of digital communication can considerably be enhanced by error detection and correction techniques, data security, data retransmission, and flow control in case there is a possibility of receiver memory being overwhelmed.

On the flip side, a digital system needs precise timing synchronization between transmitter and receiver clocks. An analog signal is at first to be digitized and the corresponding bit pattern is sent. This would entail more bandwidth than sending the analog signal. Third, since an analog signal is to be digitized before transmission and to be converted back to analog before its eventual use at the receiver, error would creep into at either end.

1.3 DATA COMMUNICATION

There are various issues involved in a data communication system that must be addressed for proper data transfer to take place between a source and destination. The following communication tasks are of importance for a multisource multidestination communication system: signal generation, synchronization, utilization of transmission facility, error detection and correction, flow control, message formatting, addressing, routing, data recovery, data security, and network management.

Signal generation refers to the signal source that must have adequate intensity and proper form such that it can propagate through the transmission medium and is understood by the receiver. A synchronized data transmission scheme is undertaken when considerable amount of data is transmitted so that each separate data packet is received correctly. When a number of devices share the same transmission medium, a multiplexing scheme is used to allocate the total transmission capacity among the devices. A data transmission scheme should include error detection and correction capabilities so that the receiver can detect and correct the actual data even in hostile transmission facilities. In flow control, the receiver lets know the transmitter when to slow down data transmission or else to totally stop transmission. Data formatting refers to particular form of the data to be transmitted. In a multisource, multidestination data transmission scheme, addressing of source and destination stations is a must, which must be included in the data frame. A data frame may reach the destination via different paths. Routing of data along a specific path is undertaken for proper data delivery to the particular destination. Recovery of data refers to some unforeseen situations in which a data transmission is interrupted. The scheme involves resuming the activity at the point at which the interruption took place, or else initiating data transmission afresh. Data security involves encryption of data at the transmitting end side such that no third party can have access to the data being sent. At the receiver, the received data is decrypted to get back the original data. The overall communication facility is managed by network management, which takes care of the present status of the facility, reports about failures, configures the system, etc.

1.3.1 Main Characteristics

The efficacy of a data transmission system is a function of four characteristics: data delivery, timeliness, accuracy, and jitter. Data delivery involves delivering data at the correct destination only. Timeliness is a quality of a transmission system that ensures timely delivery of data. In case of audio and video transmission, timeliness is the essence of transmission. Third, data must be received accurately and correctly at the receiver; otherwise, the received data would become useless. If in a system, data packets arrive at the destination with different delay times, then the system suffers from jitter. Real-time applications, such as teleconferencing, require an upper bound on jitter. The larger the delay variations allowed in a system, the larger the delay in real-time data delivery, resulting in greater size of delay buffers at the receivers.

1.4 DATA TYPES

Data can have various forms, such as numbers, text, audio, video, or images.

A number is represented by its equivalent binary. Some important number systems are as follows: decimal or base 10, binary or base 2, hexadecimal or base 16, and IP addresses or base 256.

Text is represented by different bit patterns, with each bit pattern (also known as set or code) representing a particular text symbol. The current coding system uses 32 bits to represent a bit pattern, known as Unicode. The first 127 characters of the Unicode is used by the ASCII (American Standard Code for Information Interchange) code.

Audio and video refer to recording and broadcasting of sound and picture, respectively. While audio is always continuous in nature, video can either be continuous or discrete. A TV camera beams a continuous picture; however, the concept of motion is derived by superposing discrete images.

Images are composed of pixels or picture elements. Each pixel is represented by a small dot. An image can be divided into small or huge number of pixels. The greater the number of pixels, the better is the clarity of the image. Now, each pixel is represented by a bit pattern. The size and the value of the bit pattern is a function of image quality. A colored image requires a higher value of the bit pattern compared with a black-and-white image. A colored image may be represented by RGB—a combination of primary colors: red, green, and blue. Another way of representation is by YCM—yellow, cyan, and magenta.

1.5 DATA TRANSFER CHARACTERISTICS

Digital data is converted into digital signal by means of a process called *line coding*. Digital data or data element is the smallest entity that represents a piece of information. It is represented by bit. Data rate is the number of data elements that can be sent in a second. The unit is bits per second or bps. It could be kbps (kilo bits per second) or Mbps (mega bits per second).

When a bit is converted into Manchester-coded form, then in each bit period, there is a transition halfway through the bit period of the coded data. That is, the first 50% of the bit period is 0 and the rest of the period is 1 or vice versa—depending on whether the bit is 0 or 1, respectively. The bit is called the data element, while the smallest element of the coded data is called the signal element. The number of signal elements that can be sent in a second is called *signal rate* or *pulse rate* or *modulation rate* or *baud rate*. Thus, signal rate refers to the speed of signal element after encoding and modulation on data is performed.

It should be clearly understood that in communication, it is the data that need to be sent but it is the signal that actually travels through the link. Increasing the speed of data transmission means more data rate—this increases the throughput. If signal transmission rate can be decreased, it would result in lesser bandwidth requirement of the channel.

Communication speed on the transmission link is limited by bandwidth of the link. It refers to the maximum rate at which the signal changes can be handled before attenuation degrades the quality of the signal at the receiving end so much so that it cannot be retrieved faithfully. The difference between the maximum and minimum frequencies contained in a signal is its bandwidth. The *absolute* bandwidth is the difference between the maximum and minimum frequency contained in the spectrum of the signal, while the *effective* bandwidth refers to the difference between the maximum and minimum frequency in the spectrum in which most spectral energy is contained. Bandwidth can either be expressed in Hz or bps.

The bandwidth in Hz of a channel is the range of frequencies it can pass without degradation of the signal, while bandwidth in bps is the number of bits per second that a channel can pass through it.

Latency or time delay is another very important characteristic in message transmission via a transmission link. Latency refers to the time taken by an entire message to reach the destination via various links. There are several components in latency: propagation time, transmission time, queuing time, and processing time. Thus, latency is the sum of all these delays taken together. The less the delay in transmission for a message, the better the system is. Propagation time is the time needed by a single bit to reach the destination from the source. It is measured by dividing the distance by the speed of the medium through which propagation is taking place. Transmission time is the time between the first bit leaving the sender and the last bit of the message to reach the destination. It is defined as the message size divided by the bandwidth. Queuing time is a variable one and depends on the load or traffic through which the data/message has to pass. This is akin to a car taking more time covering a distance during day time when traffic is heavy, but takes much less time during morning when traffic is expectedly less. Message from the source to destination has to pass through different nodes. The nodes themselves are to cater to traffic from other sources that pass through them. Thus, depending on the en route traffic, there is always a variable delay for the message to reach the destination.

1.6 DATA FLOW METHODS

Data flow between two communicating devices is done by using simplex (also called *receive only*, *one way only*, or *transmit only*), half duplex (also called *two-way alternate* or *either way*), full duplex (also called *two-way simultaneous*, *duplex*, or *both ways*), and full/full duplex modes. These schemes are shown in Figure 1.2.

In the simplex mode, communication is always unidirectional—from the transmitter to the receiver. Examples are radio broadcasting and computer keywords. The entire channel capacity is used for transmission purpose.

In half-duplex mode, two-way data transmission is possible—but not at the same time. Like in simplex mode, here also the entire channel capacity is utilized for data transmission since transmission is always in a single direction at any given instant of time. An example is a walkie-talkie used in traffic control.

In full duplex mode, transmission in both directions takes place between any two stations at the same time. In this mode, transmission of data can take place in either of the two ways: the link may contain two physically separate lines—one for sending and the other for receiving; or else the capacity of the channel is shared between the two signals traveling in either direction.

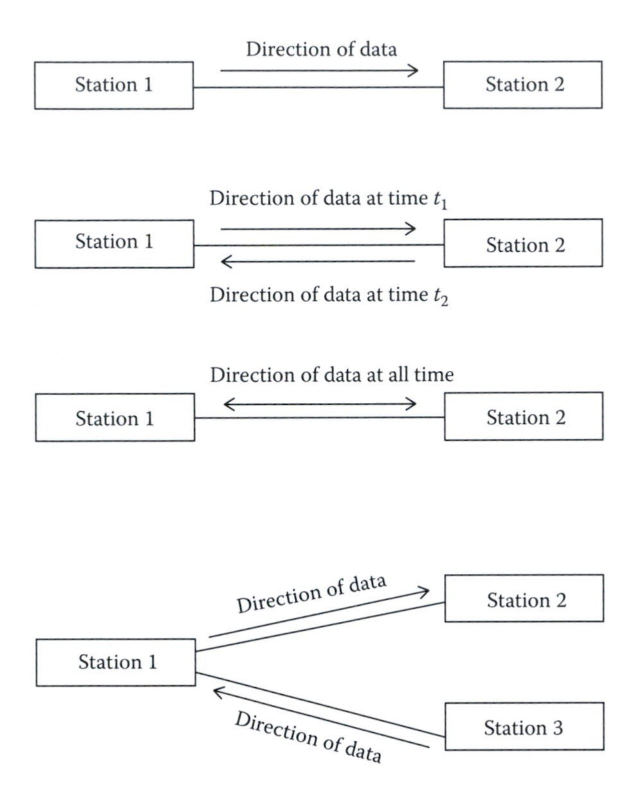

FIGURE 1.2 Different transmission modes. (From B. A. Forouzan. *Data Communications and Networking*, 4th Edition, Special Indian Edition. Tata McGraw Hill Companies Inc., New Delhi, India, p. 6, 2006.)

In full/full duplex mode, transmission in both directions is possible at the same time—like full duplex mode, but this is not limited to the same two stations. In this mode, a station is transmitting to a second station and receiving from a third station.

1.7 TRANSMISSION MODES

Data from the transmitter to the receiver can be executed in different ways. It can be parallel or serial. The serial data transmission scheme is again subdivided into asynchronous, synchronous, and isochronous. The different modes of transmission are shown in Figure 1.3.

1.7.1 Parallel

Whereas a message consisting of n bits requires n clocks for sending the same to the receiver, it requires a single clock for parallel transmission. Thus, in essence, parallel transmission is much faster than serial

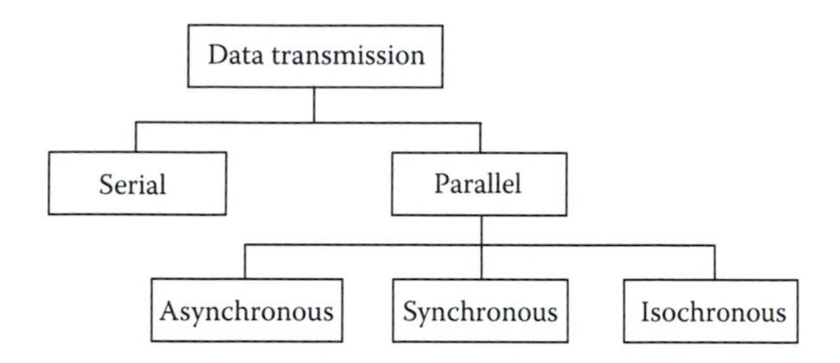

FIGURE 1.3 Different data transfer schemes. (From B. A. Forouzan. *Data Communications and Networking*, 4th Edition, Special Indian Edition. Tata McGraw Hill Companies Inc., New Delhi, India, p. 131, 2006.)

transmission; however, the cost of laying *n* wires would entail extra cost. Thus, parallel transmission is undertaken only if the distance between the transmitter and receiver is not considerable.

1.7.2 Serial

Serial transmission involves transmission of bits of a message in serial manner—one after the other, i.e., only one communication channel is required. When the distance between the transmitter and the receiver is considerable, it makes sense to transmit information serially because in such cases parallel transmission would entail considerable cost, and laying of wires may sometimes be difficult. Serial transmission, as already mentioned, can be of three types—the kind of serial transmission depends on the specific requirements.

1.7.3 Asynchronous

Although termed asynchronous communication, it is basically a synchronous type of communication where synchronization is maintained for each character—each character consisting of 5–8 data bits. The receiver resynchronizes at the beginning of each character frame. Synchronization basically means agreeing or coinciding exactly in time scale. Asynchronous communication is called *start–stop* type transmission because framing of each character is between a *start* and *stop* bit(s). A character starts with a start bit (which is always a logic 0), followed by the bits of the data (with LSB being sent first), the parity bit, and lastly one, one and a half, or two stop bits. The stop bit(s) is/are always 1 or high state. Synchronization is achieved at the receiver on receiving the high to low transition of the start pulse. The stop bit provides a minimum guard band or buffer period between two characters. An asynchronous frame format is shown in Figure 1.4.

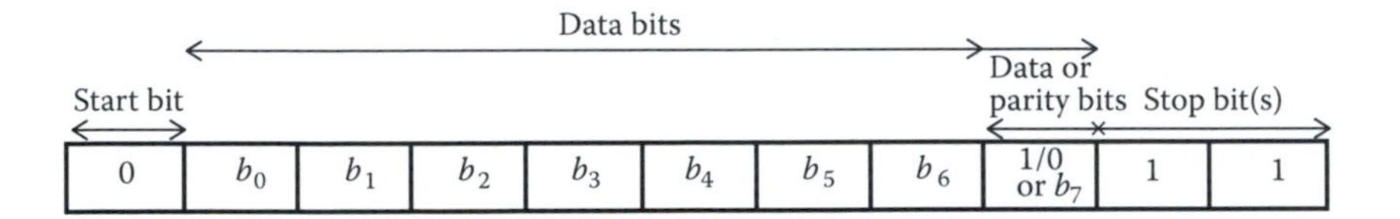

FIGURE 1.4 Asynchronous data frame format. (From W. Tomasi. *Advanced Electronic Communications Systems*, 6th Edition. Pearson Inc., New Delhi, India, p. 165, 2004.)

An idle line (i.e., no transmission) is always a stream of 1's. When a character is being sent down the line, it starts with a start bit of status 0. Thus, the start bit is identified by the receiver with a high to low transition at which time the bits of data follow, and lastly the stop bit(s). Since the stop bit is always 1 in nature and assuming a second character immediately follows the first, then again there would be high to low transition at the receiver (due to the stop bit of the first character and the start bit of the second character). If no second character is sent, the line continues in the idle line high state after completion of the stop bit.

An example of an asynchronous data transmission is an operator typing data into the computer (a real-time transmission). Since the typing speed can never remain constant, the number of idle line 1's will vary.

For reliable communication, be it asynchronous or synchronous, a pre-agreed set of rules has to be obeyed by both the transmitter and the receiver, called *protocol*. The major elements comprising a protocol are *syntax* (refers to format and signal levels), *semantic* (refers to control information for proper coordination and error handling), and *timing* (refers to speed matching and sequencing). In the present case of asynchronous communication, the protocols are clock speed, character frame length, signal level, number of start and stop bits, and type of parity bit (odd or even).

A separate clock line is normally not taken from the transmitter to receiver in asynchronous transmission. However, the receiver clock is designed to be as close to the transmitter clock value. If these two clocks differ somewhat, a *clock slippage* may occur. There can be underslipping or overslipping. The former occurs if the transmitter clock is slower than the receiver clock, while if the transmitter clock is faster than the receiver clock, overslipping would occur. Thus, for the case of overslipping, the received data is being sampled at a rate slower than the rate at which the data bits are received from the transmitter. In this case, as the samples are being analyzed and stored in the receiver memory, a time will be reached at which a data bit will be completely skipped.

Asynchronous transmission is undertaken when the data is sporadic or intermittent in nature and also the volume of data to be handled is not huge. It is simple and cheap but its overhead is somewhat quite high—about

2–3 bits for each 8 bits of data. This comes to about 20–25% of overhead bits and hence it affects throughput (it is the number of actual data bits sent in unit time) in a big way. A framing error can occur if the sudden appearance of noise causes a change of idle line condition from 1 to 0, which the receiver would assume inadvertently to be a valid start pulse.

1.7.4 Synchronous

In synchronous transmission, a huge chunk of data is transported from transmitter to receiver at a fast rate in predefined frames. While transmission is character-by-character type in asynchronous transmission, it is frame-by-frame type in synchronous transmission. In this transmission scheme, it is best to insert clock information in the data signal itself. One such example is Manchester coding. Thus, irrespective of data signal pattern, there would always be a level transition in each period. At the receiving end side, this level transition is utilized to generate the clock signal that would be totally in synchronism with the received data. Thus, even if there is any change in the data rate during transmission, no loss of synchronism would occur, which is realized by employing PLL (phase locked loop). A synchronous frame format is shown in Figure 1.5. The frame contains several fields. The first field is a synchronous pattern, also called a flag field. The synchronous character places the receiver in the character mode and conditions it to receive data bits byte-wise. In case of BSC (Binary Synchronous Communications), two synchronous characters are sent, one after the other, to avoid misinterpreting any random data byte to be a synchronous character.

Synchronous transmission becomes more and more efficient as the amount of data to be transferred increases. Efficiency is the ratio of information bits to total transmitted bits. In this case, the percentage of overhead bits is less than 1% assuming 1000 character block of data having approximately 48 overhead bits in the form of synchronous and control bits.

1.7.5 Isochronous

In synchronous transmission, frames travel down the line in fixed time slots. Here, frames are synchronized for transmission. It fails if there is an

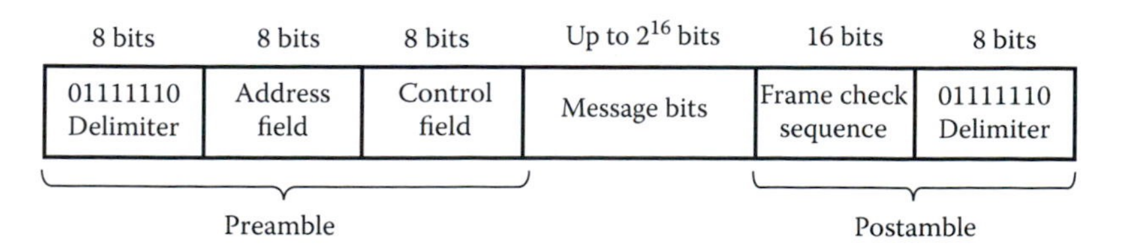

FIGURE 1.5 Synchronous data frame format.

uneven delay between frames, which occurs in real-time audio and video. TV transmissions involve sending 30 images per second. TV viewing must also have to be at the same rate. Isochronous transmission guarantees such transmissions such that images arrive at the same rate for the purpose of viewing. A major part of the bandwidth is allocated to a single or two devices for data transfer in isochronous mode. This method is used for high-speed data transfer. A packet consists of a maximum of 123 bytes in this mode, and there is no limit to the maximum number of packets that can be sent.

1.8 USE OF MODEMS

The term modem stands for modulator and demodulator. At the transmitting end, it acts as a *signal modulator* whose output is a digitally modulated analog signal, while at the receiving end, it acts as a *signal demodulator* by demodulating the received signal into binary data. A modem used for data communication purpose is called a data communication modem, dataset, data phone, or simply modem.

Normal telephone lines, used for carrying voice signals, operate in the range of 300–3300 Hz, giving a 3000 Hz bandwidth. Voice signals can accept some amount of interference and distortion to the extent that its intelligibility is not lost. Binary data requires higher data integrity for data retrieval and hence the two edges are not used. The bandwidth for data communication is 2400 Hz, with lower and upper limits at 600 and 3000 Hz, respectively. This is shown in Figure 1.6. Thus, voice band data modems are used to carry data. In this, the bits or the digital signals modulate the carrier, producing digitally modulated analog signals that modulate

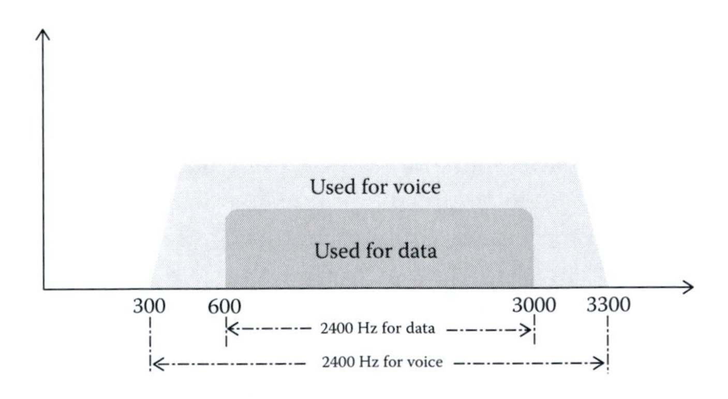

FIGURE 1.6 Bandwidth of voice and data signals. (From B. A. Forouzan. *Data Communications and Networking*, 4th Edition, Special Indian Edition. Tata McGraw Hill Companies Inc., New Delhi, India, p. 248, 2006.)

data into analog form for transmission through the telephone channel using a bandwidth of 2400 Hz of the total telephone line bandwidth.

1.9 POWER SPECTRAL DENSITY

A time-limited signal has an infinite bandwidth. However, most of the power of such signals is concentrated within some finite band. In this context, power spectral density is a very important term that shows the power content of a signal as a function of frequency. The effective bandwidth of a signal is the frequency band in which most of the power of the signal is concentrated.

If a channel has a high-frequency response between two frequencies, f_1 and f_2, then the power spectral density of the code that is chosen for transmission should have high power content in this band to avoid signal distortion.

1.10 TRANSMISSION IMPAIRMENTS

Transmission media, which are not ideal or perfect, cause transmission impairments. This means that the signal that is sent is not the one that is received. For analog signals, the impairments cause degradation in signal quality of the received signal, while for digital signals the effect is an error in bits that is received. The most significant impairments are *attenuation*, *distortion*, *limited bandwidth*, and *noise*.

Attenuation causes a loss of energy of the received signal, i.e., the strength of the signal falls off with distance as it travels down the medium. This loss is due to the resistance of the medium. To overcome this, amplifiers are used for analog signals, while for digital signals repeaters are used. For guided media, attenuation is generally logarithmic in nature, while for unguided media it is a complex function of distance.

Signals received at the receiver must have strength sufficiently higher than noise to eliminate the possibility of any error. Second, attenuation is an increasing function of frequency. Equalizers are employed to overcome this problem across a defined band of frequencies. A second method is to employ amplifiers that amplify high frequencies more than the lower ones.

Distortion changes the form or shape of a signal. A composite signal has different frequencies that travel at different speeds through the medium. Thus, the phase relationships that the different frequencies had at the transmitting end would not be maintained at the receiving end. This would give rise to delay distortion. It may result in some frequency components of one bit spilling into the next one, resulting in intersymbol interference (ISI). This limits the maximum transmission frequency.

Leaving aside other impairments, the larger the bandwidth of the medium, the more closely the received signal is to the transmitted one.

The last major cause of impairment is noise. There are different types of noise, such as induced noise, thermal noise, crosstalk, intermodulation noise, and impulse noise. Induced noise comes from operations of industrial motors and electrical appliances. Thermal noise occurs owing to random motion of electrons in a wire and is a function of temperature. Crosstalk refers to the effect that a wire causes on a neighboring wire. It is an unwanted coupling between two signal paths. Intermodulation noise may arise when two or more signals of different frequencies share the same transmission medium. It may result in an additional signal of frequency that is the sum of the two existing frequencies. Impulse noise generally comes from lightning or power lines. It comes in the form of a spike having sufficient energy and considerable magnitude but exists for a short duration of time. It is unpredictable in nature. Lastly, it can be said that as bit rate increases, error rate would also increase. This is because with increasing data rate, the bit time period decreases, thereby exposing more number of bits to noise.

1.11 DATA RATE AND BANDWIDTH RELATIONSHIP

The data communication channel is the most important facility in the whole communication process. The greater the bandwidth of a channel, the higher the cost of the facility. Thus, a given bandwidth must be used as efficiently and prudently as possible. For a given level of noise, the maximum data rate is determined by the bandwidth of the channel. A given channel has a limited bandwidth determined by the physical properties of the medium. Data rate through a channel is determined by available bandwidth, number of levels present in the signal, and also the amount of noise present, i.e., quality of the channel.

The Nyquist bit rate for a noiseless channel and the Shannon capacity for a noisy channel are the two yardsticks for determining the maximum data rate through a channel. The maximum theoretical data rate, in bps, is given by $2 \times B \times \log_2 L$, where B is the bandwidth and L is the number of levels in the signaling element. Thus, for a given bandwidth, if the number of levels is increased, the corresponding bit rate would increase. Although it is true theoretically, data retrieval at the receiver would be more and more difficult because of presence of so many levels. As per Shannon, the channel capacity, in presence of noise, is given by $B \times \log_2(1 + \text{SNR})$. SNR is the signal-to-noise ratio at the receiver where the received signal is processed and the bits are retrieved. This formula does not contain the number of levels present in the signal element. Thus, channel capacity increases with either bandwidth or signal strength. However, with increasing signal strength, nonlinearities in the system also increases as also the intermodulation noise. Again, since the noise is assumed to be white, the greater the bandwidth, more noise would get into the system. This effectively decreases the SNR with increasing bandwidth.

1.12 MULTIPLEXING

Bandwidth is one of the most precious resources in communication, and its judicious use is seemingly the main challenge to the communication engineers. If a low bandwidth (narrow bandwidth) signal occupies a link whose bandwidth is high, then the link's resources are woefully utilized. A communication engineer's task is to utilize the available bandwidth as fully as possible. Multiplexing is used to utilize the bandwidth of a link most efficiently and effectively.

1.12.1 Introduction

Multiplexing is the transmission of multiple signals simultaneously over a single link. Although they share the same medium for transfer of information, they do not necessarily occur at the same time or occupy identical bandwidth. Metallic wires, coaxial cables, satellite microwave, optical fiber, etc., may act as the transmission medium. At the receiver, demultiplexing is done to retrieve the original signals. Figure 1.7 shows the basic principle of operation of a multiplexer–demultiplexer (MUX–DEMUX) system. They are connected by a single link through which n channels transmit their information. A multiplexer combines the input signals into a single stream (many-to-one) while a demultiplexer (one-to-many) separates the signals into individual ones.

1.12.2 Types

There are three basic multiplexing schemes. These are frequency division multiplexing (FDM), wavelength division multiplexing (WDM), and time division multiplexing (TDM). Of these, the FDM and WDM techniques are used for analog signals, while TDM is used for digital signals. Apart from the above, another multiplexing technique called space division multiplexing (SDM) is sometimes used, which is rather not a very sophisticated one. In this,

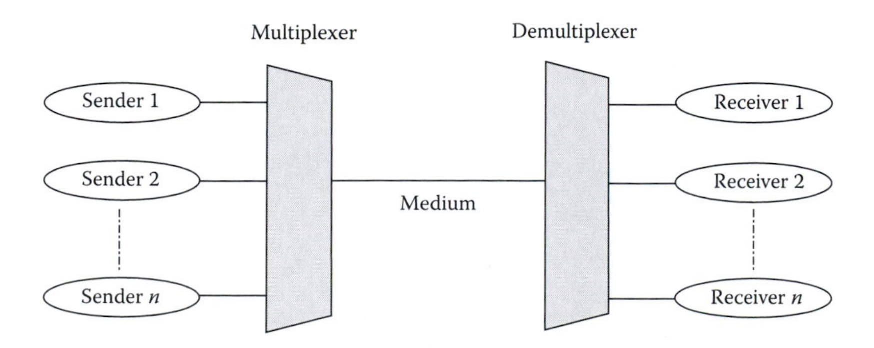

FIGURE 1.7 Schematic of MUX–DEMUX system. (Available at www.comsci.liu.edu/~jrodriguez/cs154fl08/Slides/Lecture5.pdf.)

individual cables are allocated for individual signals, which are then put within the same trench. The trench itself is considered to be the transmission medium.

An example of SDM is the local telephone system. Here, each telephone is connected to the central office by a local loop not shared by any other subscriber. The SDM technique is not considered to be a true multiplexing scheme and, hence, its demultiplexing is not necessary in the way it is done for FDM, WDM, or TDM techniques.

1.12.3 FDM

Frequency division multiplexing (FDM) is a technique in which the available bandwidth in a communication link is divided into a series of nonoverlapping frequency subbands, each of which carries the modulated version of the original signal. Cable television uses a single cable to transmit many channels for viewing using FDM technique. Other uses of FDM are handling multiple telephone calls through high-capacity trunk lines, communication satellites that transmit different channel data for both uplinking and downlinking purposes.

An FDM scheme and its subsequent demultiplexing is shown in Figure 1.8. The multiplexer section consists of a low-pass filter, modulator, and band-pass filter for each channel.

In the demultiplexer side, the same three are present in the reverse order. Each input signal is modulated by separate distinct carrier frequencies. The modulated carriers consist of a narrow band of frequencies, called the

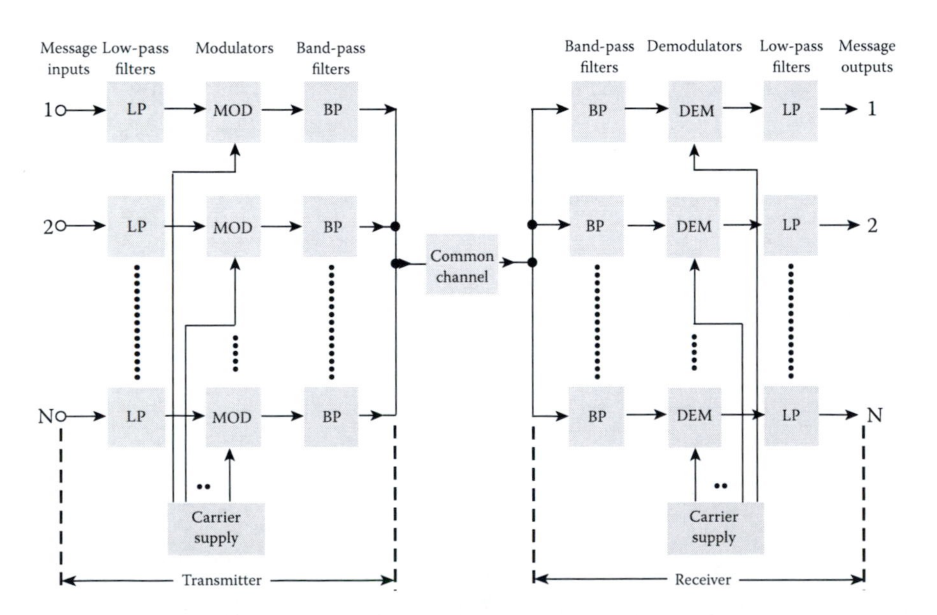

FIGURE 1.8 FDM MUX–DEMUX system.

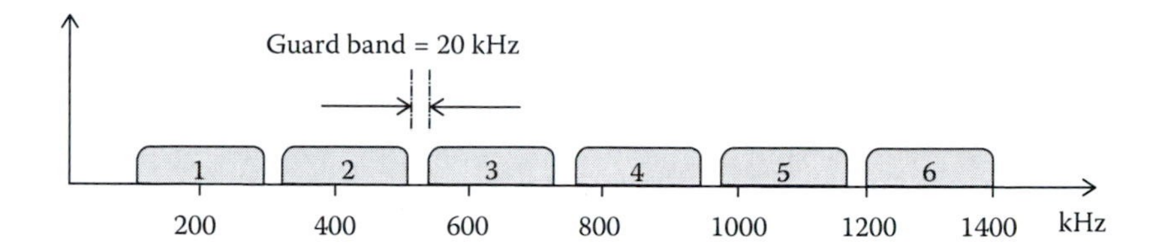

FIGURE 1.9 Different carrier frequencies with guard band in between. (Available at www.comsci.liu.edu/~jrodriguez/cs154fl08/Slides/Lecture5.pdf.)

passbands, centered around the carrier frequency of each individual channel. The input signal information is contained in these passbands. The carrier frequencies of each individual channels are so chosen that their passbands do not overlap, minimizing chances of interference. Figure 1.9 shows six channels with the carrier frequency of the first channel fixed at 200 kHz and the sixth channel at 1300 kHz with a guard band of 20 kHz between any two channels.

1.12.4 WDM

Wavelength division multiplexing (WDM) is, in a sense, identical to FDM. However, in this case, optical signals of different wavelengths (i.e., of different colors) are used and sent via a single optical cable. Optical signals have very high frequencies, unlike FDM. Multiplexing ensures that the very high bandwidth associated with optical fibers is effectively utilized. Figure 1.10 shows how a multiplexer combines several different wavelengths and the combined signal is demultiplexed at the receiving end. The methodology applied here is that a prism can bend a beam of light, which depends on the angle of incidence and also the frequency. SONET uses WDM technology in which more than one optical cable is used for MUX–DEMUX purposes. Of late, dense WDM (DWDM) is used, which uses many channels with very less spacing between them, thereby enhancing efficiency even more.

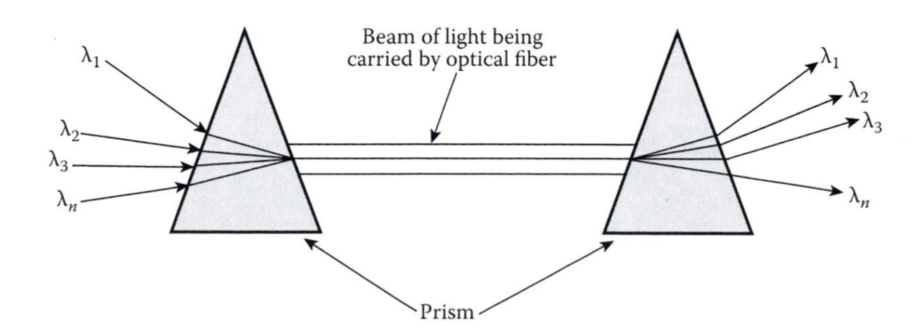

FIGURE 1.10 Schematic of WDM system. (Available at www.comsci.liu.edu/~jrodriguez/cs154fl08/Slides/Lecture5.pdf.)

1.12.5 TDM

Time division multiplexing (TDM) is a digital communication process that allows several signals from different sources to time share the resources of a link. In TDM, the whole bandwidth of the link is utilized at any instant of time by a single signal, while the total bandwidth is always shared by the communicating signals in FDM. In TDM, all signals have their own precise clocks to send data that needs proper synchronization. The different channels have their own scheduling for data transfer. The receiver can extract the channel signals by proper clocking and synchronization with the individual channels. Proper redesigning entails TDM to be adaptive in nature to any load changes. The concept of time division multiplexing is shown in Figure 1.11. It shows *n* channels that send data one after the other, utilizing the full bandwidth of the link.

1.12.5.1 Synchronous TDM

In synchronous TDM, data from different sources are divided into fixed time slots, in which a slot may contain a single bit, a byte of data, or a predefined amount of data. As shown in Figure 1.12, data from the first source is sent in the first time slot, followed by data from the second source in the second time slot. This is continued until data from the last source is sent. Then the system repeats itself. Thus, in the first four time slots, data $\mathbf{A}_1$ from source 1, data $\mathbf{B}_1$ from source 2, no data from source 3, and data $\mathbf{D}_1$ from source 4 are fed into the multiplexer. In this sequence, the next set of data from the four sources are sent, following the same logic. It should be noted here that even if some source does not have any data to be sent at any given instant of time, because of preallocation, that time is simply wasted. For Figure 1.12, four time slots are wasted—one for source 2, two for source 3, and one for source 4. In a particular case when a channel does not have any data to be sent for a considerable time, underutilization of the channel takes place, leading to a less efficient system. Statistical TDM

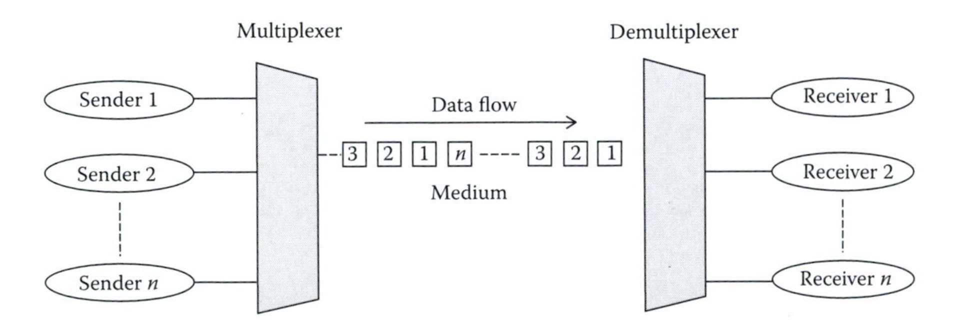

FIGURE 1.11 TDM MUX–DEMUX system. (Available at www.comsci.liu.edu/~jrodriguez/cs154fl08/Slides/Lecture5.pdf.)

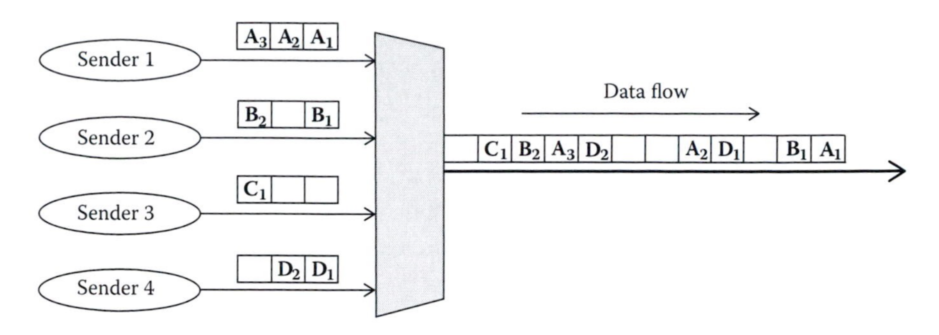

FIGURE 1.12 Synchronous TDM system. (Available at www.comsci.liu.edu/~jrodriguez/cs154fl08/Slides/Lecture5.pdf.)

addresses this problem by skipping slot allocation for a source if it does not have any data in that particular time slot.

1.12.5.2 Statistical TDM

To overcome the shortcomings of synchronous TDM, statistical TDM utilizes only those slots that do have data, skipping those slots that do not have data at the times they are to send their data on the channel. Referring to Figure 1.13, one slot for source 2, two slots for source 3, and one slot for source 4 are skipped because they are empty. The scheme must have an identifier to indicate and identify the particular receiver, since the traditional multiplexing scheme is not used here for the sake of improved channel utilization.

1.12.6 Variable Data Rate

Thus far, the discussions on TDM centered on the assumption that all the sources involved in data transmission has the same data rate—which is not always the case. There are different approaches to overcome such data rate

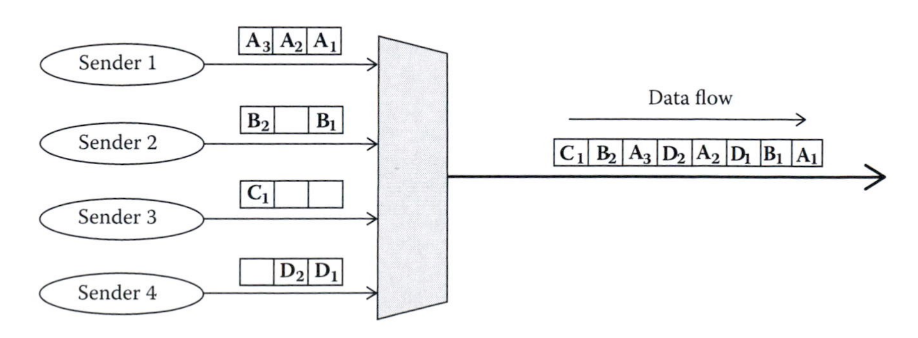

FIGURE 1.13 Statistical TDM system. (Available at www.comsci.liu.edu/~jrodriguez/cs154fl08/Slides/Lecture5.pdf.)

variations. These are multilevel multiplexing, multislot multiplexing, and pulse stuffing technique, which are discussed below.

1.12.7 Multilevel Multiplexing

Multilevel multiplexing is employed when most of the lines have higher data rates and fewer ones slower data rates, with the data rates of the latter an integral multiple of the former. The slower ones are multiplexed so that the data rate of the multiplexed output is equal to the other ones. Figure 1.14 shows a two-level multiplexing scheme with the two 20 kbps lines combined by the first multiplexer. The second multiplexer effectively combines four input lines. The output frame is a 160 kbps line. It should be noted that demultiplexing at two levels are needed to get back the original five lines—two of 20 kbps and three of 40 kbps.

1.12.8 Multislot Multiplexing

Multislot multiplexing is employed when most of the lines have slower data rates and fewer ones faster data rates, with the data rates of the former an integral multiple of the latter. Figure 1.15 shows the 50 kbps line divided

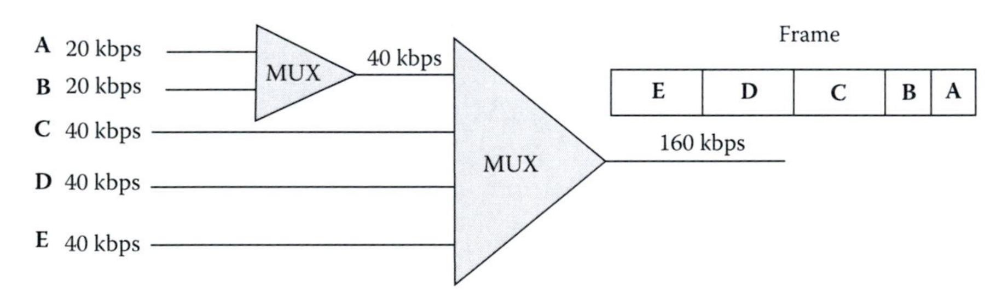

FIGURE 1.14 Multilevel multiplexing scheme. (From B. A. Forouzan. *Data Communications and Networking*, 4th Edition, Special Indian Edition. Tata McGraw Hill Companies Inc., New Delhi, India, p. 174, 2006.)

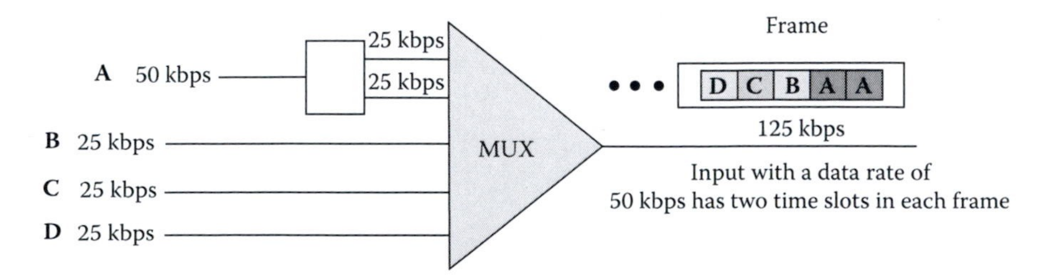

FIGURE 1.15 Multislot multiplexing scheme. (From B. A. Forouzan. *Data Communications and Networking*, 4th Edition, Special Indian Edition. Tata McGraw Hill Companies Inc., New Delhi, India, p. 174, 2006.)

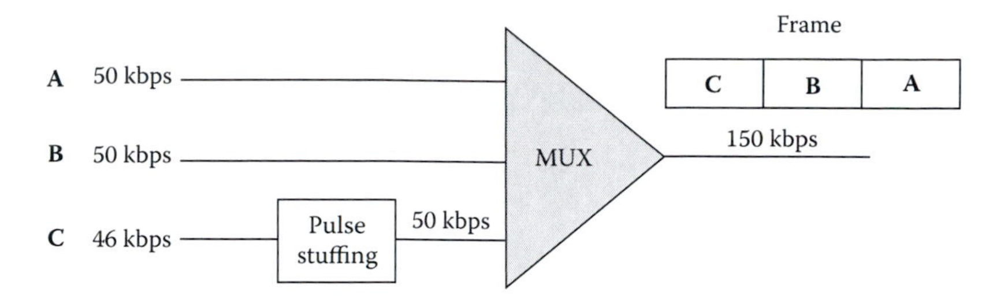

FIGURE 1.16 Pulse stuffing scheme. (From B. A. Forouzan. *Data Communications and Networking*, 4th Edition, Special Indian Edition. Tata McGraw Hill Companies Inc., New Delhi, India, p. 175, 2006.)

into two 25 kbps lines so that the reduced rate is equal to the other three. The 50 kbps line is allocated two time slots in the output data frame of 125 kbps rate.

1.12.9 Pulse Stuffing Multiplexing

The above two schemes fail when data rates of the sources are not an integral multiple of each other. In this scheme, extra bits are added to the slower data rate sources so that ultimately their data rates become equal to the faster ones. This scheme is also called bit padding or bit stuffing and is shown in Figure 1.16.

1.13 SPREAD SPECTRUM

The spread spectrum technique is used to spread the spectrum of the original signal to camouflage the original signal from possible eavesdropper or being subjected to jamming by some malicious intruder. Spread spectrum was developed initially for military or intelligence applications. It is also used in cellular and cordless telephones. These secrecy requirements far outweigh the greater bandwidth requirements in this case because of spread of the signal over a larger bandwidth. It is designed for wireless applications such as local area networks (LANs) and wide area networks (WANs). Two considerations must be borne in mind in the spread spectrum technique: spreading allows *redundancy* in the system and the larger bandwidth after spreading is *independent* of the original signal.

In FDM, several signals are allocated different frequency bands in the available bandwidth: thus, the bandwidth is utilized in a more effective and efficient way. In spread spectrum technique, different signals from different sources are combined to fit into a larger bandwidth. The latter addresses the concerns of intentional eavesdropping and line jamming. In spread spectrum, wide-band, noise-like signals are used during transmission,

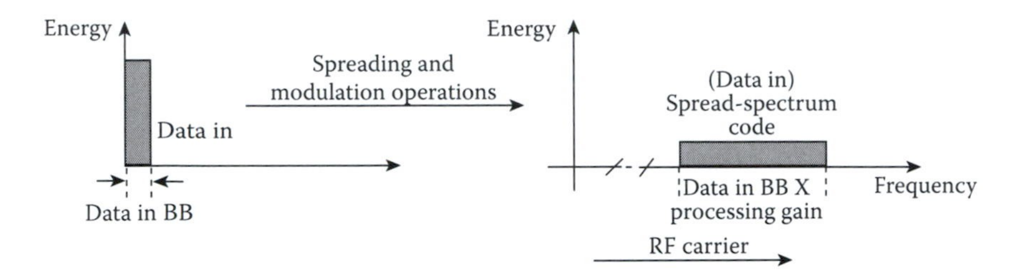

FIGURE 1.17 Energy vs. frequency characteristic of an information signal and its spread spectrum signal.

which are difficult to detect, intercept, and demodulate. This spread-out transmitted spectrum has a power level that may be less than the noise power. Because of this, an unauthorized listener cannot sense anything and the coded message remains invisible. The energy vs. frequency characteristics of a spread spectrum signal and the original baseband signal are shown in Figure 1.17.

1.13.1 Introduction

The general scheme of a spread spectrum communication scheme is shown in Figure 1.18. It includes encoder/decoder, modulator/demodulator, and a pseudorandom bit sequence (PRBS) generator. The encoder receives the input and produces a relatively narrow bandwidth analog signal. This signal is further modulated by the modulator with the help of a PRBS generator. The result is a considerable spreading of the bandwidth, which is fed into the communication channel. On the receiving end side, the received signal is demodulated by the same PRBS generator in a synchronous manner. The channel decoder finally decodes the signal to retrieve the data to get back the original signal.

If the bandwidth of a channel is B and that of the spread spectrum is B_{ss}, then $B_{ss} \gg B$.

Two techniques are employed to spread the spectrum: frequency hopping spread spectrum (FHSS) and direct sequence spread spectrum (DSSS).

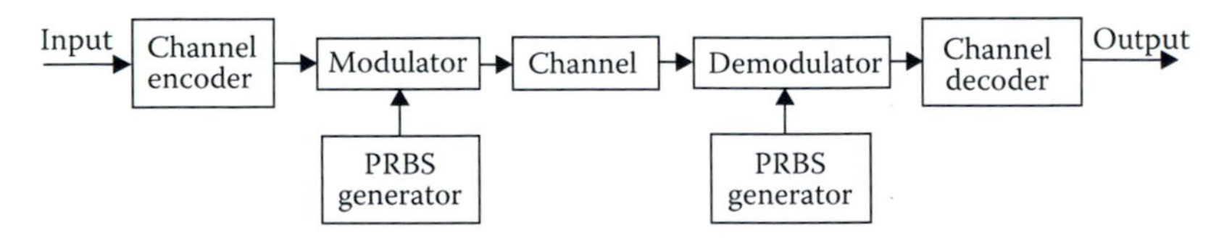

FIGURE 1.18 General scheme of spread spectrum communication technique. (From W. Stallings. *Data & Computer Communications*, 6th Edition. Pearson Inc., New Delhi, India, p. 163, 2000.)

1.13.2 FHSS

In this scheme, FHSS-modulated signals hop from one frequency to another in the RF range. The number of hopping frequencies equals the number of channels that are to be multiplexed. At any given instant of time, a signal modulates only one carrier frequency—the one chosen from the frequency or channel table. The channel usage pattern is called hopping sequence.

A scheme for implementation of FHSS is shown in Figure 1.19. NRZ (nonreturn-to-zero) data is initially modulated into either FSK (frequency shift keying) or BPSK (binary phase shift keying) form. A PRBS generates a p-bit pseudosequence where each sequence corresponds to a hopping period. Each sequence in the p-bit pattern selects a frequency from a table, which is fed to a frequency synthesizer. It generates the frequency selected from the table and acts as the carrier frequency, which is modulated by the

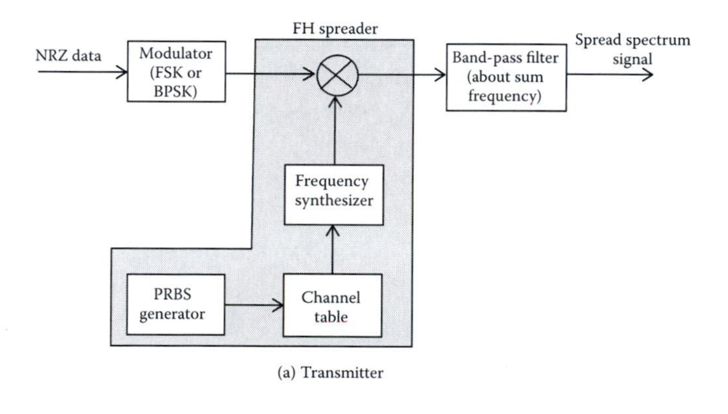

(a) Transmitter

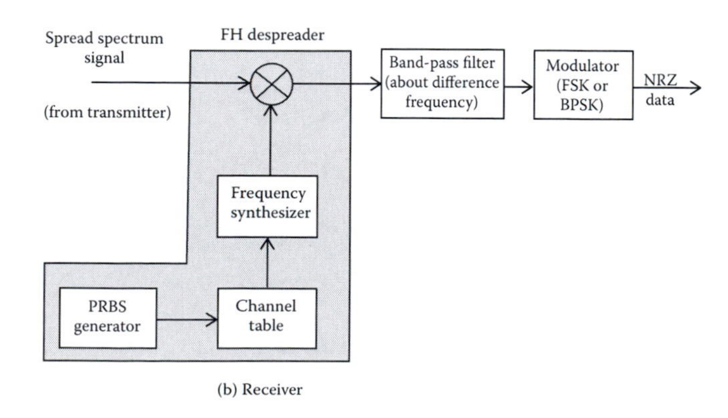

(b) Receiver

FIGURE 1.19 FHSS system (a) transmitter and (b) receiver. (From W. Stallings. *Data & Computer Communications*, 6th Edition. Pearson Inc., New Delhi, India, p. 164, 2000.)

first modulated output. This modulated output is passed through a band-pass filter to get the final FHSS.

The frequency selection process needs a bit of explaining and shown in Figure 1.20. The PRBS outputs eight pseudosequences as per the design. It can be obtained by employing a 3-bit (p = 3) LFSR (linear feedback shift register). As per Figure 1.20, the first 3-bit pattern from the PRBS is 100, which selects a frequency of 1800 kHz from the table, generated by the frequency synthesizer. Thus, it is apparent that in the first hop period, the frequency selected is 1800 kHz and in the eighth or last period it is 1000 kHz, after which the PRBS generator repeats itself.

The receiver hops between the frequencies in synchronism with the transmitter, as per the agreed protocol. In practice, the PRBS length sequence is much more than the example cited here (p = 3). If the hopping period is kept very small, the modulated output hops from one frequency to another very fast, and thus any would-be eavesdropper would not be able to decipher the modulated signal. Malicious jamming by inserting noise may jam the signal for one hopping period, but not the whole period. Thus, only one channel may get jammed but not the other ones.

FHSS is similar to FDM in that both use multiplexing techniques. However, while in FDM the individual channels occupy separate bands in the overall bandwidth at the same time, in FHSS, any one channel occupies a particular portion of the overall bandwidth at any given time.

Two types of hopping are possible in FHSS: slow hopping and fast hopping. The transmitter uses only one frequency for a few bit periods for the former but changes frequency several times in a single bit period for the latter. *Dwell time* is known as the time spent on a channel with a fixed frequency. If the frequency is changed thrice in a bit period, then dwell time is one third of bit period. Figure 1.21 shows how frequency changes for slow and fast hopping cases.

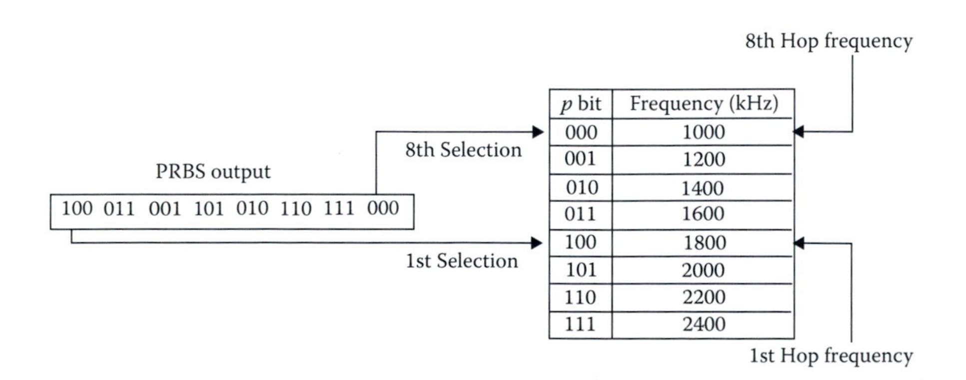

FIGURE 1.20 Frequency selection technique in FHSS.

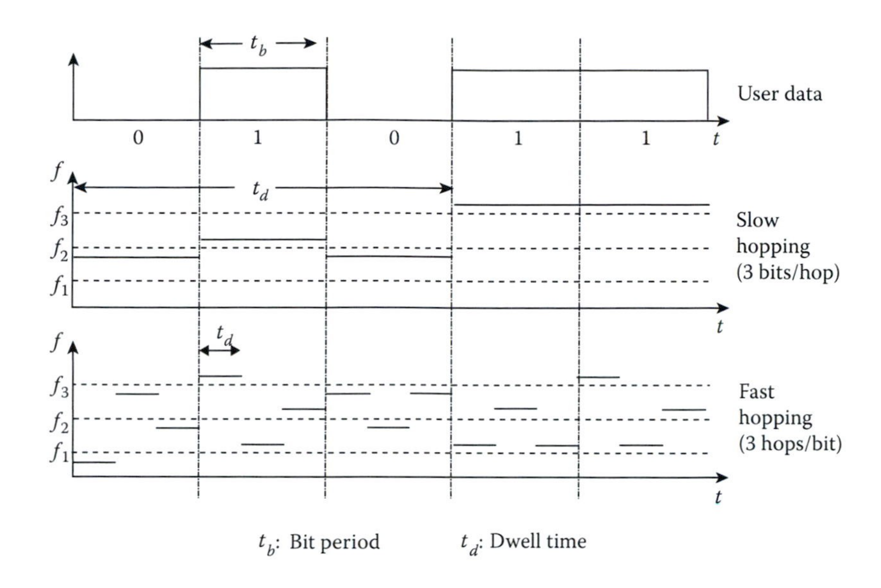

FIGURE 1.21 Frequency changes in slow and fast hopping.

Slow hopping is cheaper, less complex with relaxed tolerances, but less immune to narrow band interferences, while the opposites are true for fast hopping.

1.13.3 DSSS

Like FHSS, this scheme also enhances the bandwidth of the original signal, but in a different manner. Here, each bit in the original message or signal is represented by multiple bits in the transmitted signal—also called *chipping code* (or *spreading code*). This code spreads the signal in a wider frequency band in direct proportion to the number of bits used in the chipping code.

In wireless LAN, the chipping code used is 11 bits in length, also called *Barker sequence*. The schematic of the transmitter and the receiver used in DSSS technique is shown in Figure 1.22. The input NRZ data is modulated by a BPSK code, while the PRBS output is modulated by a carrier frequency. After this, these two modulated signals are combined to get the transmitted signal.

If the original message signal rate is n and the chip code has N bits, the rate of the spread signal is $n \times N$. Thus, a higher value of N would entail a higher spreading of the transmitted signal (spread signal). Immunity to interference is enhanced if each station uses a different chip code.

Bandwidth in DSSS cannot be shared if a spreading code is used that spreads the spread signals from different stations in such a manner that

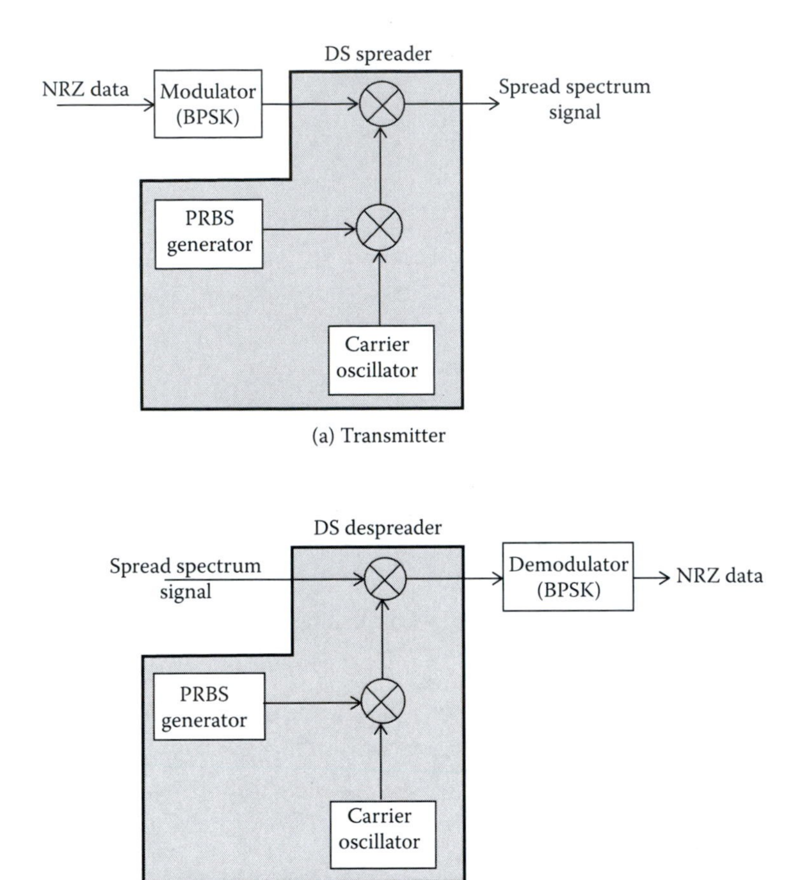

FIGURE 1.22 Schematic of DSSS (a) transmitter and (b) receiver. (From W. Stallings. *Data & Computer Communications*, 6th Edition. Pearson Inc., New Delhi, India, p. 166, 2000.)

they cannot be combined and separated. If it is otherwise, then the bandwidth can be shared.

1.13.4 Comparison between FHSS and DSSS

Although the basic idea remains the same, they have some characteristic differences that distinguish them. These are now discussed.

The carrier of the direct sequence spread spectrum always remains the same; however, the frequency hopping spread spectrum technique uses different frequencies at different times. The carrier frequency in case of FHSS hops around the designed band in a predefined manner. The FHSS system does not provide any processing gain. It is the increase in power density when the signal is despread. Higher processing gain provides for better SNR at the receiving end side.

Compared with DSSS, an FHSS system is much more difficult to synchronize—in FHSS, both time and frequency need to be synchronized at either end, while in DSSS only the timing of the chips needs synchronization. In FHSS, to make initial synchronization possible, it parks at a fixed frequency before initiating communication. In the uneventful case of the jammer frequency equaling the parking frequency, there would be no hopping and communication would not be possible. In another instance when the communication is going on, it is very difficult to resynchronize once the receiver loses synchronism. In FHSS, there is more latency time because the system needs to search the signal in order to lock on to it, while in case of DSSS it needs a few bits to lock on to the chip sequence.

FHSS deals with multipath issues in a better manner than DSSS, because in the former the frequency hops from one frequency to another at the end of each hopping period. FHSS is not as robust as DSSS, but simpler to detect.

1.13.5 Advantages of Spread Spectrum

There are several advantages that have led to the wide acceptability of spread spectrum techniques in communications. These are now discussed.

A spread spectrum system, if it operates in the ISM band, would be allowed to have more power because of its inherent noninterfering nature. Higher transmit power entails higher operating distance than a traditional analog communication system.

By judiciously choosing pseudorandom binary sequences, multiple spread spectrum systems can *coexist* without interfering with one another. This thus would enhance the bandwidth utility of the system.

Reduced crosstalk interference is another advantage of this system. This is achieved because the spread spectrum technique employs processing gain in the receiver. The effect of crosstalk interference can be eliminated if noise below a threshold is rejected altogether. This is particularly so in voice communication where error in several bits would not significantly influence the quality of a voice signal.

Because of atmospheric reflection and refraction, a receiver may receive a transmitted signal in several paths—multipath. This is shown in Figure 1.23. Multipath fading occurs because of interference between direct path (D) and reflected path (R). However, since the despreading process involves synchronization to direct path signal, the reflected signal is rejected.

A spread spectrum system is inherently secure in nature compared with both conventional analog and digital communications. Irrespective of FHSS or DSSS, the transmitted signal is totally random in nature, only to

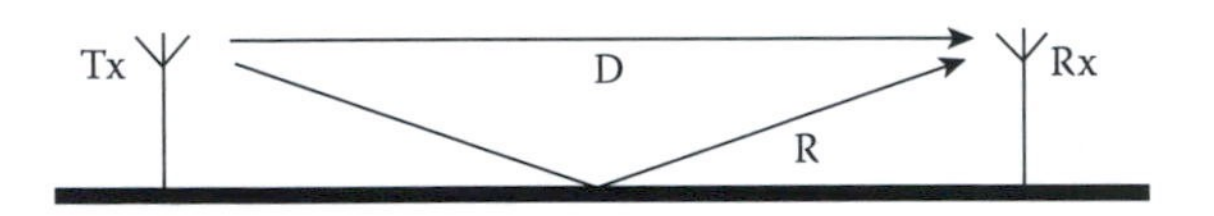

FIGURE 1.23 Signal may reach receiver in multipaths.

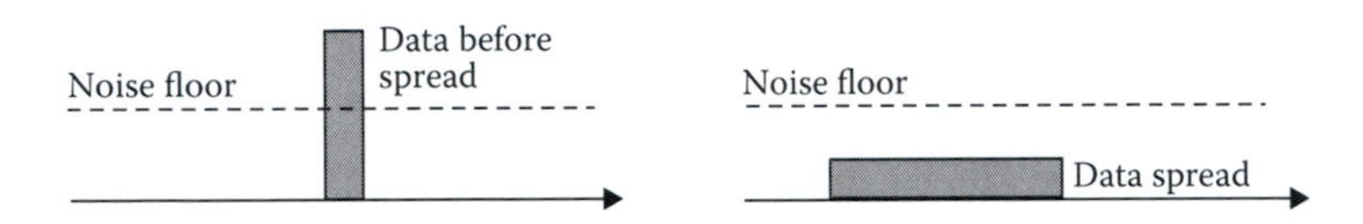

FIGURE 1.24 Spread signal level lies below noise level.

be recognized by the receiver by proper synchronization. Thus, eavesdropping by any malicious intruder is eliminated.

Because of the "pseudo" nature of the code, the number of carrier frequencies used, and also their values, it is very difficult to decipher the transmitted signal and thus any intruder would be unable to retrieve the original signal.

It is very difficult for an unauthorized listener to decode the transmitted signal because the pseudocode used in the spreading process is unknown to him. Sometimes the spread signal spectral density lies below the noise floor, which is particularly so in case of DSSS. This is shown in Figure 1.24. An intruder into the system cannot "see" the coded signal because it lies much below the noise floor.

1.14 DATA CODING

An error-free message received at the receiver is central to any communication scheme. A message, sent from the transmitter, may get corrupted along its passage to the receiver. Many factors play into corrupting message bits—the magnitude of which depends on the type of communication. For audio or video transmissions, some level of error may be tolerated; however, in some cases, say monetary transactions must have to have zero level of tolerance. Various schemes are employed to detect and correct transmission errors. Redundant bits, which do not form message bits, are sent along with data bits to achieve the same. Redundancy is realized by various coding schemes that are performed on message bits to ensure error-free message reception. There are several reasons for coding a message before it is sent to the receiver: (a) to simplify transmitter and receiver design; (b) to achieve higher transmission efficiency; (c) to decrease error; and (d) to provide secrecy in transmission, i.e., eavesdropping by a third party is eliminated.

1.14.1 Introduction

It is very simple to comprehend that error correction is more difficult to realize than simple error detection. Error detection involves the detection of corrupted bit(s). On the other hand, error correction involves, first, detecting the places of error, and subsequently, correcting them. Error correction involves two schemes: forward error correction and retransmission. In the former, the received bits are corrected with the help of redundant bits following some established methodology. The retransmission scheme is straightforward in which retransmission of message is requested when the receiver detects any error in data bits.

Errors that corrupt messages may be single bit or multibit in nature, although multibit errors are more common. Multibit error is known as burst error. The number of bits affected in burst error depends on the duration of noise and data transmission rate.

1.14.2 Characteristics of a Line Code

There are various line codes employed to decrease potential error at the receiving section. A line code should have the ability to detect and correct errors. Second, the line code should have adequate timing content, i.e., it should be possible to extract clock timing from the coded data, which is essential for data retrieval. Third, the code should be transparent in nature. There are situations in which long strings of 1's and 0's are transmitted that form part of data. If it is possible to retrieve such long strings 1's and 0's from the coded data, then the code is said to be transparent. Fourth, the code should be immune from channel noise and ISI. Lastly, the frequency response of a coded signal should match the frequency response of the channel through which the coded data is being transmitted. In essence, their power spectral densities should match each other.

Some other characteristics that a line code should have are baseline wandering and DC components. Data, after it is decoded at the receiver, has an average power in its spectrum. This average is called *baseline*. A data element is evaluated for its value after the received signal element power is compared with this average. If there are long strings of 1's and 0's in the data element, there is a possibility of a drift in this baseline value—called *baseline wandering*. Such a phenomenon makes determination of data element value very difficult. Thus, a data element should be coded such that baseline wandering can be avoided. An example of such a code is Manchester coding. When a digital data do not change its status for a few number of bits, there is a possibility of low-frequency existence in the spectrum—called DC components. Since AC coupling is used in the repeaters, it is expected that the spectrum should be devoid of any DC components.

1.14.3 Types

There are various line codes that are used for sending messages to the receiver in an effort to achieve error-free reception. The Manchester code is a very good example of a transparent code for which there is always a change in signal level at half the bit period. This code is particularly suitable for data having long strings of 1's and 0's. Huffman codes utilize the probability of occurrence of different characters and assign shorter codes to characters that are transmitted most often. It should be ensured that no shorter code word should form the initial bits of any longer code word—this is necessary for the receiver to recognize the character boundaries.

Coding schemes can broadly be categorized into block coding and convolution coding. Messages are divided into several blocks—each such block is called a data word. To each such data word, redundant bits are added to get code words. In block coding scheme, each identical data word gives rise to the same codeword—i.e., it is a one-to-one coding scheme. Block coding can be employed for both error detection and correction.

Linear block codes are a form of block codes that are employed for both error detection and correction. For such a linear block code, the XORing of two valid code words would give rise to another valid codeword. Parity check code and Hamming code are some examples of linear block codes. Parity check code is an error-checking code and it has its own drawbacks and is not used in noisy environments. A Hamming code can correct a single error or detect a double error. It can also detect a burst error. A cyclic code is a special type of linear block code in which if a codeword is cyclically shifted, it would result in another codeword. Cyclic redundancy code (CRC) is a type of cyclic code. If automatic repeat request (ARQ) protocol is used in conjunction with CRC code, then it becomes very effective in reducing the BER of a message. A CRC-16 code may attain one undetected error in every 10^{14} bits.

2 Networking

2.1 INTRODUCTION

The field of communication has expanded at a very fast pace over the last several decades. Initially, only voice communication was in vogue, but availability of microprocessors, microcomputers, and peripheral equipments have drastically changed the whole scenario. Today, a multitude of information, mostly in digital form, is transported worldwide in a fraction of a second, almost without any kind of error. Very high volumes of digital data along with faster, more reliable, and secure communications are the need of the hour.

The original information from the source can be in either analog or digital form. If it is analog, it is first to be digitized and then sent. A data communication system involves transmission, reception, and extraction of the original data after processing. Analog information is obtained at the receiving end after proper digital-to-analog conversion. Examples of analog information may be voice or simply audio music, while digital information may consist of binary coded data, alphanumeric codes.

A network is a set of devices (often termed as nodes or stations) interconnected by communication links. Examples of a node can be a computer, a printer, or any device that is capable of sending/receiving data to/from other nodes connected with the network. A data communications network essentially does not virtually have any limit with regard to its capacity. Data communication networks connect a workstation with mainframe computers, ATMs to bank systems, airline and hotel bookings, electronic mail transfers, information highways, media and news centers, etc.

Figure 2.1 shows the structure of a computer communications network. The black circles are switching computers that are connected by high-speed lines. There are also computers connected to different kinds of terminals, and some do not have any connections associated with them.

Again, there are three types of terminals: local, remote, and orphan—the names are apparent from their positions Figure 2.1. The local computers operate at moderate bandwidth for data transmission, while the high-speed switching computers need higher bandwidth for their very fast nature of operation.

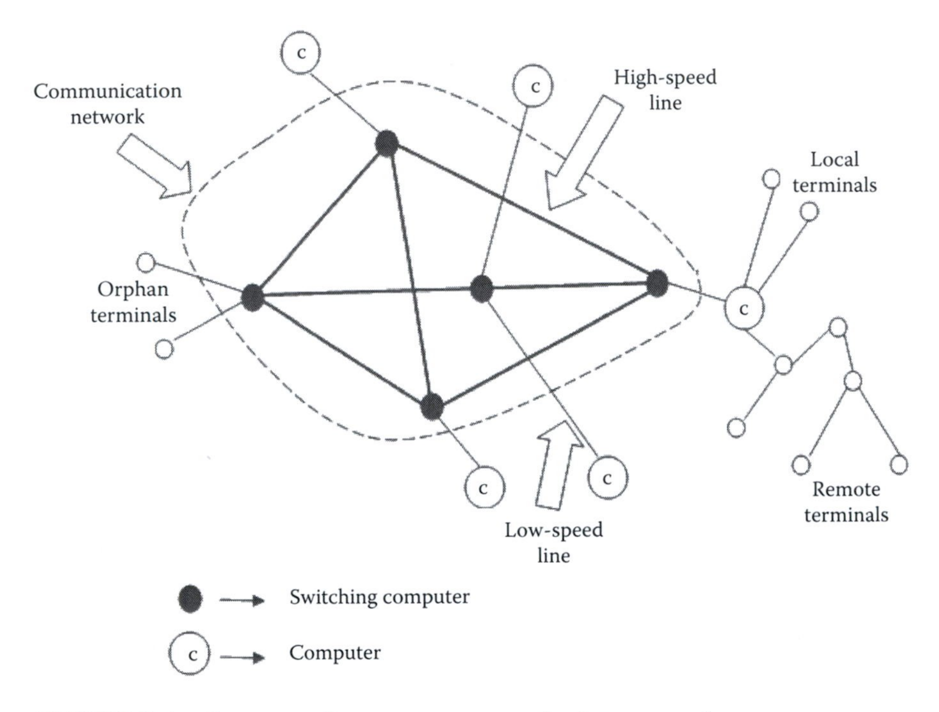

FIGURE 2.1 Structure of computer communications network.

2.2 CHARACTERISTICS

A data communications network must fulfill the following criteria or attributes for an effective and reliable network:

1. Fastness: Data transmission from a host must be as fast as possible. Throughput is a measure of fastness. As network traffic increases, throughput decreases and delay increases.
2. Transmit and response times: They must be kept as low as possible. Transmit time is the time for a message to travel from one node to another, while response time is the time between an inquiry and the corresponding response.
3. Cost: The cost of establishing, maintaining, and operating the network must be a minimum.
4. Efficiency: A major part of the network must not remain idle for a long time.
5. Secure: Any malicious intruder must not have access to the network.
6. Error: The system must be as error free as possible. One parameter that is used for this purpose is bit error rate or BER.

2.3 CONNECTION TYPES

A network consists of two or more nodes connected by a communication link. A link is a path through which information travels from on node to another. There are two types of connections for transfer of information from node to node: point-to-point and multipoint or multidrop.

A dedicated link is a point-to-point connection for data transfer from one node to another. Figure 2.2 shows such a connection. A physical wire may connect two personal computers or a personal computer and a printer for data transfer. The remote control of a TV with the set is executed without wire by a point-to-point connection method. Another example is the high-speed data transmission between one mainframe computer to another.

A multipoint or multidrop connection is shown in Figure 2.3. In this, several nodes share the same link. Either a single node shares the link at a time or several nodes share the link at the same time.

2.4 DATA COMMUNICATION STANDARDS AND ORGANIZATIONS

Standards are essential for data and telecommunications industry so that equipment manufacturers abide by these agreed standards. This would ensure national and international interoperability between devices from

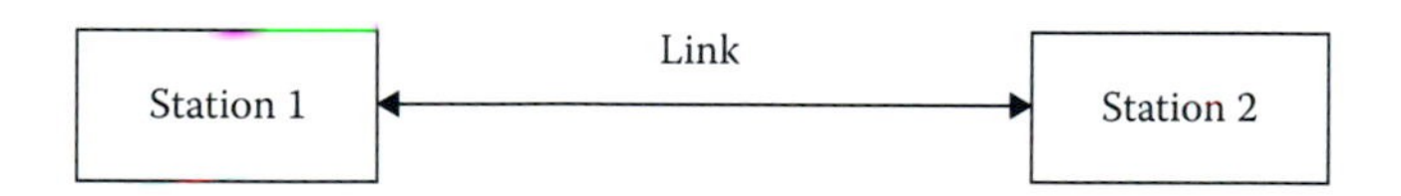

FIGURE 2.2 Point-to-point connection. (From B. A. Forouzan. *Data Communications and Networking*, 4th Edition, Special Indian Edition. Tata McGraw Hill Companies Inc., New Delhi, India, p. 9, 2006.)

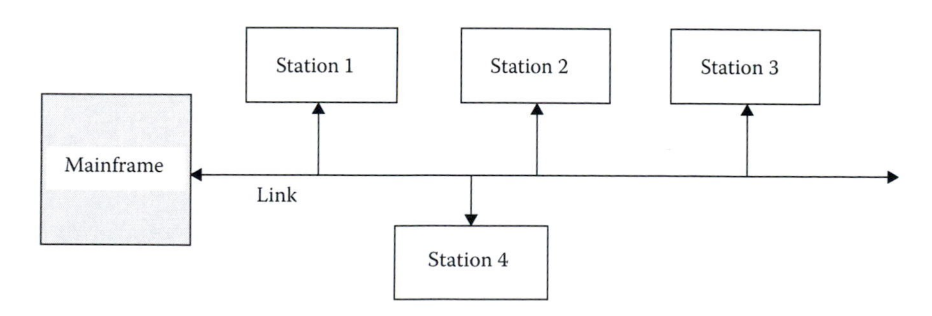

FIGURE 2.3 Multipoint connection. (From B. A. Forouzan. *Data Communications and Networking*, 4th Edition, Special Indian Edition. Tata McGraw Hill Companies Inc., New Delhi, India, p. 9, 2006.)

different manufacturers. Standards, guidelines, and frameworks are established by a consortium of organizations, government agencies, manufacturers, and users of equipments in the communication industry. The standard organizations in the data, telecommunications, and networking industries are shown in Figure 2.4. These are now discussed.

The **International Standards Organization (ISO)** was formed in 1946. It is a voluntary organization that aims at creating the set of rules and regulations for graphics and document exchange and works for equipment and system compatibility, reduced cost, improved performance and productivity, better quality products, etc.

The **International Telecommunications Union-Telecommunications Sector (ITU-T)** is based in Geneva, Switzerland. Previously it was known as the Consultative Committee for International Telegraphy and Telephony. Sets of rules and standards for data communications and telephone are recommended by the ITU-T. It has developed three sets of standards: the V series for modem interfacing and data transmission over telephone lines; X series for data transmission for public digital networks, e-mails, etc.; and the I and Q series for Integrated Services Digital Network (ISDN) and Broadband ISDN.

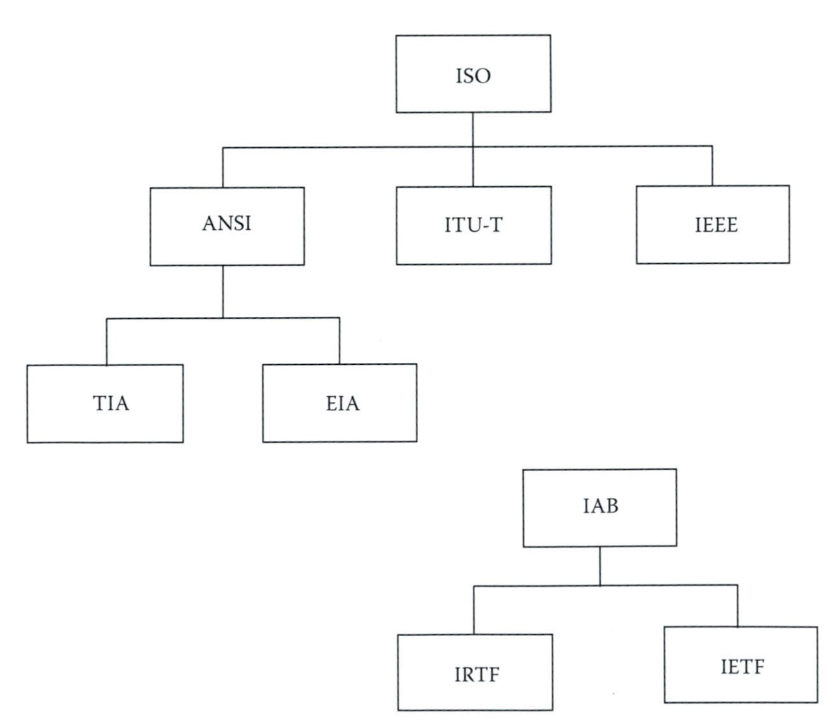

FIGURE 2.4 Different standard organizations in data and networking. (From W. Tomasi. *Advanced Electronic Communications Systems*, 6th Edition. Pearson Inc., New Delhi, India, p. 115, 2004.)

The **Institute of Electrical and Electronics Engineers (IEEE)** is the world's largest professional body in the fields of electronics, communication, and computer. It aims at enhancing product quality, advancing theory, etc., in the above fields. It oversees the current developments and their adoption as standards in the fields of communications and computing.

The **American National Standards Institute (ANSI)** is the official standards agency and has the voting right in ISO on behalf of the US government. It coordinates in the fields of computing and communications in the United States and is a nonprofit voluntary organization. People from communication industries, regulatory bodies, professional societies, are among its members.

The **Electronics Industries Association (EIA)** is also a nonprofit organization that recommends industrial standards. It has considerable contributions in the fields of physical connection interfaces and electronic signaling specifications for data communication. It is responsible for RS (recommended standard) in the above field.

The **Telecommunications Industry Associations (TIA)** is a leading trade association in the fields of communications and information technology. Convergence of new communication networks is facilitated by TIA. It provides its competencies for manufacturers of communication and information technology products to market their devices.

The **Internet Architecture Board (IAB)** has its name derived from Internet Activities Board. Advanced Research Projects Agency (ARPA), the research wing of the Department of Defense of the United States, was established in 1957. Its purpose was to take forward the technological advancements for military use. ARPANET was formed in the late 1970s to oversee such activities. It was renamed as Internet Activities Board in 1983. IAB is a technical advisory group of the Internet Society whose responsibilities include the following: (a) advises and guides the board of trustees and officers of the Internet Society with regard to technical, procedural, architectural policy-making processes concerning the Internet and its enabling technologies; (b) oversees the architectural protocols and procedures used by the Internet; (c) acts as representative of the Internet Society in close association with organizations that are responsible for standards, technical, and other issues pertaining to worldwide internet activities; (d) acts as an appeal board to remedy the improper utilization of internet standards; and (e) is responsible for the administration of various Internet-assigned numbers.

The **Internet Engineering Task Force (IETF)** has network designers, vendors, operators, and researchers in its fold and is responsible for the evolution of Internet architecture and the consequent smooth operation of the Internet.

The **Internet Research Task Force (IRTF)** promotes and oversees future short- and long-term researches in the areas of Internet protocols, architecture, applications, etc.

2.5 NETWORK TOPOLOGY

Network topology refers to the manner of interconnecting computers, cables, and components, both physically and logically. Physical connection refers to the actual laying of the network, while logical connection refers to how data actually passes from one node (station) to another.

The different topologies used for networking purposes are mesh, star, bus, ring, and hybrid. All these are examples of multidrop topology.

2.5.1 Mesh

In a mesh topology, each station or node is connected, via a point-to-point link, with every other node connected to form the mesh network. Here, a particular link carries data only between the two nodes with which the link is connected. To realize the above, each node in an n mesh network must be connected to the rest $(n - 1)$ nodes. Thus, a total of $n(n - 1)$ physical links are needed. However, for a full duplex mode, where communication takes place in either direction at the same time, a total of $n(n - 1)/2$ duplex mode links are needed. Figure 2.5 shows a mesh network consisting of five nodes.

There are several advantages associated with employing a mesh network. First, because of dedicated links, data can move from one link to another irrespective of system load. Second, because of dedicated links, data security/privacy is ensured. Third, the system is robust in the sense that a fault in a link does not affect the other parts of the system. Fourth, a faulty node can very easily be detected. Fifth, a traffic diversion can be done if a faulty node is detected.

Because of point-to-point connection between each and every node, considerable cable length is required and it increases largely as the number of nodes increases. On the one hand, it increases cost and, on the other hand, huge amount of cabling may lead to space problems.

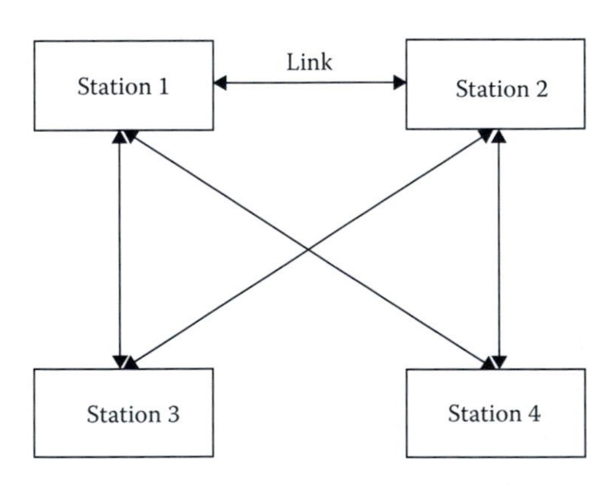

FIGURE 2.5 Mesh network.

Mesh topology is implemented in telephone regional offices where each such office is connected to each other office connected to the system.

2.5.2 Star

Unlike mesh topology, no point-to-point connection exists between any two nodes in the case of a star network. Instead, each station or node is connected to a central computer, called a hub. It acts as a multipoint connector. Thus, in this case, the point-to-point connection is between each station and the central computer only. Any data exchange between two nodes takes place only via the central computer. This is shown in Figure 2.6. The central computer or the hub has store-and-forward facility—this enables it to handle more than one information at a time.

Some advantages are associated with a star network. First, the system is robust in nature. That is, if a link fails, data exchange between that particular node and the hub fails—without affecting any other node connected to the system. Fault detection and isolation is very easy in this case. Second, less cabling leads to lesser cost. Third, reconfiguration is very easy in case addition/alteration/deletion of any node to the system.

If the central computer or the hub fails, the total star network fails. Again, depending on certain situations, more cable is needed instead of lesser cable length.

Star topology is normally used in local area networks (LANs).

2.5.3 Bus

In this case, there is a long cable—acting as the backbone from which small drop cables (lines) are connected to individual stations or nodes.

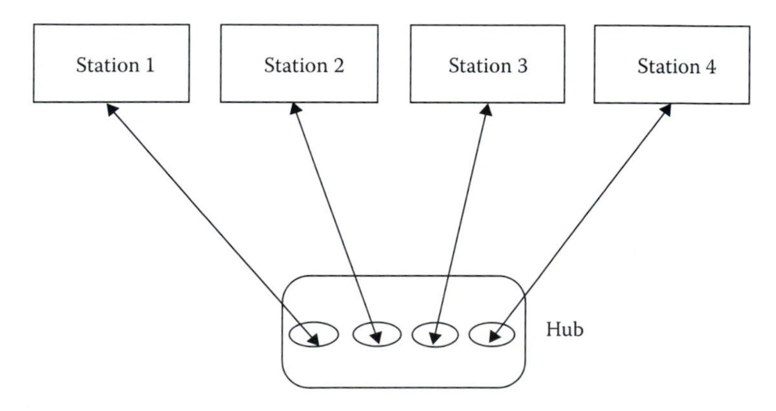

FIGURE 2.6 Star network. (From B. A. Forouzan. *Data Communications and Networking*, 4th Edition, Special Indian Edition. Tata McGraw Hill Companies Inc., New Delhi, India, p. 11, 2006.)

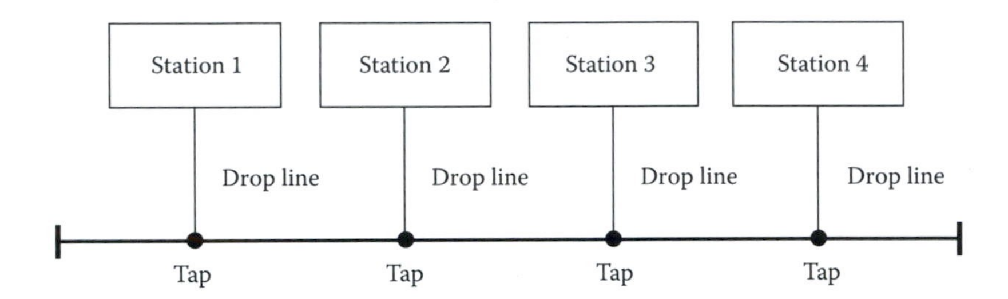

FIGURE 2.7 Bus network. (From B. A. Forouzan. *Data Communications and Networking*, 4th Edition, Special Indian Edition. Tata McGraw Hill Companies Inc., New Delhi, India, p. 11, 2006.)

This connection is made by the means of taps. As a signal travels along the backbone, it becomes weaker because of the presence of taps. This puts a limit to the number of taps that the bus technology can support and the distance involved. The scheme is shown in Figure 2.7. Bus topology is sometimes referred to as *linear* or *horizontal* bus. Bus topology normally involves a centrally placed host computer that controls data/information flow to and from other computers.

Bus topology involves deploying less cable lengths than those in star topology. Installation is very easy in this case. Fault detection/isolation/reconnection is a difficult proposition in case of bus topology. It is also difficult to add extra nodes to an existing system. There exists also a possibility of signal quality degradation at the taps because of signal reflection. Any fault or break in the trunk or bus cable leads to a total breakdown of the whole system. Bus topology used to be dominated in LANs and Ethernet LANs earlier, but is used much less today.

2.5.4 Ring

Ring topology has a dedicated point-to-point connection between any two successive devices. Stations are connected in series to form a closed loop. It is shown in Figure 2.8. Traffic in the ring is unidirectional. Information, meant for a particular station, is passed along the ring in one direction. This is not accepted by stations along the route for which it is not meant. Instead, they regenerate the signal with the help of the respective repeaters until it reaches the destination station or node. Ring topology, like bus and star topologies, normally has a centrally located host computer that controls data/information flow to other stations in the network.

It is advantageous to use ring topology because it offers easy installation and reconfiguration. Fault isolation is very easy. Whenever any addition/deletion of nodes are required, maximum ring length along with the maximum number of devices used in the ring topology are the only issues to be borne in mind.

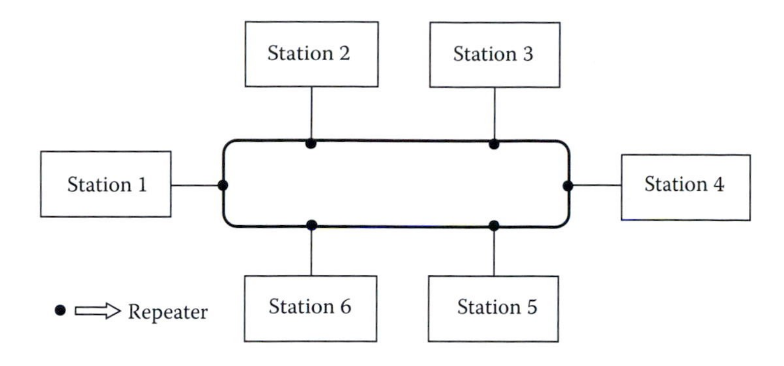

FIGURE 2.8 Ring network. (From B. A. Forouzan. *Data Communications and Networking*, 4th Edition, Special Indian Edition. Tata McGraw Hill Companies Inc., New Delhi, India, p. 12, 2006.)

Unidirectional traffic in a ring network is a major advantage of this topology. Any break at any place in the network will bring the entire network down. This can be overcome by employing dual cables (rings) or switches that have the capability of closing off a break.

In IBM's LAN networks, ring topology was used, but it is less prevalent because of high-speed LANs that are used today.

2.5.5 Hybrid

It is normally a mix of the topologies discussed already. A combination of star and bus topology is shown in Figure 2.9. It shows a network in which

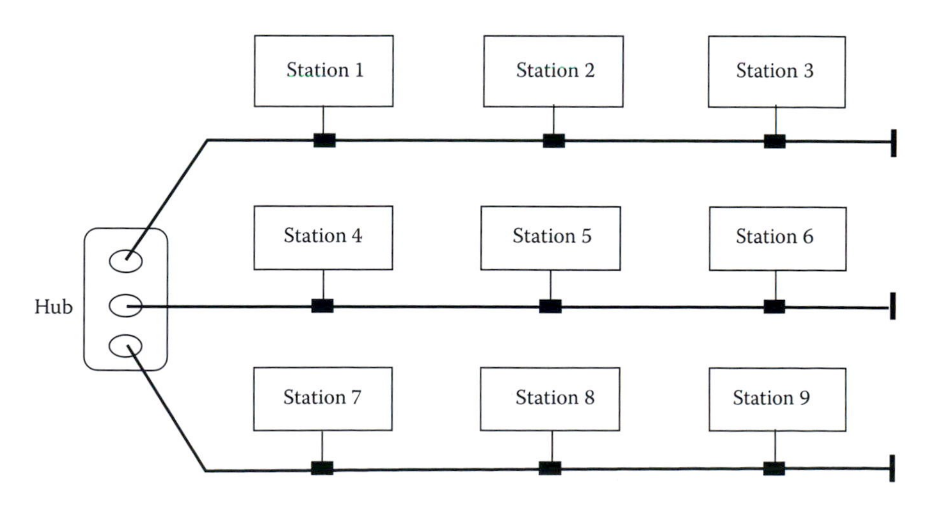

FIGURE 2.9 Hybrid network. (From B. A. Forouzan. *Data Communications and Networking*, 4th Edition, Special Indian Edition. Tata McGraw Hill Companies Inc., New Delhi, India, p. 13, 2006.)

three lines are star connected to the hub. Stations or nodes are connected to each of these three lines. A hybrid topology combines the advantages of both star and bus topologies.

2.6 NETWORK APPLICATIONS

Networking is all about sharing resources and data between computers over the communication links. The important parameters of a communication network are reliability of transmission, speed, data security, and performance.

A network must be designed with the specific application in mind. A network designed for a specific application would work nicely for that application category, but may not work so if the application profile changes. Thus, proper design of the network as per the application needs is of prime importance for expected performance to be obtained from the same. Table 2.1 shows the various network applications that have been categorized as per the different needs.

Network applications are categorized into connection-oriented and connectionless protocols. Considerable amounts of overhead in the form of encryption, error checking, and handshaking slows down the speed of data transmission rate in a major way for connection-oriented protocols while in the case of connectionless protocols, these reliability features are dispensed with, thereby increasing data transmission rate in a big way.

2.7 NETWORK COMPONENTS

Various network components help in proper operation of the network. These are servers, transmission media, shared data, clients, shared printers

TABLE 2.1
Different Networking Applications

Standard office applications	E-mail, file transfer, data storing, printing, etc.
High-end office applications	CAD, software development, video imaging, high-volume data transfer, etc.
Manufacturing automation	Process and discrete control, CNC machines, etc.
Mainframe connectivity	PCs, workstations, servers, etc.
Multimedia applications	Live interactive video, etc.

Source: W. Tomasi. *Advanced Electronic Communications Systems*, 6th Edition. Pearson Inc., New Delhi, India, p. 127, 2004.

and other peripherals, network interface cards (NIC), hardware/software resources, local operating system (LOS), and network operating system (NOS).

A server is a computer processor that provides specific service to the network. Users can access network resources by means of server. There are different kinds of servers that provide different services. Some of them are routing servers, gateway servers, terminal servers, web servers, printer servers, file servers, mail servers, communication servers, database servers, and a host of others.

A routing server connects nodes and networks having like architectures. Gateway servers connect nodes and networks having different architectures with the help of protocol conversions. A file server provides a copy of the file from the server to the client (user) who has requested the same.

Transmission media are also called links, channels, or lines. They can be coaxial cables, twisted wire pairs, or optical fiber, and are used to interconnect computers to the network.

Shared data are data used by the end users (computers) and provided by file servers. Examples include e-mail and printer access programs.

Computers stationed at either end of the network are the clients/users/customers who take the network facilities for proper end-to-end process data delivery. Clients request the requisite services from the servers and receive the same and execute data transfers.

Printers and peripherals are the hardware resources whose services are used by the clients at the behest of the servers. Printer software and data files are among the resources provided to the clients.

Each computer on the network has a NIC, whose purpose is to format, process, control, transmit, and receive data to and from the network. At the transmitter, the NIC passes the formatted data to the physical layer and then it is passed on to the physical layer. At the receiver, the NIC receives the formatted data from the physical link, strips the overhead from it, and processes the same as per its contents. NICs must have a driver to operate on the network to which they are connected. NICs manufactured by different manufacturers vary in speed, cost, and complexity. Each NIC has a 6-byte media access control (MAC) address, which is permanently embedded during manufacture. The MAC address is also known by node (station), hardware, physical, Ethernet, or LAN address.

LOSs permit computers to access files, CDs, and prints as per the needs. Some examples are MS-DOS, PC-DOS, Windows 95, Windows 98, and Windows 2000.

NOS is a software that allows computers to interact with the network. Thus, it is run on both computers and servers. A NOS provides password authentication, network administration functions, data file and printer access, etc. Some examples of NOS are IBM LAN Server, UNIX, Microsoft Windows NT Server, and Novell NetWare.

2.8 CLASSIFICATION OF NETWORKS

Networks are classified by their sizes. It encompasses the geographical area, speed, number of computers, media, and the physical architecture comprising the network. Initial primary classification includes LANs, metropolitan area networks (MANs), wide-area networks (WANs), and global area networks (GANs). Some more primary types of networks are building backbone, campus backbone, and enterprise network. In addition, personal area networks (PANs) and power line area networks (PLANs) are fast catching up in the area of networking. In PANs, data transfer takes place via the human body by simply touching them, while PLANs use existing alternating current distribution networks to transmit data from one place to another.

2.8.1 LANs

The IEEE defines LAN as "a datacom system allowing a number of independent devices to communicate directly with each other, within a moderately sized geographic area over a physical communications channel of moderate data rates." LANs are privately owned data communication networks in which around 10–50 computers share their resources with the help of file servers and are typically housed in a room or a building or a multistoried complex. Two PCs and a printer can make a LAN system. Geographically, a LAN can spread to a few kilometers with data rates in the range of 10 to 100 Mbps. Because of size limitations, transmission time on the LAN is known and can be used for its configuration in an effective way.

One of the computers comprising the LAN may serve as a server with software loaded in it, which can be downloaded by other computers as needed. Common topologies used for LANs are bus, ring, and star.

Some characteristics of a LAN are as follows: fully administered by the owner; the nodes on the network can share the information in a peer-to-peer manner, i.e., any station or node can send data to any other station; all stations must have full connectivity; it runs over a shared transmission medium—normally cabling.

2.8.2 MANs

The physical size of a MAN lies inbetween a LAN and a WAN. Normally, MANs cover an entire city geographically. Speeds range from 1.5 to 10 Mbps, and support both voice and data and sometimes also video.

A MAN could be seen as combining several LANs together to form larger network. It enables resource sharing between LANs or between stations lying on the same LAN network. An example of MAN is the cable TV distribution network. Large corporations use MANs to coordinate and share data between their different units housed at different places in the same city.

2.8.3 WANs

Wide-area networks (WANs) have a geographical area that covers an entire country or an entire continent. They operate at bit rates of 1.5 Mbps to 2.4 Gbps; cover a distance in excess of 1500 km; and support data, voice, and video.

WANs are generally implemented through service providers and follow circuit-switching technology. Services to systems connected to the WAN at different locations are made via routers. Examples of WANs are ISDN, ATM, X.25, frame relay, and T1 and T3 digital carrier systems.

2.8.4 GANs

The GAN provides data transfer between countries around the globe and Internet is an example of this. It is obviously a network of networks that encompasses the entire globe and operates between 1.5 Mbps to 100 Gbps.

2.8.5 Building and Campus Backbone and Enterprise Network

A building backbone carries data from different LANs from a single enterprise. It is forwarded to MANs, WANs, etc., via switches and routers.

A university campus normally has several buildings that have their own LANs. A campus backbone carries traffic from such individual LANs. The campus backbone operates relatively at a high speed to accommodate the high traffic volume generated from individual buildings, and they may be interconnected by fiber optic cables supported by switches and routers.

An enterprise network is an assortment of some or all of the previously mentioned networking schemes that are connected in such a manner that the enterprise network operates faithfully and reliably.

2.9 INTERCONNECTION OF NETWORKS

An internet is a few networks connected together that can communicate and share their resources and data traffic. On the other hand, the Internet is a public data communication network consisting of hundreds and thousands of networks, used by millions of users. Private individuals, public agencies, governmental organizations, university and research organizations, schools, colleges, airlines, railways, and libraries worldwide use the Internet to share data and information. Internet/intranet is executed by such browsers as Microsoft Explorer and Netscape Navigator. The Internet came into being only in 1969 at ARPA. Internet, intranet, and World Wide Web (WWW) have together created a virtual explosion in communication arena for data and information transfer at a very high speed with higher and higher traffic density. WWW is a server-based application that allows subscribers to have access services offered by the web.

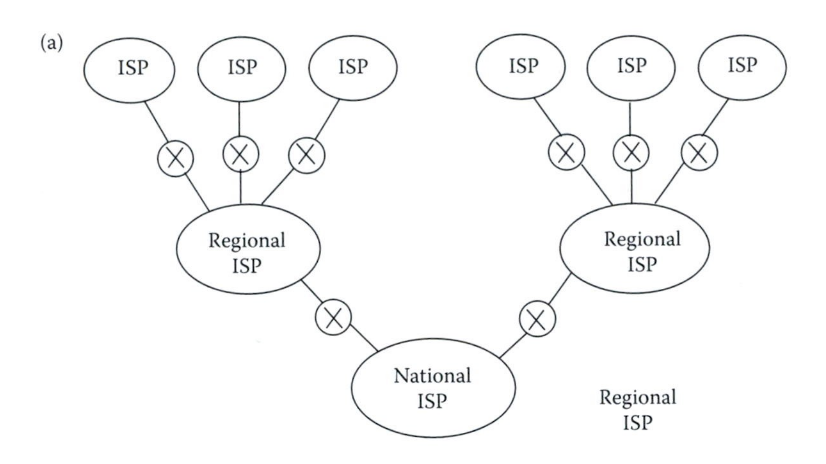

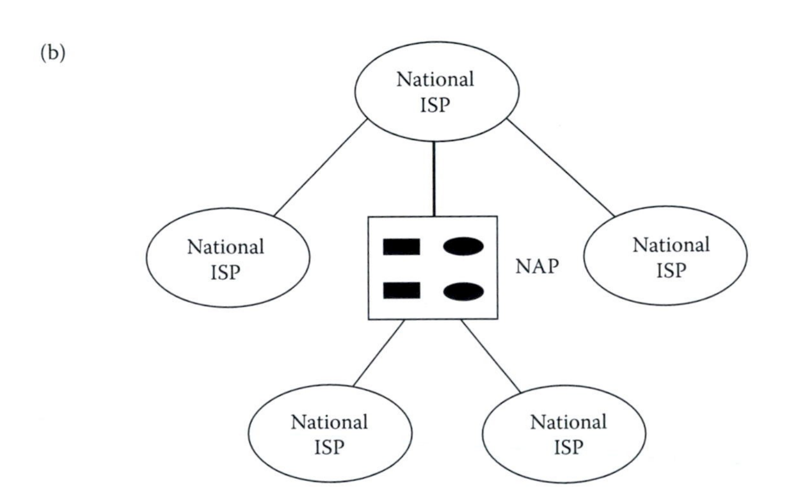

FIGURE 2.10 Hierarchical organization of the Internet. (a) Hierarchical structure of a national ISP and (b) interconnection of a national ISP. (From B. A. Forouzan. *Data Communications and Networking*, 4th Edition, Special Indian Edition. Tata McGraw Hill Companies Inc., New Delhi, India, p. 18, 2006.)

An accurate representation of the Internet is very difficult because of its continual change—some networks are always added/deleted to/from the Internet. The Internet, as today, consists of a complex hierarchical structure. At the lowest rung of this structure lie the LANs to which the end users are connected. The end users are connected to the Internet via the local Internet service providers (ISPs); successively comes the regional, national, and international service providers. The local ISP may be an agency providing local Internet connectivity or a nonprofit organization such as a university that runs, owns, provides, and maintains its own network. Figure 2.10 shows the hierarchical organization of the Internet.

3 Network Models

3.1 INTRODUCTION

A network is all about interconnecting two or more devices together so that data/information can flow either way. When the devices are from the same manufacturer, it is rather easy to interconnect them because they follow the same set of rules, specifications, and guidelines. Problems begin to surface when devices from different manufacturers are interconnected. Apart from hardware problems, software compatibility must be present for proper and reliable communication to take place.

"Closed" systems are those in which communication occurs when the devices, adapters, and others are from the same suppliers but fail miserably when devices from different manufacturers are interconnected because of incompatibilities in both hardware and software between different interconnecting devices. These are termed as "proprietary." On the other hand, an "open" system is one in which devices from different manufacturers can exchange information faithfully without encountering any kind of problem whatsoever. Thus, for such systems, the specifications and guidelines are "open" to all devices connected to the network. The set of protocols underlined in an open system allows any two different systems to communicate with each other.

All networks, be it standard, proprietary, or open, are defined by the "International Standard ISO/IEC 7498-1:1994 Information Technology—Open Systems Interconnection—Basic Reference Model: The Basic Model." This was first introduced in 1986. This model can be applied to all communication systems—from PCs to satellite systems.

3.2 THREE-LAYER MODEL

Before the introduction of the seven-layer OSI model, a three-layer model was conceptualized, involving an application layer, a transport layer, and a network access layer. Files existing in a computer, which may be transported to another computer, belong to the application layer.

Two levels of addressing are needed for an application residing in a computer to reach correctly to another computer. First, applications in a computer must have their own separate addresses (called the *service access point* [SAP]), which would enable the transport layer to support multiple applications in a computer. Again, each computer connected to the network must have a *unique address*. This enables the network to deliver data at the proper destination computer.

The application layer supports different applications with the help of software dedicated for these applications. A three-layer model involving four computers is shown in Figure 3.1. Each computer in the network has its own software to support application, transport, and network layers.

The transport layer helps in exchanging data from one computer to another reliably—irrespective of the nature of the application. It ensures delivery of data at the exact destination application. Control information existing at the transport layer ensures proper data delivery.

The network access layer, or simply network layer, helps in exchanging data between the computer and the network. This layer provides the address of the destination computer to which data is to be ultimately delivered. It is the responsibility of the network to route the data traffic on the network properly so that it reaches its destination. This means that the other two layers are not concerned about the specifics of the control software inherent in the network layer. Again the network layer does not know the SAP at which the data is to be finally delivered at the destination computer. Different softwares are used at this layer depending on the type of the network—circuit switching, packet switching, local area networks (LANs), etc.

Figure 3.2 shows how the application data passes through the transport and network layers by adding control information at the respective layers

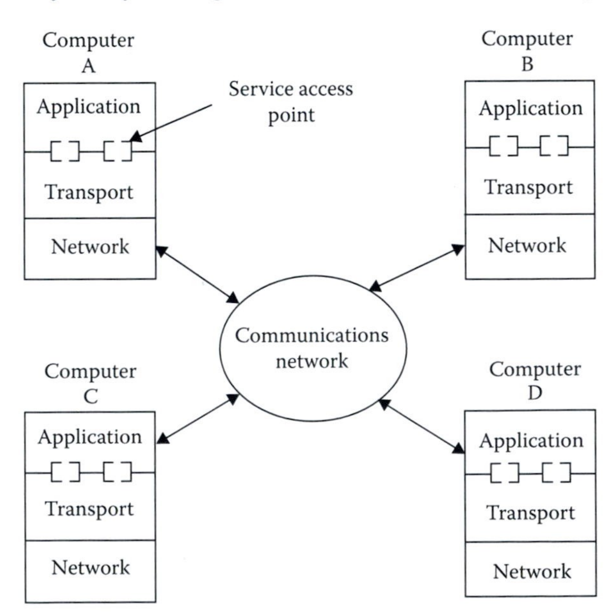

FIGURE 3.1 Three-layer communication model. (From W. Stallings. *Data & Computer Communications*, 6th Edition. Pearson Inc., New Delhi, India, p. 16, 2000.)

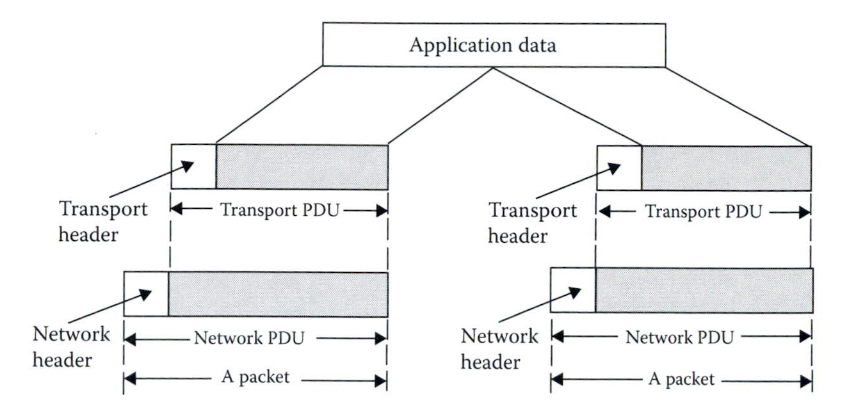

FIGURE 3.2 Transport and network PDUs. (From W. Stallings. *Data & Computer Communications*, 6th Edition. Pearson Inc., New Delhi, India, p. 17, 2000.)

(called headers). The protocol data unit (PDU) of a layer is the combination of control information (also called header) of the layer and the total block from just its upper layer. In Figure 3.2, a network PDU and a transport PDU are shown. Header information in the transport PDU includes a destination SAP, sequence number, and error detection mechanism. With the help of the destination SAP, the receiver computer can direct the received data at the correct application file. A sequence number is essential when the transport PDU is sending a series of the same. It helps in rearranging the received data in correct order in case they are received otherwise. Finally, a transport PDU includes a code that helps in detecting whether the received data has been received correctly or not. The receiver can then take action accordingly. Again the network PDU would include a network header that contains the destination computer address and some extra facilities, if needed. The network PDU, with the help of its header, guides the received data to the destination computer. The network header may include some extra facility that would help in prioritizing some message to be transported ahead of others.

3.3 OSI MODEL

An Open Systems Interconnection (OSI) reference model provides a common basis that aids in the development of systems interconnection standards. The model covers all aspects of network communication as envisaged in the International Standards Organization (ISO). ISO is an organization and not a model, whereas the OSI reference model is not a protocol or a set of rules but is an overall framework that forms the basis to define protocols. OSI has a layered architecture that facilitates design of network systems allowing communication between all types of computer systems. It consists of seven layers and is shown in Figure 3.3.

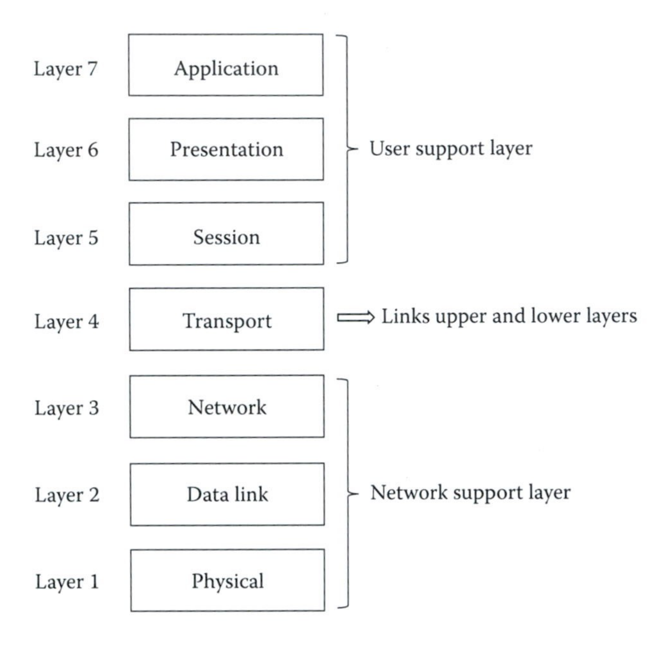

FIGURE 3.3 Seven-layer OSI model. (From B. A. Forouzan. *Data Communication and Networking*, 4th Edition, Special Indian Edition. Tata McGraw Hill Companies Inc., New Delhi, India, p. 30, 2006.)

The seven layers can be visualized to be divided into three subgroups. Layers 1, 2, and 3 are known as *network support layers*, while layers 5, 6, and 7 are called *user support layers*. The layer lying in between, i.e., layer 4, links the two subgroups. The network support layers help in moving data from one device to the other by taking care of physical connections, physical addressing, electrical specifications, transport timings, etc. The user support layers help in interoperability among software systems that are different from each other. The bottommost layer, i.e., layer 1, is mostly hardware based; other lower layers are a combination of hardware and software; while the upper layers comprise mostly of software.

Each layer of the seven-layer OSI model performs the following tasks, shown in Figure 3.4. While developing the model, similar types of networking functions were clubbed together and put into a specific layer. This way, the different layers were assigned different functions and architecture developed that is comprehensive and at the same time flexible. Since the functionalities of each layer are separate and well defined, standards can be developed independently and simultaneously of each other—thus speeding up the whole process of standardization. Again, since the layers are independent of each other, any change in standard in one layer would not affect the existing software in another layer.

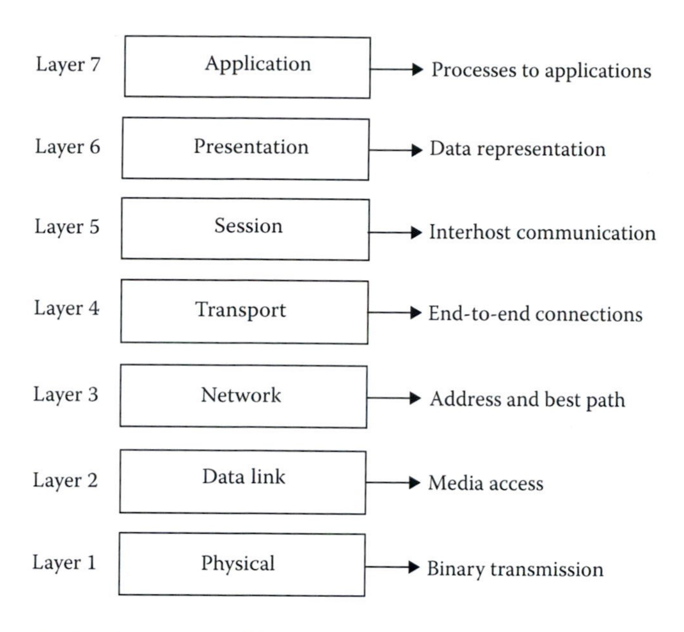

FIGURE 3.4 Tasks performed by seven-layer OSI model.

Source data is encapsulated in packet form, which starts at the upper layer and moves down the successive layers, adding control information in the form of header at each layer and also trailer at the data link layer. When the packet (with headers added at each layer and trailer) reaches layer 1, i.e., the physical layer, it is sent across a physical communication link that passes through the intervening nodes before finally reaching the destination station. This is shown in Figure 3.5. A layer on a device communicates with the same layer in another device. Such communications are based on a set of rules and conventions already agreed on. Such communications are termed peer-to-peer communications and shown in Figure 3.5. Figure 3.6 shows data exchange between two computers with headers and trailers placed at appropriate places in each layer.

3.3.1 Physical Layer

Figure 3.7 shows the mechanism of transporting data from the physical layer onto the physical medium. The physical layer receives data from the data link layer. It deals with the physical and electrical specifications of the interface and the medium, as also the functions and procedures that the physical devices and interfaces have to perform for transmission to take place.

Characteristics associated with the physical layer are as follows:

- It transforms the bits into signal, i.e., how the 0's and 1's are to be encoded for transmission over the physical medium.

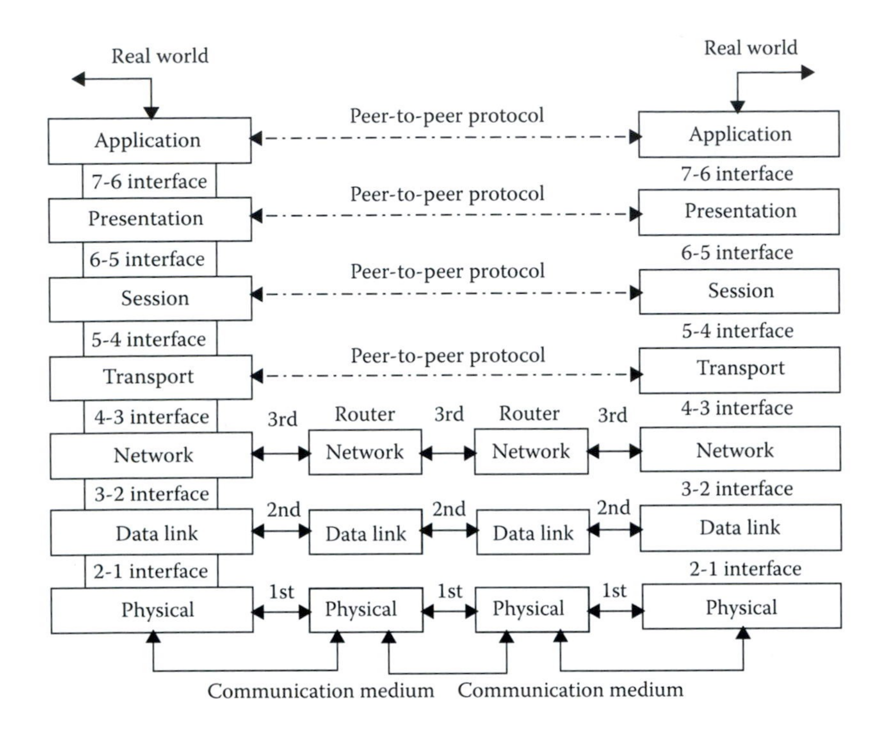

FIGURE 3.5 Data transfer between devices and peer-to-peer processes.

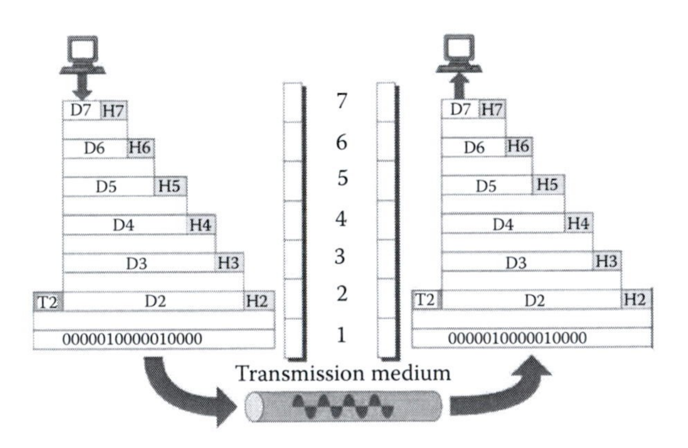

FIGURE 3.6 Data exchange using OSI model. (Available at www.eazynotes.com/notes/computer-networks/slides/osi-model.pdf.)

- It defines the data rate or transmission rate of the bits.
- It is the duty of the physical layer to synchronize the transmitter and receiver clocks.
- It defines the physical topology—i.e., how the devices are connected, viz., mesh or star or ring or hybrid.

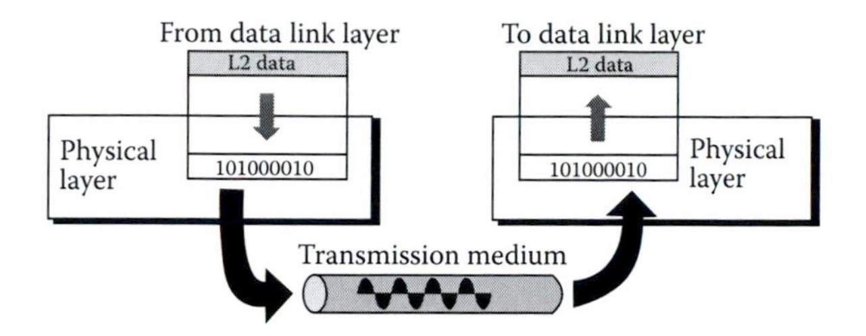

FIGURE 3.7 Operation of physical layer. (Available at www.eazynotes.com/notes/computer-networks/slides/osi-model.pdf.)

- Line configuration, i.e., either point-to-point or multipoint configuration of the devices is the responsibility of the physical layer.
- The physical layer is concerned with the mode of transmission, i.e., simplex, half duplex, or full duplex.
- The physical layer defines the characteristics of the interface between the devices and the transmission medium.
- The physical layer provides the necessary specifications for different types of hardware such as cabling, connectors and transreceivers, network interface cards (NICs), and hubs.

3.3.2 Data Link Layer

The data link layer is shown in Figure 3.8. Characteristics associated with the data link layer are as follows:

- It divides the whole message received from the network layer into smaller manageable data units—termed *frame*.
- It is the responsibility of the data link layer to move the frames from one node to another node (hop-to-hop).

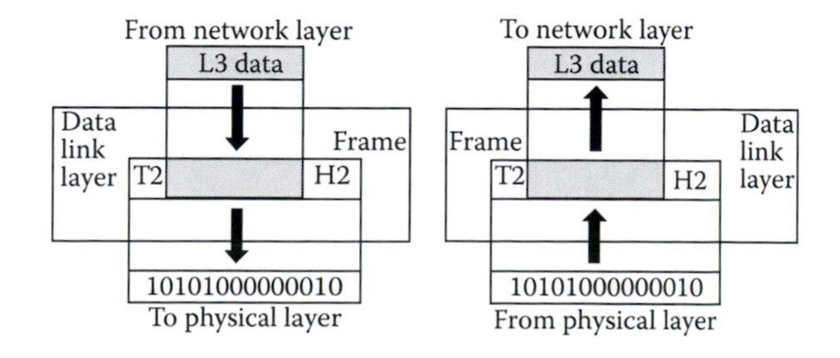

FIGURE 3.8 Operation of data link layer. (Available at www.eazynotes.com/notes/computer-networks/slides/osi-model.pdf.)

- It imposes a *flow control* mechanism when data produced by the sender is at a rate higher than the rate at which data can be absorbed by the receiver.
- The layer has an *error control* mechanism by virtue of which it can detect and retransmit damaged or lost frames. It also can recognize duplicate frames. It is achieved by adding a trailer at the end of each frame.
- The data link layer is subdivided into an upper sublayer called logical link control (LLC) and a lower sublayer called media access control (MAC). LLC is responsible for flow and error control. LLC ensures that protocols such as Internet Protocol (IP) can function regardless of the type of physical technology used. Multipoint access is resolved by MAC, i.e., MAC acts as a mediator. Technologies that are used to achieve the above are Carrier Sense Multiple Access with Collision Detection for Ethernet and token for token ring systems.
- The data link layer adds a header to a frame that needs to be distributed to different systems. Then the header includes the sender and receiver addresses. This is known as *physical addressing.*

3.3.3 Network Layer

The mechanism of data flow through the network layer is shown in Figure 3.9. The responsibilities of the network layer include the following:

- The network layer takes the responsibility for the source to destination delivery of a packet across multiple networks.

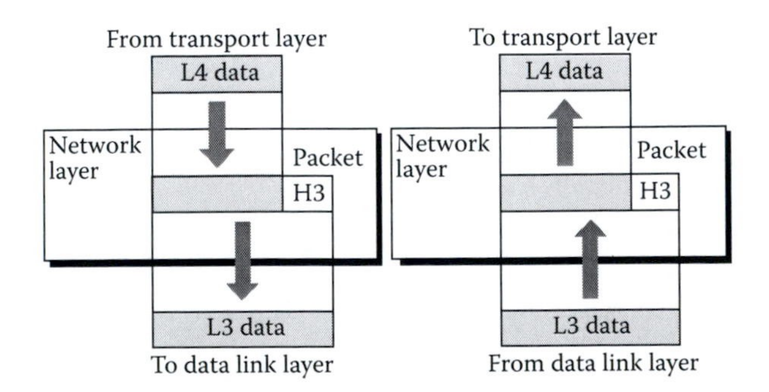

FIGURE 3.9 Operation of network layer. (Available at www.eazynotes.com/notes/computer-networks/slides/osi-model.pdf.)

- If a packet, residing in a network, is to be sent to another network, the network layer then adds the logical (network) address of the sender and receiver to each packet.
- Such addresses are assigned to local devices by a network administrator. This is assigned dynamically by a special server called Dynamic Host Configuration Protocol (DHCP).
- Several networks are connected by routers and switches to form a large network. The network layer identifies the best path to route the packets to the final destination.

Network layer protocols are IP and Novell's Internetwork Packet Exchange, although the latter is almost deprecated.

3.3.4 Transport Layer

The transport layer is shown in Figure 3.10. Responsibilities carried out by the transport layer are as follows:

- It ensures process-to-process delivery of an entire message.
- While the network layer treats each packet independently, the transport layer treats the entire message en masse and sees to it that all the packets are in order.
- The transport layer can be either connectionless or connection oriented. In connection-oriented transmissions, the receiving device sends an acknowledgment back to the source after a packet is received. This is not so for a connectionless transmission; thus, while the former is a slower transmission method, the latter is relatively faster.

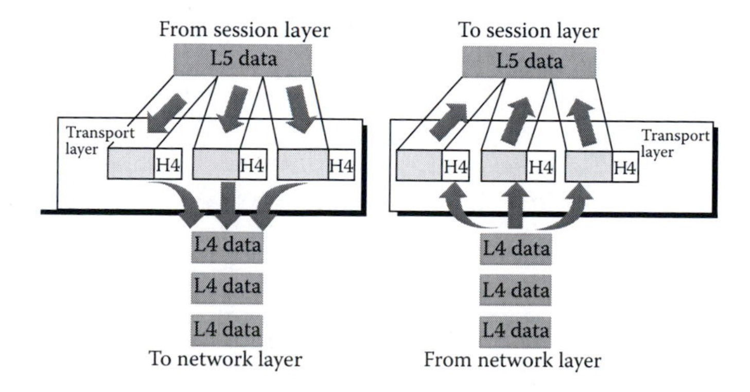

FIGURE 3.10 Operation of transport layer. (Available at www.eazynotes.com/notes/computer-networks/slides/osi-model.pdf.)

- A computer may run several processes at the same time. A transport layer header assigns a port address to each such process.
- The transport layer divides a message into segments, with each segment containing a sequence number. The sequence numbers enable the assembly of the message at the receiver. It also identifies and replaces packets lost in transmission.

Flow control in the transport layer is end to end rather than a single link.

Error control in the transport layer is process-to-process rather than across a single link.

Transport layer protocols include Transmission Control Protocol (TCP) and User Datagram Protocol (UDP). The former is connection oriented while the latter is connectionless.

3.3.5 Session Layer

Figure 3.11 shows the operation of the session layer. It acts as the *dialog controller* for the network. Its job includes establishing, maintaining, synchronizing, and lastly terminating the interaction among the devices that communicate with each other. If a session is broken, it attempts to retrieve the session.

Responsibilities carried out by the session layer include the following:

- Dialog control involves determining which of the two devices are to communicate data between themselves. Data sharing may be simplex, half duplex, or full duplex.

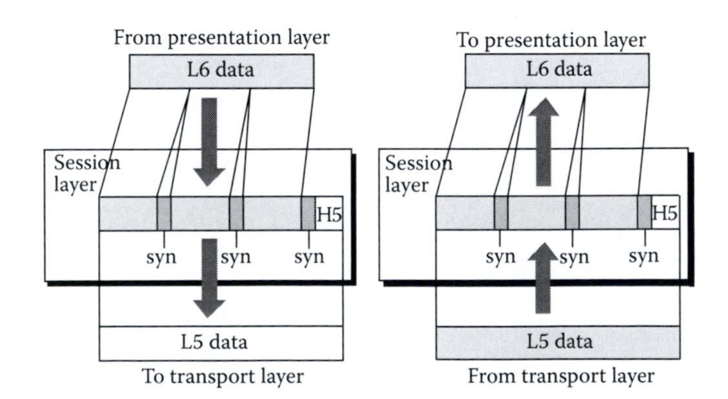

FIGURE 3.11 Operation of session layer. (Available at www.eazynotes.com/notes/computer-networks/slides/osi-model.pdf.)

- The session layer adds checkpoints, also called synchronization points, to a stream of data. For a huge amount of data, it adds checkpoints in between at predetermined intervals to ensure that data up to each consecutive checkpoint are received and acknowledged properly. The process of adding checkpoints and markers to a stream of data is called *dialog separation*.

3.3.6 Presentation Layer

Syntax and semantics of a data message are taken care of by the presentation layer. Figure 3.12 shows the operation of a presentation layer. Its responsibilities include the following:

- It is the responsibility of the presentation layer to ensure that data encoded differently by different computers are interoperable.
- Sensitive information, to be exchanged between the sender and the receiver, must be kept away from possible eavesdroppers. Data is encrypted in a manner that hides the information from such malafide data poachers. Decryption is done to transform the message back to its original form at the receiver.
- Data compression is a method that reduces the bit numbers contained in a data stream, without losing vital information. It is particularly applied in multimedia systems.

Presentation layer formats include text (ASCII, EBCDIC, RTF), images (JPG, TIF, GIF), audio (MP3, WAV), and movies (MPEG, AVI, MOV).

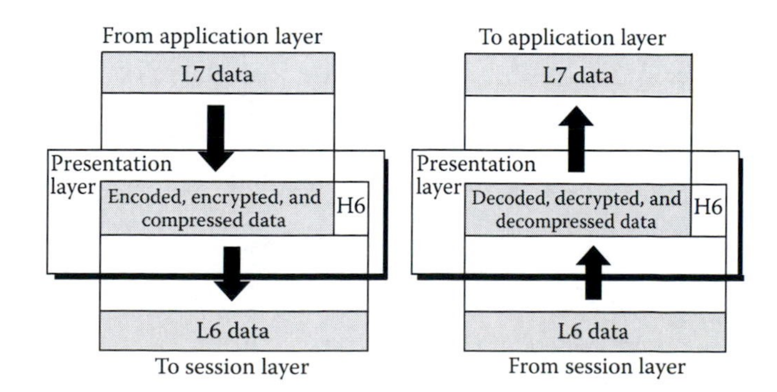

FIGURE 3.12 Operation of presentation layer. (Available at www.eazynotes.com/notes/computer-networks/slides/osi-model.pdf.)

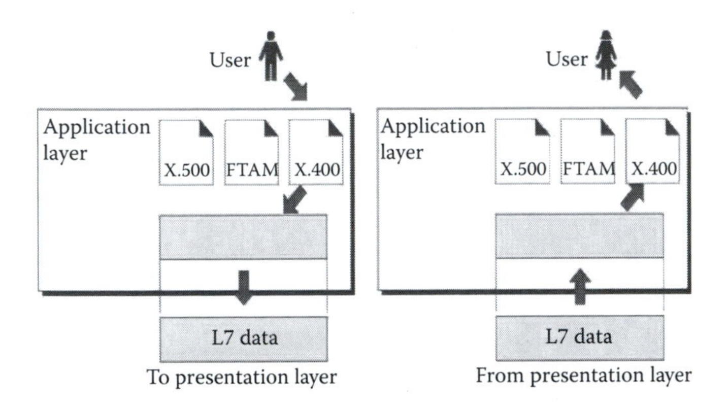

FIGURE 3.13 Operation of application layer. (Available at www.eazynotes.com/notes/computer-networks/slides/osi-model.pdf.)

3.3.7 Application Layer

Figure 3.13 shows the operation of the application layer whose main characteristics are given below:

- It provides user interface and supports various services such as e-mail, file transfer, and access to the World Wide Web.
- It supports e-mail service to any part of the world.
- It allows users to locate data from a remote location, retrieve the same, and use the same at the user's place.
- A user can log into a remote computer and use its resources.

3.4 TCP/IP PROTOCOL SUITE

The TCP/IP protocol suite was in vogue before the OSI model. The original TCP/IP had four layers and obviously did not match the layers of the OSI model. When the OSI model was introduced, there was a belief that it would overwhelm TCP/IP commercially, which it never did. Ultimately, during the 1990s, it became apparent that TCP/IP would dominate the commercial protocol architecture on which further developments would be based.

3.4.1 Introduction

TCP/IP was developed by the Department of Defense before the introduction of the seven-layer OSI model. It is the *de facto* global standard for the Internet. The Internet (earlier known as ARPANET) was a part of a military project of the Advanced Research Projects Agency (ARPA) and the communication model based thereon is known as the ARPA model.

ARPA was developed in the United States before the OSI model was developed in Europe by the ISO. Whereas the OSI model specifies exactly what function(s) each layer has to perform, TCP/IP comprises several relatively independent protocols that can be combined in several ways. Although TCP/IP and OSI were developed at different times by different bodies, they form the basis for data communications having different types and of different complexities.

It is not mandatory to use all the layers in the TCP/IP model; for example, some application-level protocols operate directly on top of IP. TCP/IP does not include the bottom network interface layer, but it depends on the same for access to the medium.

3.4.2 Protocol Architecture

The five-layer TCP/IP reference model is shown in Table 3.1.

The application layer corresponds to the upper three layers of the OSI model, i.e., application, presentation, and session layers. TCP residing in the transport layer ensures data delivery to the proper process. The network layer routes data from the host to the destination node via one or more networks, with the help of the IP addresses. The data link layer interfaces an end system with the network, while the lowest or physical layer is concerned with signal rate, signal encoding, etc.

Figure 3.14 shows the protocols available at different layers in the TCP/IP protocol suite.

At the different layers, different protocols are available. Among them, TCP and UDP belonging to the transport layer and IP in the network layer are the ones that form the basis for data delivery from one computer stationed at one end of the globe to another computer housed at the other end.

TABLE 3.1
Five-Layer TCP/IP Model

Layer No.	Layer Name	Purpose
5	Application	Specifies communication methodology between different processes/applications residing on different hosts
4	Transport	Provides end-to-end reliable data transfer
3	Network	Routes data between host and destination nodes through one or more networks connected by routers
2	Data link	Concerned with logical interface between an end system and a network
1	Physical	Concerned with physical transmission medium, signal encoding scheme, signal transmission rate, etc.

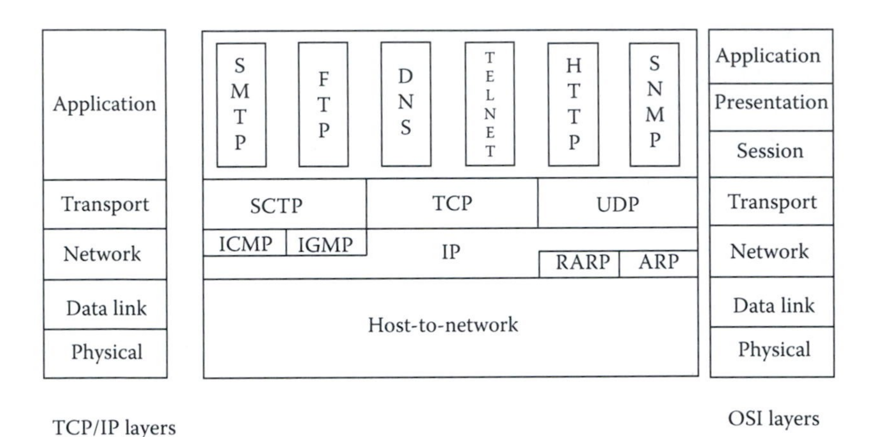

FIGURE 3.14 Protocols in TCP/IP protocol suite.

3.4.2.1 TCP

TCP is a connection-oriented transport layer protocol. Functionalities include reliable data delivery, congestion control, duplicate data suppression, and flow control. Most of the user application protocols, such as FTP and Telnet, use TCP. Two processes can communicate with each other through the TCP connection with the help of IP datagrams. This is shown in Figure 3.15.

TCP establishes a session between the transmitting process and the receiving process before initiating transmission. There are facilities to check that all packets have been received and arranged for retransmission, in case of packet loss. These involve additional overhead and lead to higher

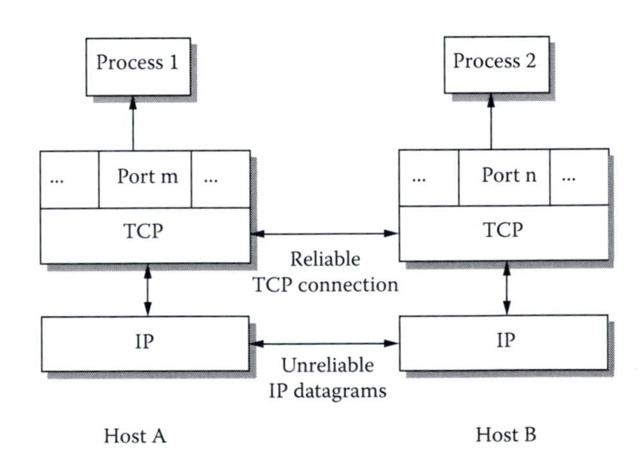

FIGURE 3.15 Connection establishment between two processes by TCP. (Available at www.redbooks.ibm.com/redbooks/pdfs/gg243376.pdf.)

processing time and header size. However, at the same time, it makes the system more reliable.

TCP fragments a large chunk of data into smaller segments when necessary, numbers the segments, reassembles the whole message, detects and arranges for retransmission in case of failure, issues acknowledgments for data received, and provides socket services for multiple connections to ports on remote hosts. The segmented messages may be received out of order at the receiver, which the TCP reassembles in correct order.

Each and every TCP segment has a header that comprises all the necessary information for proper data delivery and retrieval. The TCP header format is shown in Figure 3.16.

The sending and destination ports are each 16 bits. While the former identifies the sending end host, the latter identifies the receiving host. A destination host is identified by its IP number, and the process on the host is identified by its port number. The IP number in conjunction with port number is called socket.

The sequence number, consisting of 32 bits, ensures the sequentiality of data stream that is sent by the sending end host. It specifies the first byte of the user data in the segment. During the initial setup, it represents the Initial Sequence Number (ISN) that identifies the first byte of data in every segment of data sent by TCP. For a particular connection, ISN ensures that data is reassembled in correct order at the receiving host.

<table>
<tr><td colspan="3">Source port (16)</td><td>Destination port (16)</td></tr>
<tr><td colspan="4">Sequence number (32)</td></tr>
<tr><td colspan="4">Acknowledgment number (32)</td></tr>
<tr><td>Offset (4)</td><td>Reserved</td><td>Control bits (6)</td><td>Window (16)</td></tr>
<tr><td colspan="3">Checksum (16)</td><td>Urgent pointer (16)</td></tr>
<tr><td colspan="4">Options and padding</td></tr>
<tr><td colspan="4">Data</td></tr>
</table>

FIGURE 3.16 TCP header format. (From N. Mathivanan. *PC Based Instrumentation: Concepts and Practice*. Prentice Hall of India, New Delhi, India, p. 549, 2007.)

Acknowledgment number is 32 bits in length and is "sequence number + 1" of the last successfully received data byte. The sending host, while transmitting, sets a timer and if the acknowledgments are not received within the specified time, an error is assumed and the data is again retransmitted.

Data offset is 4 bits in length and indicates where the data begins. The six flags, each of 1 bit, control connection and data transfer. These are as follows:

1. URG: Indicates that the header contains the valid urgent pointer.
2. ACK: Indicates that the header contains a valid acknowledgment number.
3. PSH: It is push function forcing TCP to promptly deliver data.
4. RST: Exercising this causes the connection to be reset.
5. SYN: It ensures that sequence numbers are synchronized and handshaking operations occur.
6. FIN: Indicates that there is no more data from the sender.

The window size is 16 bits and provides flow control. The checksum field is used for error control, while the urgent pointer field, in association with URG flag, can insert a block of "urgent" data at the beginning of a segment.

3.4.2.1.1 *Window Principle*

In TCP, sequence number is assigned to each byte in the data stream, giving rise to a segment and they are sent one after the other. The window principle is applied at the byte level in case of TCP, the segments sent, and the ACKs received. The window size is expressed in terms of number of bytes instead of number of packets.

The receiver determines the window size, and it is a variable one during data transfer. An ACK sent by the receiver is representative of the window size the receiver is capable of handling at that instant of time.

A sender's data stream at any given instant can look like that shown in Figure 3.17.

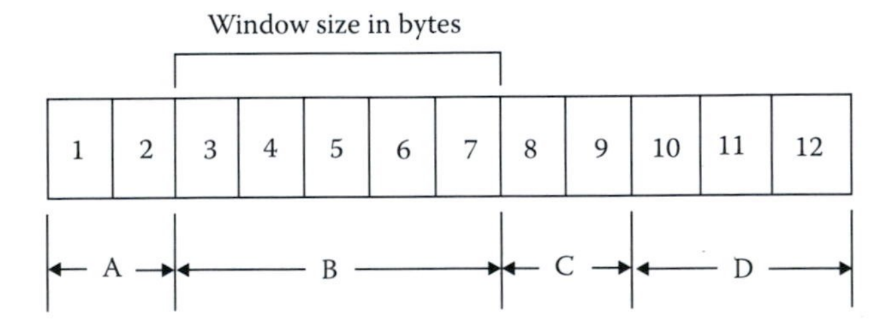

FIGURE 3.17 TCP window principle. (Available at www.redbooks.ibm.com/redbooks/pdfs/gg243376.pdf.)

In the figure

- A represents the number of bytes transmitted and acknowledged.
- B represents the number of bytes sent but yet to be acknowledged.
- C represents the number of bytes that can be sent without waiting for acknowledgment.
- D represents the number of bytes that cannot be sent at present.

3.4.2.1.2 Congestion Control

TCP uses a very helpful congestion control mechanism that can overcome any possible overwhelming of the receiver by the sender. This is a possibility for slow WAN links. TCP congestion control algorithms can adapt the sender to the network capacity at any point of time and thus avoid any potential congestion situation. TCP follows a variety of congestion control algorithms to avoid congestion: *slow start*, *congestion avoidance*, *fast retransmit*, and *fast recovery*.

Slow start is a process in which, once the connection is established, a sender can send multiple segments depending on the window size as advertised by the receiver. This is fine when two hosts are on the same LAN, but poses a serious problem if there are routers and slow links in between. The consequences are that packets are dropped, performance is degraded, and the situation calls for retransmission.

The slow start algorithm avoids such a situation. It operates on the principle that packets can be injected into the network only at a rate acknowledgments are received from the sender. The slow start algorithm adds another window to the sender's TCP. This is called congestion window, abbreviated *cwnd*, and is initialized to one segment only. Each time an ACK is received by the sender, it increases its congestion window by one. The sender has the option to send the lower value as imposed by the congestion window or the advertised window as advertised by the receiving host. The former is controlled by the sender and the latter by the receiver.

The congestion avoidance algorithm is based on the assumption that packet loss due to congestion is very small. Packet loss may be ascertained from the following two conditions: if a time out occurs or duplicate ACKs are received. Although the slow start and congestion avoidance algorithms are independent of each other, loss of packet leads to implementation of both algorithms at the same time. Thus, another algorithm, called a slow start threshold size, abbreviated *ssthresh*, is used along with *cwnd*.

In fast retransmit algorithm, TCP does not wait for time out to resend the lost segments.

Fast recovery algorithm is performed by fast retransmit. It is undertaken to allow for high throughput under moderate congestion—particularly for large windows. The first retransmit and fast recovery algorithms are normally implemented together.

3.4.2.2 UDP

UDP is a connectionless protocol. It does not require any connection establishment before data transmission. UDP requires no sequence numbers, no timers, no synchronization parameters, no retransmission of data packets, and no priority options. Thus, it has very little overhead. Its major drawback is that it does not guarantee delivery. UDP is normally used for broadcasting, general network announcements, real-time data, etc.

The UDP header is shown in Figure 3.18. It has only four fields.

The source port is an optional one. When it is used, it indicates the port address of the sending process. When not used, a value of zero is inserted for this field. The destination port indicates the process to which the data is to be delivered. The "length" is the length in bytes of the used datagram, including the header. The checksum is an optional 16-bit field, used for validation purposes.

3.4.2.3 IP

The internetworking layer in TCP/IP has some very important protocols: IP, Internet Control Message Protocol (ICMP), Address Resolution Protocol (ARP), and Dynamic Host Configuration Protocol (DHCP). They together perform datagram addressing, routing, delivery, dynamic address configuration, and resolving between internetwork layer addresses and network interface layer addresses.

IP is an unreliable, connectionless, and best-effort packet delivery protocol. Best-effort delivery means that packets sent by IP might be lost, may reach out of order, or even may get duplicated. It is the responsibility of the higher layer to address these concerns. Connectionless network protocol is

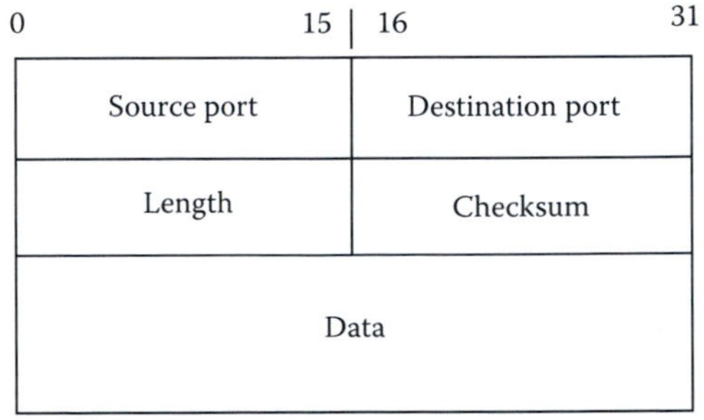

0 — 15	16 — 31
Source port	Destination port
Length	Checksum
Data	

FIGURE 3.18 UDP header format. (From S. Mackay et al., *Practical Industrial Data Networks Design, Installation and Troubleshooting*. Newnes An Imprint of Elsevier, U.K., p. 272, 2004.)

used to minimize the dependence on specific computing centers that uses hierarchical connection-oriented networks.

IP addressing is a must to identify a host on the Internet. Thus each host is assigned an *IP address* or an *Internet address*. A host is recognized by this IP address. A host may be connected to more than one network, called *multihomed*, in which case the host must have a separate address for each network interface.

IP addresses are represented by a 32-bit unsigned binary value and are expressed in a dotted decimal format. Each IP address consists of *a network number* and a *host number*. The network number is administered by one of the three Regional Internet Registries: American Registry for Internet Numbers, Reseaux IP Europeans, and Asia Pacific Network Information Centre. For example, 128.3.7.8, is an IP address, where 128.3 represents the network number while 7.8 represents the host number. Sometimes terms such as *network address* or *netID* are used instead of network number, while *host address* or *hosteID* are used for host number.

An IP datagram (the basic data packet exchanged between hosts) contains a *source IP address* and a *destination IP address*. For a datagram to be sent to a destination IP address, it must be translated or mapped into a physical address. For example, in LANs, the IP address is translated into a physical MAC address by the ARP.

There are five classes of IP addresses: A, B, C, D, and E. Class A address is suitable for an extremely large number of hosts, class B for a small number of hosts, and class C for medium-sized networks with moderate number of hosts. Class D is used for multicasting, while class E is for future or experimental use. Delivery of a datagram using IP addresses can be any of the following types: unicast, broadcast, multicast, or any cast. This is shown in Figure 3.19. A connectionless protocol can send a message in any

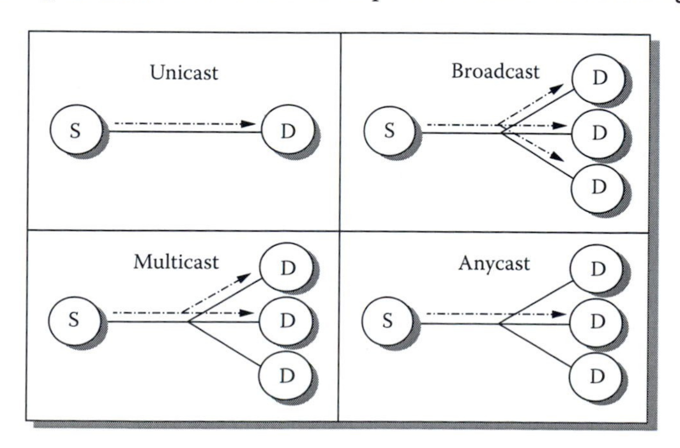

FIGURE 3.19 Different packet delivery modes. (Available at www.redbooks.ibm.com/redbooks/pdfs/gg243376.pdf.)

of the above four methods, while a connection-oriented protocol can only use unicast addresses.

IP version 4 (IPv4) is responsible for delivery of data packets (datagrams) between the sending host to the receiving host. Different systems, for example, Ethernet, can handle 1500 bytes while X.25 can handle 576 bytes. Because of this frame size limitation, a message is broken up into fragments, called datagrams. Each datagram from an entire message is given an IP header, which is then sent from the sending host. The receiving host rebuilds the message from the datagrams received. The IPv4 header consists of at least five 32 bit "long words," i.e., 20 bytes in all, and is shown in Figure 3.20. This IP header is appended to the information that it receives from higher-level protocols.

The "ver" field is 4 bits long and indicates the IP protocol version in use (in this case, it is four). The 4-bit IHL (Internet header length) indicates the length of the IP header. This header length is not a fixed one. The ToS (type of service) field is 8 bits long and its different bit positions correspond to minimizing delay and monetary cost and maximizing throughput and reliability. The "total length" corresponds to the length of the datagram, measured in bytes. This field, along with IHL, determines where

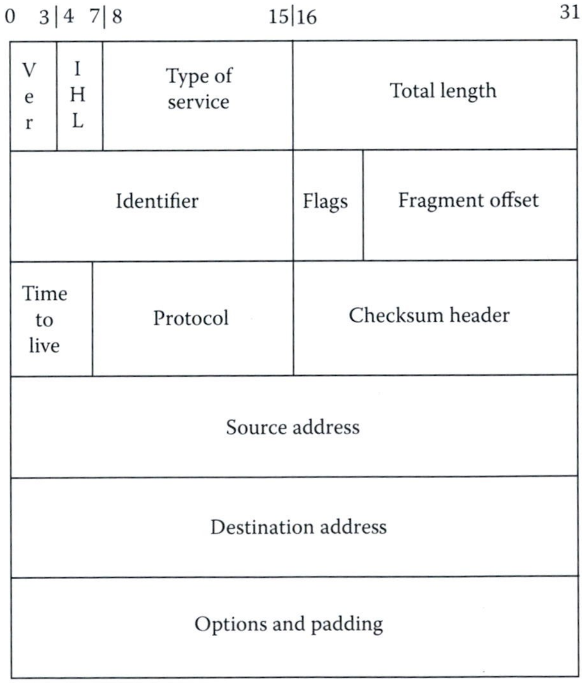

0 3	4 7	8 15	16	31
Ver	IHL	Type of service	Total length	
Identifier			Flags	Fragment offset
Time to live	Protocol		Checksum header	
Source address				
Destination address				
Options and padding				

FIGURE 3.20 IPv4 header. (From S. Mackay et al., *Practical Industrial Data Networks Design, Installation and Troubleshooting*. Newnes An Imprint of Elsevier, U.K., p. 260, 2004.)

data starts and ends. Maximum datagram length can be $2^{16} = 65,536$ bytes, although such long datagram length is impractical. Datagram lengths up to 576 octets are allowed.

The 16-bit identifier uniquely identifies each datagram sent by the host with its value incremented by one each time a datagram is sent. When fragmentation of a message is necessary, the identifier is appended to each successive fragment in order to retrieve the datagram correctly at the receiver. The 3-bit flag field follows the identifier field and has two flags. These two flags are used in the fragmentation process with DF (do not fragment) set to 1 by the higher-level protocol, in case IP is not allowed to fragment a message. When fragmentation is done, MF = 1 indicates that more fragments follow. The last fragment has MF = 0 in its flag field.

The fragment offset is 13 bits long and indicates where in the original datagram a particular datagram belongs. The TTL (time to live) sees to it that datagrams that cannot be delivered are ultimately destroyed. The protocol field is 8 bits long and indicates the next higher-level protocol header present in the data portion of the datagram. The checksum header is a checksum on the header portion only. Both source and destination addresses are 32 bits long each and represent the origin host and target host, respectively.

3.4.3 Operation

The TCP/IP protocol suite helps in sending a message from a process associated with a port residing at a host to another process associated with a port at a second host. The receiving host may reside on the same network or on another network. For the latter case, the message has to pass through several routers along its passage to the final destination. It should be kept in mind that IP is implemented in all the end systems and the routers, while TCP is implemented only in the end systems.

As already mentioned in Section 3.2, two levels of addressing are needed for a process data in one host to be sent to another process in another host. A local port address is needed, which would ensure correct data delivery at the process at the receiving host. Again a network address is needed, which would enable the message to be delivered to the receiving host.

Let's say a process data residing at a port belonging to a host is to be delivered at another process having its own port address and belonging to another host. The sending end process hands the message down to TCP. It has instructions to send the same to the second host at the particular port. TCP hands over the message to IP with instructions to deliver the same to the second host. IP is remaining to be totally transparent about the port address of the destination host. All these are managed by control information appended to the message at each layer of the TCP/IP protocol suite.

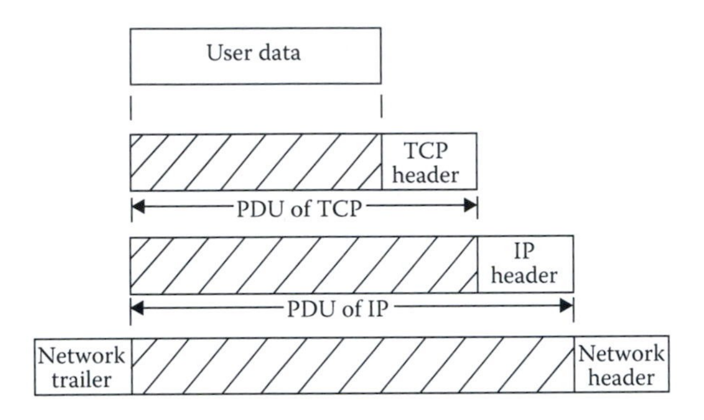

FIGURE 3.21 Data encapsulation and PDUs in TCP/IP architecture.

3.4.4 PDUs in Architecture

Control information, in the form of headers and trailers, is appended to the message at the different layers to ensure proper data delivery at the proper destination with utmost reliability. Figure 3.21 shows the data encapsulation and PDUs in the TCP/IP architecture. At the TCP layer, control information, in the form of TCP header, is appended forming a TCP segment or a PDU of TCP. Control information that is included in the TCP header includes the destination port address, sequence number, and checksum. Similarly, at the network layer, an IP header is added, giving rise to PDU of IP. The IP header includes the destination network address. It may include some other control information, such as priority in data delivery.

3.4.5 Addressing

Addressing through the TCP/IP protocol involves sending data from one process to another via the Internet. The addressing involves the following categories: *physical* or *link*, *logical* or *IP*, *port* and *specific*. The addresses refer to specific layers in the TCP/IP model and shown in Figure 3.22.

3.4.5.1 Physical

The physical or link address is the lowest-level address. It is the address of a station or node specified in its frame by LAN or WAN. Depending on the type of the network, the size and the format of the address vary. Ethernet uses a 6-byte address that is embedded in its NIC.

3.4.5.2 Logical

Logical address corresponds to the network layer in the TCP/IP model. The physical address may vary depending on the type of the network. Logical address overcomes this difficulty by recognizing a host irrespective of the

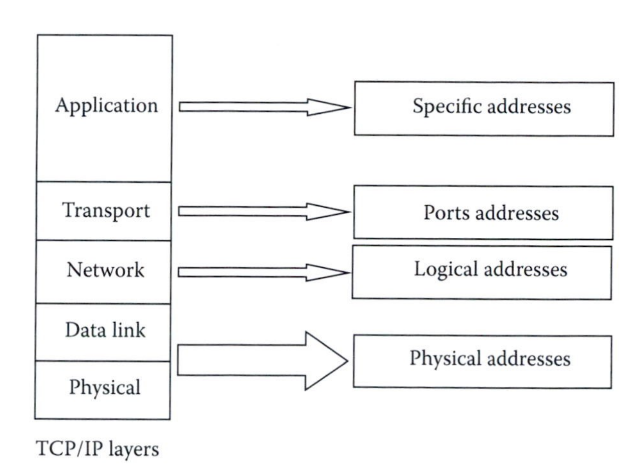

FIGURE 3.22 Addresses and their corresponding layers in TCP/IP model.

physical address type. At present, a 32-bit logical address can uniquely recognize a host connected to the Internet. No two IP addresses can be same so that two different hosts can be differentiated and recognized with their logical addresses.

3.4.5.3 Port

A computer may run several processes at the same time. It may communicate with a second computer via a file transfer protocol, message handling services, or TELNET. Thus, these processes residing on a computer must have individual separate addresses for them to receive data from other computers simultaneously. This is taken care of by port addresses. A port address is 2 bytes in length. Thus, it is the port address on a computer that helps in exact data/message delivery meant for a particular process once it has reached its destination host.

3.4.5.4 Specific

Specific addresses are user-friendly addresses such as an e-mail address or a URL (Universal Resource Locator). An e-mail address locates a particular recipient in any part of the world, while a URL helps in locating some document/writing/information available from the World Wide Web.

4 Networks in Process Automation

4.1 INTRODUCTION

Over the last two decades, the processing speed of computers has increased many folds and, along with it, the associated networking technology has improved beyond comprehension. Initially, it was local networking with only educational institutions and large business conglomerates sharing information among themselves. The Internet came in handy to share data/information globally in a reliable, secure, and fast manner. Industrial communication is all about transporting field information to the control room and controlling a process reliably and effectively. There are several network protocols that include both hardware and software to ensure robust, reliable, and sometimes real-time operations depending on the situational needs. Generically, industrial networks are referred to as a fieldbus, which includes sensor bus and device bus—the reference to a particular bus depends on the data size that is handled in one go.

There is a growing demand among the users of industrial automation systems for an industrial protocol that is vendor independent. Some key factors determine the acceptability of an industrial network protocol. Some of the key issues are as follows: an open system, reduction in the cost of wiring, and increased information availability from the field devices.

4.2 COMMUNICATION HIERARCHY IN FACTORY AUTOMATION

In an industry, be it a manufacturing or process one, data or information flows from field to the upper layers up to the management/enterprise level and vice versa. For an orderly flow of information and to optimize the performance of the process, the whole setup is divided into several hierarchical levels. This is shown in Figure 4.1.

The lowest or the field level comprises sensors, positioners, and actuators. The sensors give an indication about the process conditions, such as temperature, pressure, and flow of the process—it can be digital, analog, or hybrid as in Highway Addressable Remote Transducer (HART). The second

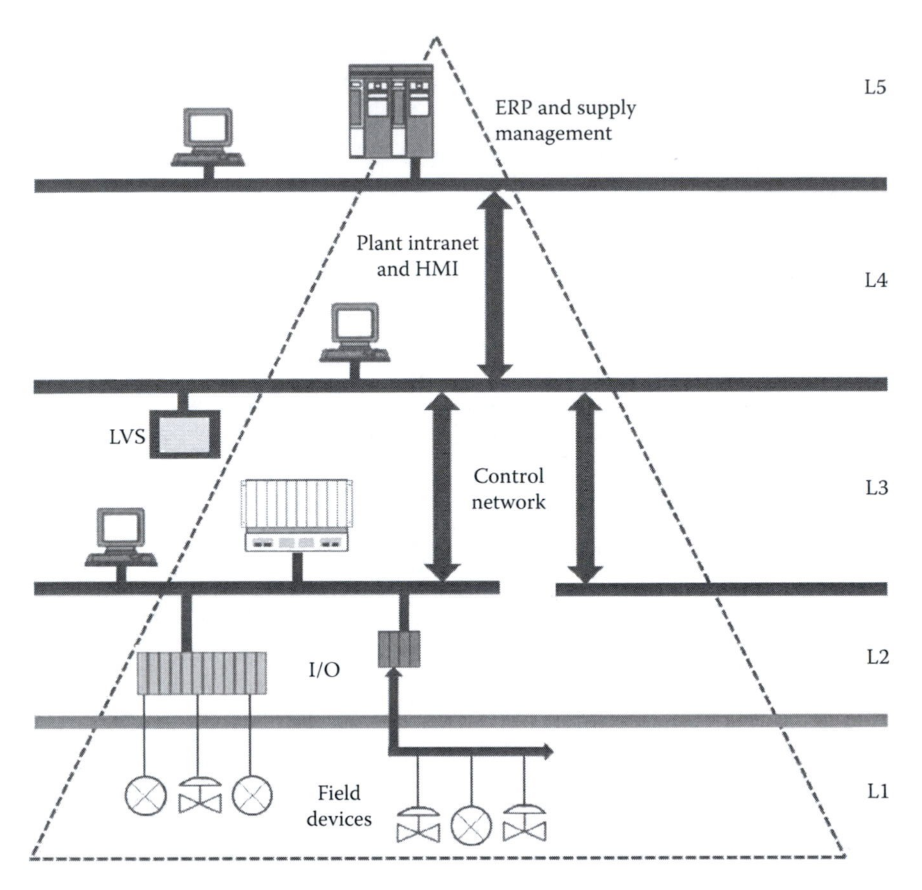

FIGURE 4.1 Hierarchical levels in process industry.

or the I/O level marshals the inputs and the outputs together. The sensor output, via the I/O level, is directed to the controller, which then generates the appropriate control signal, and it is then fed back to the actuator for control of the process. At the control level, normally programmable logic controllers (PLCs) or distributed control systems (DCSs) are used, which generate the control signals. These signals are sent to the sensor/actuator level where as per the signal received, the control action is taken—may be opening or closing of valve, starting or stopping of motors, etc. Network engineering tools, plant maintenance schedule programs, and asset management are among the jobs of the plant intranet/human–machine interface (HMI) level. The HMI level ensures that any process variable value from any place in the plant can be displayed on the operator console, can warn the operator in case any process value undershoots/overshoots predefined set limits, can change the device configurations, etc. At the topmost enterprise/supply chain management level, the information flows into the office environment for record keeping; billing via a service access point; ordering of raw materials; keeping a record of quality and quantity of finished products, etc.; and keeping a record of present stock position of different raw materials to keep the production going.

The different levels have to handle different requirements peculiar to that particular level concerned. For instance, the enterprise level has to handle a large volume of data that are neither time critical nor in constant use.

The lower three levels have some different characteristics, such as constant use, short data packet length and response time, and deterministic communication. Of course there are differences between manufacturing and process automation in these three layers. Whereas in process automation, speed is less and the signal contains more status information, the high speed of transmission and short data lengths characterize the manufacturing industries.

Because of characteristic requirements existing at different levels in the hierarchy, it is obvious that no single protocol can address all these. Speed of data transmission, data volume, and data security are the major concerns of these levels. Irrespective of whether it is manufacturing or process industry, data must be made available at the HMI and enterprise levels. For efficient and reliable data exchange between the different levels, standardization in the form of the International Standards Organization–Open Systems Interconnection (ISO–OSI) reference model is used to overcome the above issues and concerns.

4.3 I/O BUS NETWORKS

The decentralized control and distributed intelligence in the smart field devices have given rise to I/O bus networks. Data and status information are communicated through the I/O bus, for example, the on/off state of a switch, and the condition of a machine or process data are sent via the I/O bus network—be it manufacturing or process industry. Figure 4.2 shows the interconnections between a supervisory PLC, local PLCs, a local area network (LAN), and an I/O bus network.

The process bus network and the device bus network are connected to a PLC. Remote I/Os are connected to other PLCs, as shown. These PLCs are connected to a LAN, which, in its turn, is connected to a supervisory PLC and then taken to the other higher layers such as HMI and enterprise levels.

Field devices connected to the I/O bus networks must be intelligent in nature and such a field device has, within it, a sensor along with its processing circuitry, a microcontroller, transmitter/receiver network, and a power module. This is shown in Figure 4.3.

4.3.1 Types

I/O bus networks are divided into two categories: device bus networks and process bus networks. Device bus networks interface with low-level information

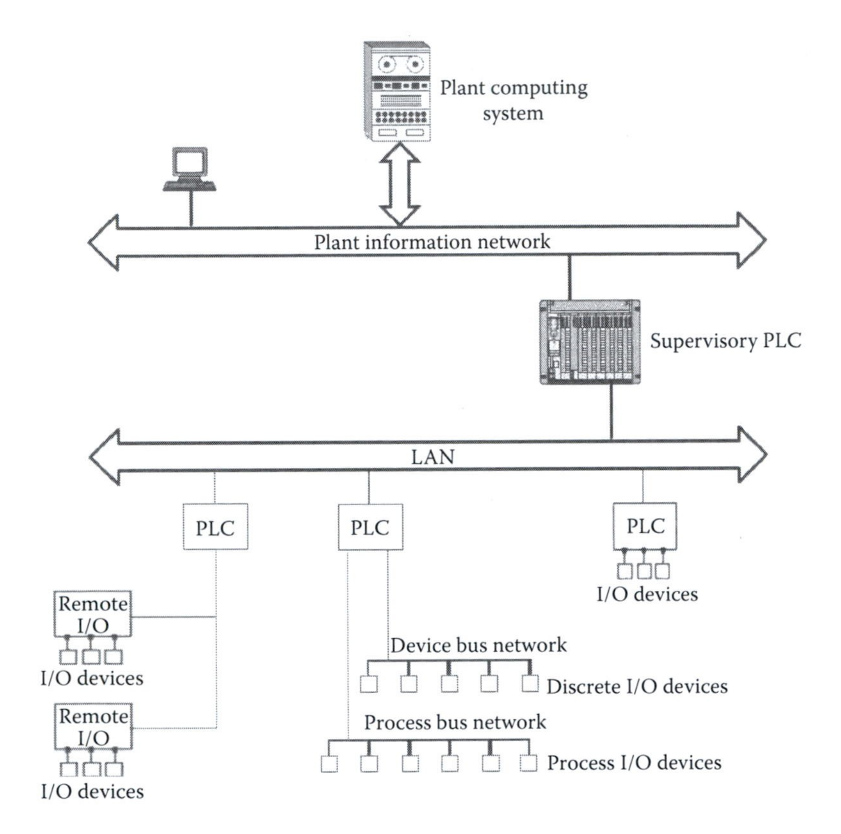

FIGURE 4.2 Interconnections between PLCs, LAN, and I/O bus network. (From Industrial Text & Video Company. *I/O Bus Networks—Including Device Net*, 1999. Available at www.idc-online.com/technical_references/pdfs/instrumentation/I_O_BusNetworks.pdf.)

devices (which are usually discrete devices such as limit switches and push buttons), while process bus networks interface with high-level information devices (which are usually smart sensors, control valves, pressure meters, etc.). Device bus networks handle from a few bits to several bytes of data, while process bus networks handle considerable amount of data—a few hundred bytes at a time. Figure 4.4 shows the classification of I/O bus networks.

Process bus networks mostly deal with analog devices, while device bus networks mostly deal with discrete devices. Examples of device bus network dealing with analog devices are thermocouples and variable speed drives—devices that transfer a few bytes of information at a time.

Bit-wide data bus networks handle up to 1 byte of data at a time, while byte-wide data bus networks handle between 1 and 50 bytes or more of data at a time. Bit-wide bus networks are also called sensor bus networks.

Process bus networks and device bus networks have differing data transmission requirements. Discrete devices transmit only a small amount of

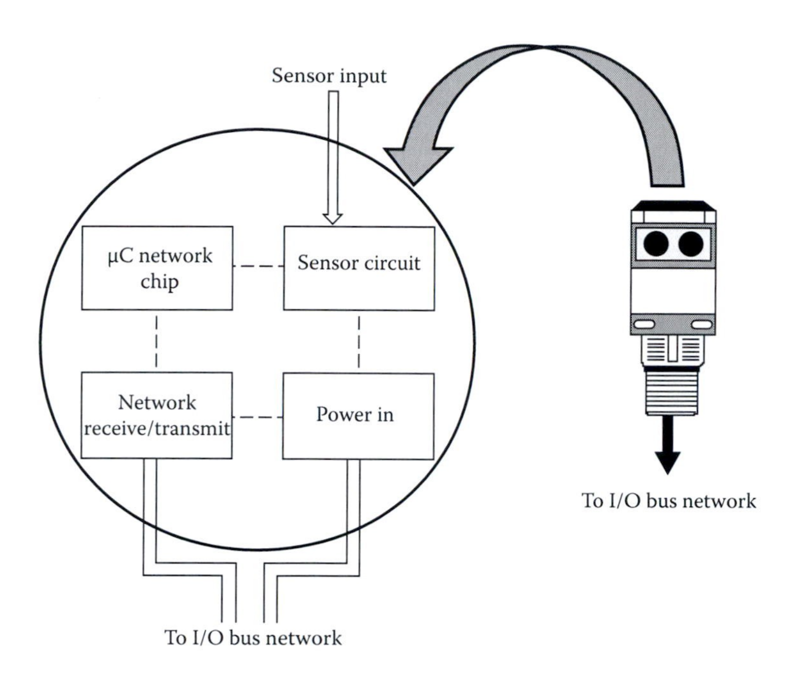

FIGURE 4.3 Intelligent field device used in I/O bus network. (From Industrial Text & Video Company. *I/O Bus Networks—Including Device Net*, 1999. Available at www.idc-online.com/technical_references/pdfs/instrumentation/I_O_BusNetworks.pdf.)

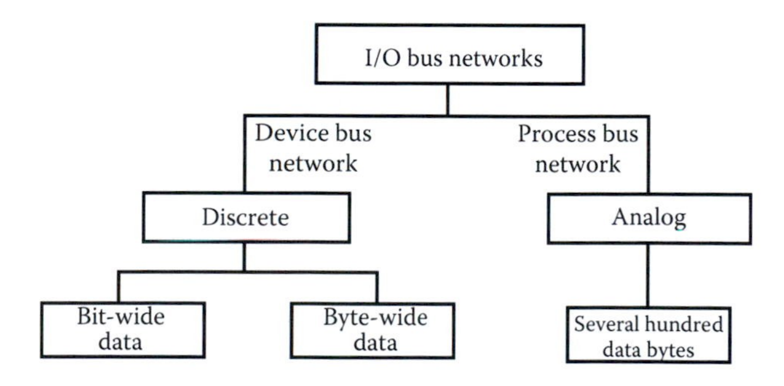

FIGURE 4.4 Classification of I/O bus networks. (From Industrial Text & Video Company. *I/O Bus Networks—Including Device Net*, 1999. Available at www.idc-online.com/technical_references/pdfs/instrumentation/I_O_BusNetworks.pdf.)

data at a time and thus meet the high speed requirements of such devices. Process bus networks are relatively slow because their data packets contain substantial amounts of information that are transmitted at a time.

4.3.2 Network and Protocol Standards

In the sphere of I/O bus networks, several organizations are working together for establishing network and protocol standards. Organizations

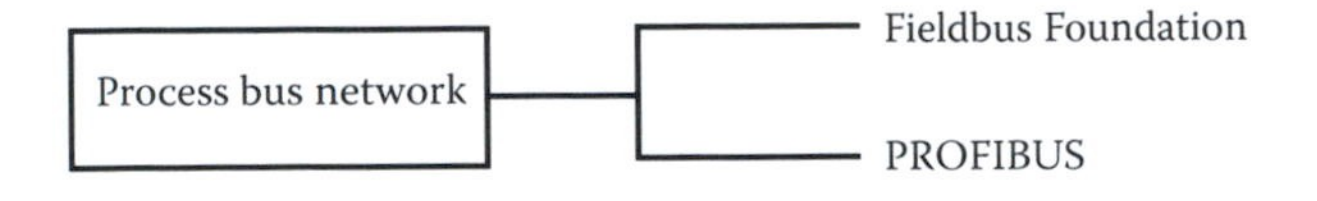

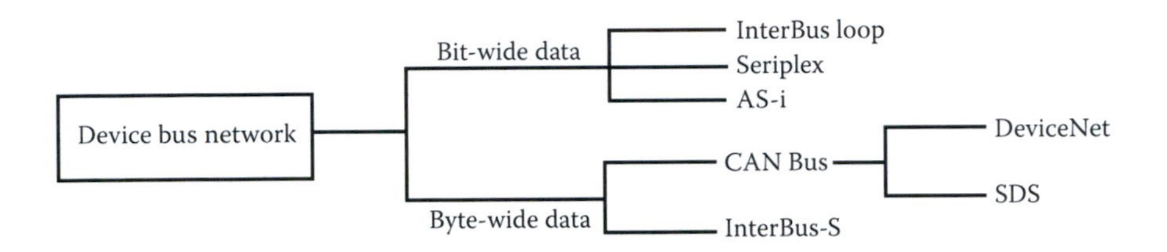

FIGURE 4.5 Network and protocol standards. (From Industrial Text & Video Company. *I/O Bus Networks—Including Device Net*, 1999. Available at www.idc-online.com/technical_references/pdfs/instrumentation/I_O_BusNetworks.pdf.)

such as the Instrument Society of America and the European International Electronics Committee (IEC) are involved in developing the standards. In the process bus area, two main organizations are working toward establishing protocol standards: the Fieldbus Foundation (established after a merger of Interoperable Systems Project and the WorldFIP North American Group) and PROFIBUS (PROcess FIeldBUS). Figure 4.5 shows the different protocol standards under I/O bus network.

In the device bus network, several bus protocol standards exist with Seriplex, AS-i, and InterBus Loop under the ambit of bit-wide data network and CAN Bus, DeviceNet, SDS, and InterBus-S under the byte-wide data network. The bit-wide data bus is also known as sensor bus.

4.3.3 Advantages

The advent of digital communication has ushered in the widespread acceptability of I/O bus networks in process automation. Digital communication allows more than one fieldbus device to communicate their data over a wire. This is due to the addressing capabilities inherent in the smart and intelligent devices that provide their process data value, device status information, device health, etc. Digital transmission has its own advantages because it is less susceptible to electromagnetic interference. Again it is very easy to restore a degraded digital signal to its original value, unlike its analog counterpart.

A second major advantage of digital communication is that such an intelligent field device can pass a digital value directly proportional to the process value—thus eliminating the need to either linearize or scale the process data.

Other established advantages of I/O bus networks are considerable reduction in wiring, the easiness with which plant expansion or else update is possible, considerable cost savings, reduced downtime, and very easy fault identification and isolation—if necessary.

4.4 OSI REFERENCE MODEL

OSI stands for Open Systems Interconnection and was first published in 1984 by the ISO. It is a benchmark on which current and future communication standards can be evaluated. An open system is a set of protocols that would allow two different communication systems to interact with each other regardless of architectural differences. The OSI model is divided into seven different functional layers or levels having clearly different tasks. Each layer has a predefined task. This ensures compatibility among the

TABLE 4.1
OSI Model from Fieldbus Perspective

Layer	Function	Task	Standards/Realizations for Fieldbuses
7	Application	Provides the user with specific network commands	DHCP, SNTP, SNMP
6	Presentation	Encodes layer 7 data before forwarding it to layer 5 and decodes layer 5 data before forwarding it to layer 7	Not relevant for fieldbuses
5	Session	Synchronizes communication sessions between two applications	Not relevant for fieldbuses
4	Transport	Prepares data string for transmission and ensures that it is reliably exchanged	TCP, UDP
3	Network	Selects data routes and ensures that the network is not overloaded	IP
2	Data link	Establishes and maintains connections between two participants	IEEE 802.2, Token passing, IEEE 802.3, CSMA/CD IEC 61158/ PROFIBUS-PA: Master/slave IEC 61158/FF: LAS
1	Physical	Pushes data into the physical medium and takes it off at the destination	EIA RS-232, EIA RS-422, EIA RS-485, IEC 61158-2 100Base T (IEC 800.3u)

Source: B. G. Liptak. *Instrument Engineers' Handbook, Process Software Digital Network*, 3rd Edition. CRC Press, Boca Raton, FL, p. 433, 2002.

TABLE 4.2
Characteristics of Various Fieldbus Standards

Attributes	RS-232 C	RS-422	RS-485	IEC-61158-2	100BaseTX
Max. no. of devices	1 Transmitter, 1 receiver	1 Transmitter, 16 receivers	32 Transmitters/ receivers	32 Transmitters/ receivers	Limited only by address range[a]
Signal coding	±12 V pulses	Differential	Differential	Manchester II	8B6T
Topology	Star	Bus	Bus	Bus	Star
Max. line length	15 m	1200 m	1200 m	1900 m	100 m
No. of lines	Min 3	4	2	2	4
Max. rate of transmission	19.2 kbps	10 Mbps	10 Mbps	37.5 kbps	100 Mbps
Comm. mode	Duplex	Duplex	Half-duplex	Half-duplex	Duplex
Transmission type	Asynchronous	Asynchronous	Synchronous	Synchronous	Synchronous
Connectors	9 or 25 pin connectors	Not specified	Not specified	Not specified	RJ 45 connectors

Source: B. G. Liptak. *Instrument Engineers' Handbook*, *Process Software Digital Network*, 3rd Edition. CRC Press, Boca Raton, FL, p. 434, 2002.

[a] Practical limit determined by response time of the system.

devices using a particular protocol and also between devices using different protocols.

The seven layers are again divided into two parts: Layers 1 to 4 serve the network and are responsible for transporting data from one layer to another. Layers 5 to 7 serve the application portion and are in a form that is understood by the user. OSI is not a protocol but a model for designing a network architecture that is interoperable, robust, and flexible. Table 4.1 gives a detail of the OSI layers from the perspective of fieldbus.

Data or message from the source travel down from layer 7, or the application layer, accepting header information from each layer. The physical layer determines the speed, transmission medium, etc., from which the information packet travels through the medium and after that it is accepted by the receiving medium. Table 4.2 shows the different interface standards so that the encapsulated data can travel through the medium.

4.5 NETWORKING AT I/O AND FIELD LEVELS

With a bidirectional control reaching up to the level of the field devices in case of fieldbus technology, process control has undergone a sea change in the last decade. Control and I/Os are now embedded into the field devices. In the Field Control System, there are only two levels existing: field level network and host level network. In the current scenario and in the existing plants, 4–20 mA analog communication is still in vogue and would continue to be so for some more time.

Binary field devices, such as limit switches, solenoids, and two-position open–close valves, are integrated via the binary I/O cards of PLCs by a point-to-point connection. Within a fieldbus controller, remote I/Os will be there to integrate these signals. For MODBUS and AS-i, an appropriate interface must be present in the controller for integrating such binary field devices. The 4–20 mA signals are connected through the analog I/O cards in the PLCs in a point-to-point manner. Figure 4.6 shows the connection to the PLC for simple 4–20 mA analog signals along with 4–20 mA hybrid, i.e., HART, signals. Legacy field devices are connected to the control network via intelligent I/Os.

In Foundation Fieldbus, the field devices are attached to the Foundation Fieldbus H1 segment as shown in Figure 4.7. A linking device connects this bus segment to the High-Speed Ethernet (HSE) or H2. The linking device performs only protocol or baud rate conversion. In case of PROFIBUS, devices are connected via a PROFIBUS segment (PROFIBUS-PA), which is then taken to the higher-speed network (PROFIBUS-DP) via a segment coupler or link. The link acts as an interface between the two segments. This is shown in Figure 4.8.

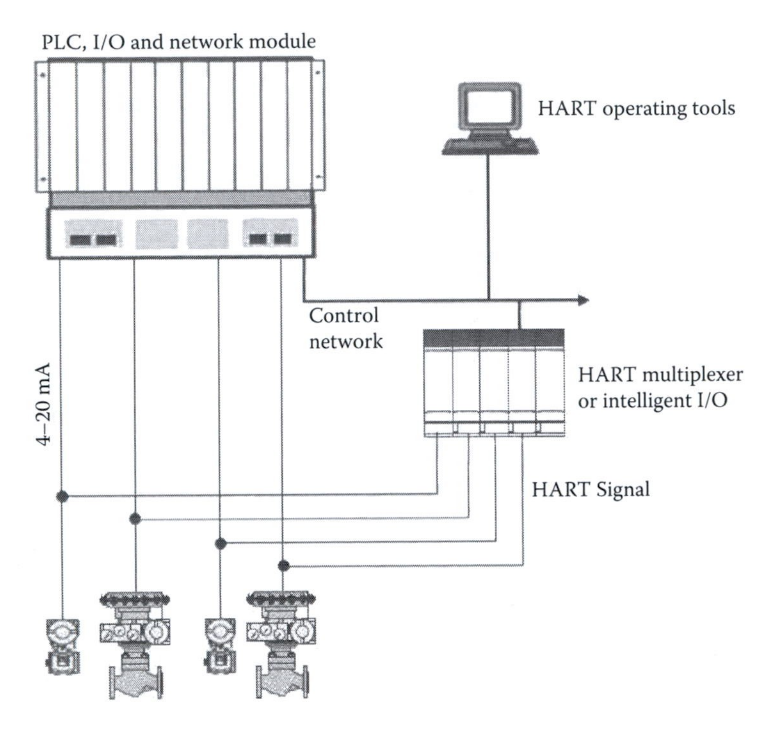

FIGURE 4.6 Connections of analog and hybrid signals to a PLC. (From B. G. Liptak. *Instrument Engineers' Handbook, Process Software Digital Network*, 3rd Edition. CRC Press, Boca Raton, FL, p. 438, 2002.)

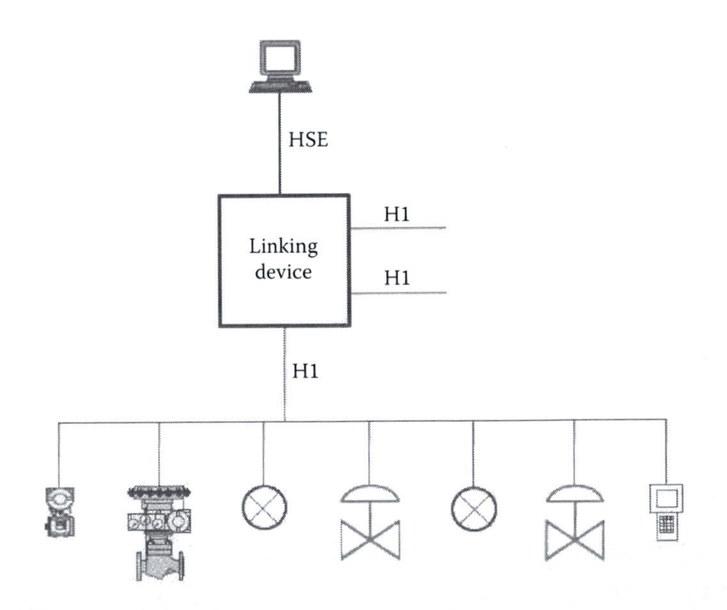

FIGURE 4.7 Connections of foundation fieldbus segment to controller.

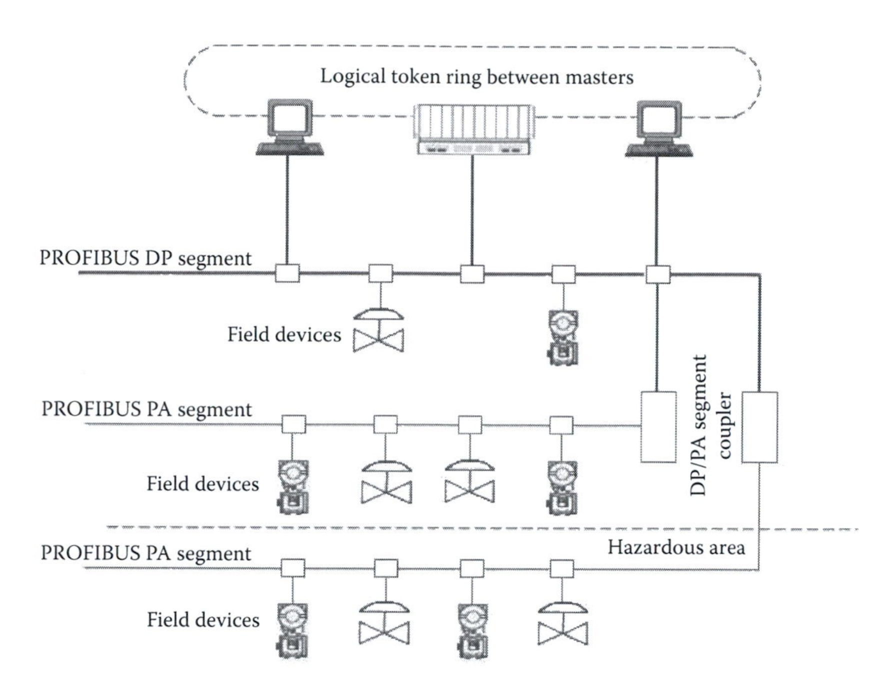

FIGURE 4.8 Connections of PROFIBUS segment to controller.

4.6 NETWORKING AT CONTROL LEVEL

In process automation, the two competing fieldbuses—Fieldbus HSE and PROFIBUS-DP—along with ControlNet govern the control level in the automation hierarchy.

Foundation Fieldbus is specially designed for process automation in which the lower-speed H1 is connected to the higher-speed HSE via a linking device. In the case of PROFIBUS, a segment coupler couples the PA network to the higher-speed DP. There are some other proprietary protocols that enjoy considerable market share, apart from InterBus, which is a manufacturing protocol used in bottling and packaging industries.

4.7 NETWORKING AT ENTERPRISE/MANAGEMENT LEVEL

Ethernet TCP/IP network is used at the enterprise/management level for internetworking and sharing data with the outside world. For this, hubs, switches, routers, etc., are used. Standard data exchange formats are used for intercompany data sharing, e-business applications, etc., for which XML text and CMG drawing formats are quite often used. Sometimes SAP or Microsoft Office programs are employed for data exchange between enterprise and management levels. For data/information exchange between management and control levels, quite often PLCs are used to which an Ethernet port is attached for connection to a supervisory station.

5 Fieldbuses

5.1 WHAT IS A FIELDBUS?

Fieldbus is a digital two-way multidrop communication link between intelligent field devices. It is a local area network dedicated to industrial automation. It replaces centralized control networks with distributed control networks and links the isolated field devices such as smart sensors/transducers/actuators/controllers. Foundation Fieldbus H1 and PROFIBUS-PA are the two fieldbus technologies used in process control.

In this two-way communication, it is possible to read data from the smart sensor and also write data into it. The multidrop communication facility in fieldbus results in enormous cable savings and resultant cost reduction. A fieldbus device must have a fieldbus interface unit for proper communication to take place, and is shown in Figure 5.1.

A comparison between a 4–20 mA system and a fieldbus system is shown in Table 5.1.

A conventional 4–20 mA current transmission system has two wires each for each of the individual field devices employed. Compared with this, a fieldbus system has two wires running for many devices that belong to the same segment. A segment may consist of 32 devices. Figure 5.2 shows a conventional point-to-point communication system and its fieldbus counterpart. There are as many wire pairs as the number of field devices for a point-to-point communication system, while it can be only a single wire pair for a fieldbus system.

5.1.1 Evolution

With digital communication making its way into process automation systems several decades back, different vendors started developing their own protocols—independent of each other. In the initial stages of fieldbus introduction, design engineers were confronted with several problems. First, a particular vendor could not provide all the parts/components needed for a plant and that a particular manufacturer cannot make all the devices better than others. This led to either choosing the less-than-the-best devices from a single manufacturer or else settling for choosing the best devices from different manufacturers. The latter option would give rise to an integrability problem and lead to isolated islands of automation and consequent interoperability difficulty with devices from different manufacturers.

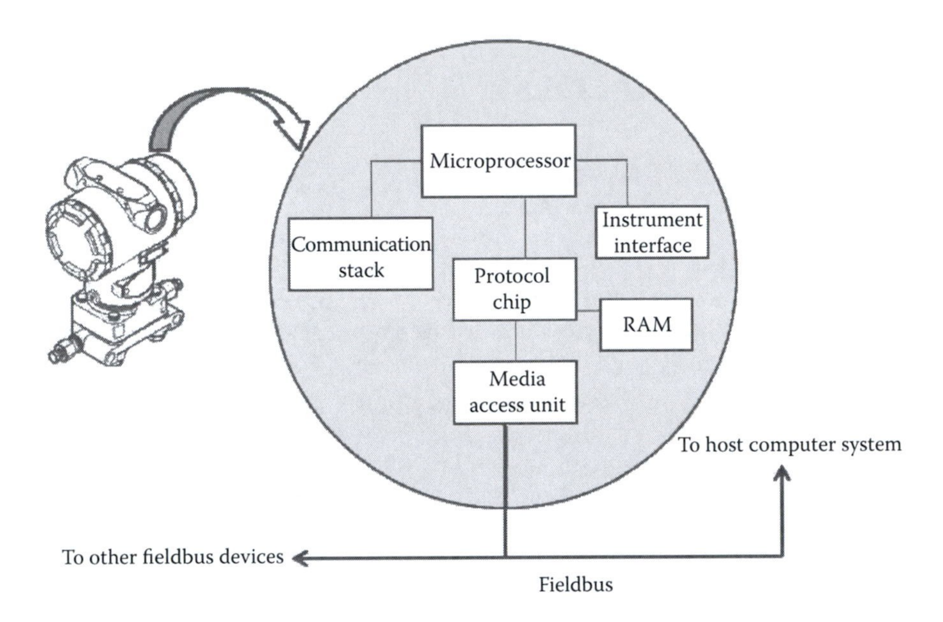

FIGURE 5.1 Fieldbus interface unit in a fieldbus transmitter. (Rosemount Inc. *The Basics of Fieldbus*, Technical Data Sheet. Chanhassen, MN, 1998. Available at http://www.rosemount.com.)

Initially when fieldbus was introduced, it suffered from numerous problems such as proprietary protocols, slow transmission speed, and different data formats. Improvements in field signal transmission technology resulted in increasing levels of decentralization.

In 1985, industry experts in the field sat together to work out a vendor-independent fieldbus standard—i.e., it would be interoperable. The bus standard would provide bus power, intrinsic safety, and the ability to communicate long distances over existing wires—the basic requirements for a process plant automation system.

Partly due to the myriad complexities of instrumentation automation systems and mostly due to the reluctance on the part of the manufacturers, a single standard protocol architecture is yet to be established. Foundation Fieldbus and PROFIBUS are now the two most dominant fieldbus technologies that are ruling the process automation field. Devices embracing these two technologies cannot communicate with each other because of protocol mismatch and thus seamless interoperability is yet to be achieved.

5.1.2 Architectural Progress

Over the years, control system architecture and field signaling developed in tandem. As field signal transmission technology continued to evolve,

TABLE 5.1
Comparison between a 4–20 mA System with a Fieldbus System

Serial Number	Variable	4–20 mA System	Fieldbus System
1	Number of field devices per wire	1	Max. 32
2	Signal/data	1	Up to thousands
3	Power supply over the wires	Yes	Yes
4	Signal degradation	Yes	Managed with terminators
5	Failure analysis	By human intervention	Reported at HMI
6	Max. run length	2000 m with proper cables and power supply	1900 m, extendable to 5700 m with repeaters
7	Field device interchangeability	Yes	Yes
8	Intrinsic safety	Yes, more barriers needed	Yes, less barriers needed
9	Control in the field	No	Yes
10	Device failure notification	Very limited	Extensive
11	Networking of field devices	No	Yes

increasing levels of system decentralization resulted. Initially it was pneumatic transmission, followed by individual 4–20 mA analog current transmission lines to the control room. The control signals, emanating from the control room, were again taken to the actuators in the field for controlling the process variables.

In direct digital control (DDC), the central computer housed in the control room executed all the controls for each and every process variable situated far off in the field. In case of failure of the central computer, there would be total failure in the control strategy and the whole plant operations would be at grave risk. This led to decentralization of control strategy and gave rise to a more decentralized distributed control system (DCS) and programmable logic controllers (PLCs) in the early 1970s. Unlike the case of a single computer controlling the whole plant in the case of a DDC, in DCS several computers share between themselves the same task. In DCS, some 20–30 numbers of loops are controlled by a single computer. A single computer failure thus affects the loops assigned to that computer

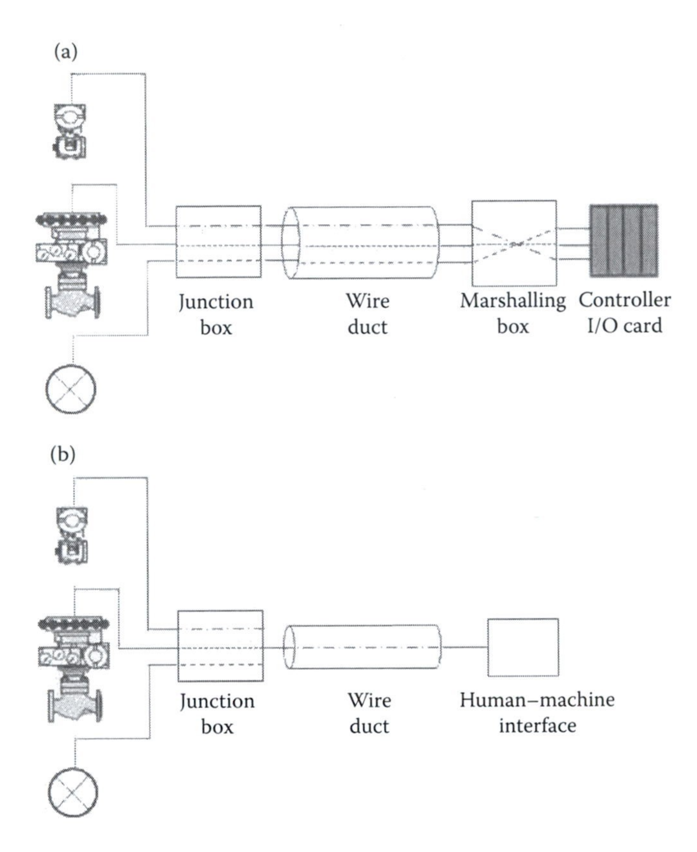

FIGURE 5.2 Schematic of (a) conventional point-to-point and (b) fieldbus communication system.

only. Traditional DCS and PLC architectures have multiple network levels, as shown in Figure 5.3.

DCSs and PLCs emerged strongly with the advent of digital communication, but their architectures were based on 4–20 mA current output for field transmitters and valve positioners. In the Field Control System (FCS), there are only two networking layers—field level network and host level network. Thus, the FCS architecture has taken the control much more into the field and it is much more decentralized than a DCS-based control system. The FCS architecture is shown in Figure 5.4.

5.1.3 Types

There are many types of fieldbuses in use today; the particular type to be used depends on the type of industry—discrete or manufacturing automation. Different types of fieldbuses include: Foundation Fieldbus, PROFIBUS, DeviceNet, ControlNet, InterBus, HART, AS-i, MODBUS, CAN Bus, Ethernet, LonWorks, and WorldFIP.

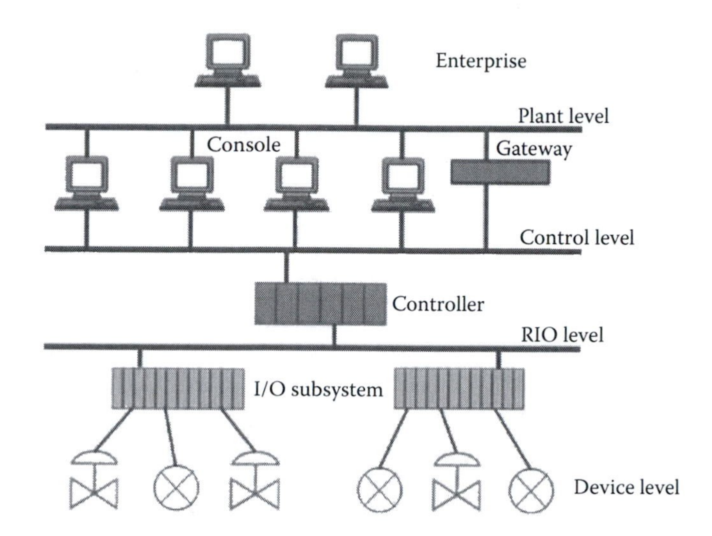

FIGURE 5.3 Multiple network architecture levels for DCS and PLC. (From J. Berge. *Fieldbuses for Process Control: Engineering, Operation and Maintenance*. ISA, p. 20, USA, 2004.)

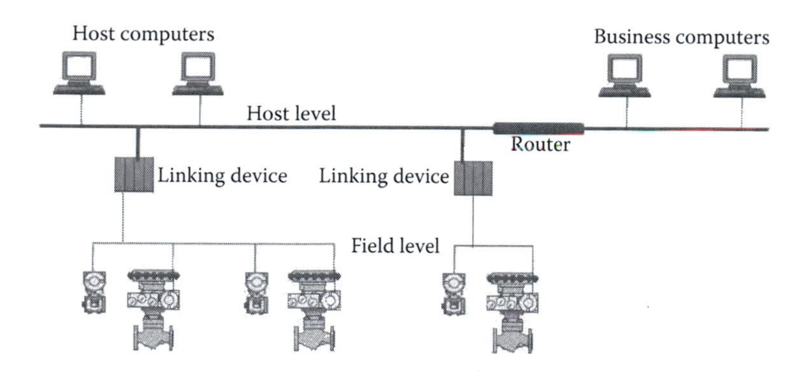

FIGURE 5.4 FCS architecture with control in the field devices. (From J. Berge. *Fieldbuses for Process Control: Engineering, Operation and Maintenance*. ISA, p. 22, USA, 2004.)

5.1.4 Expanded Network View

Application of fieldbus in process automation has its own numerous advantages. Digital communication in case of fieldbus results in better control of the process along with higher product yield. Figure 5.5 shows a traditional process control with 4–20 mA current transmission scheme and a fieldbus-based process control. For traditional control, a separate current transmission line is required for every process variable and it is unidirectional, while for fieldbus, many variables can be taken care of by a single transmission and it is bidirectional.

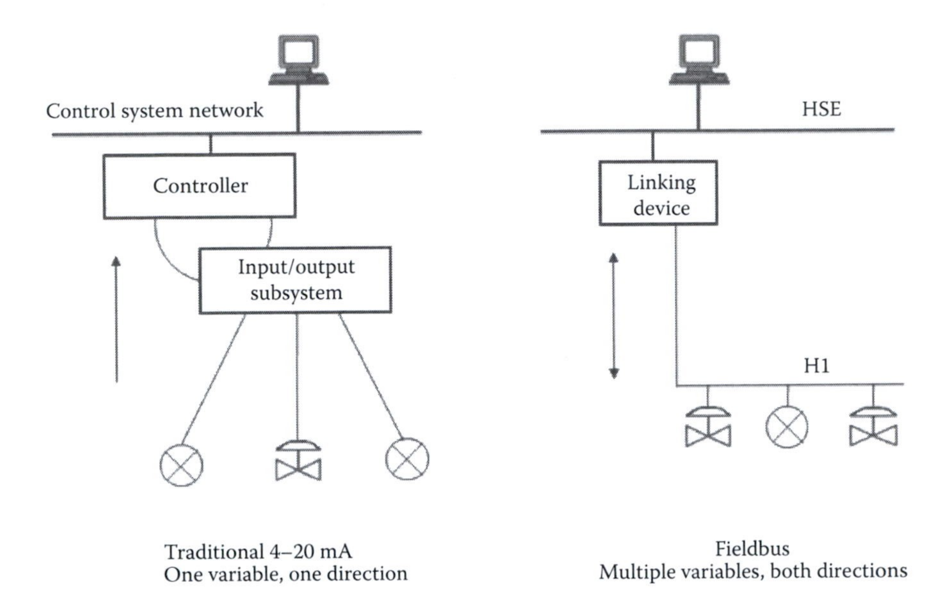

FIGURE 5.5 Traditional vs. fieldbus.

Figure 5.6 shows a comparison between a traditional and a fieldbus system in which the view is stopped at the remote input/output (I/O) for the traditional one, while it extends up to the sensors for the fieldbus system.

The I/O and the control are all housed in the smart sensor in the field in the case of a fieldbus-based system, while a separate subsystem is required

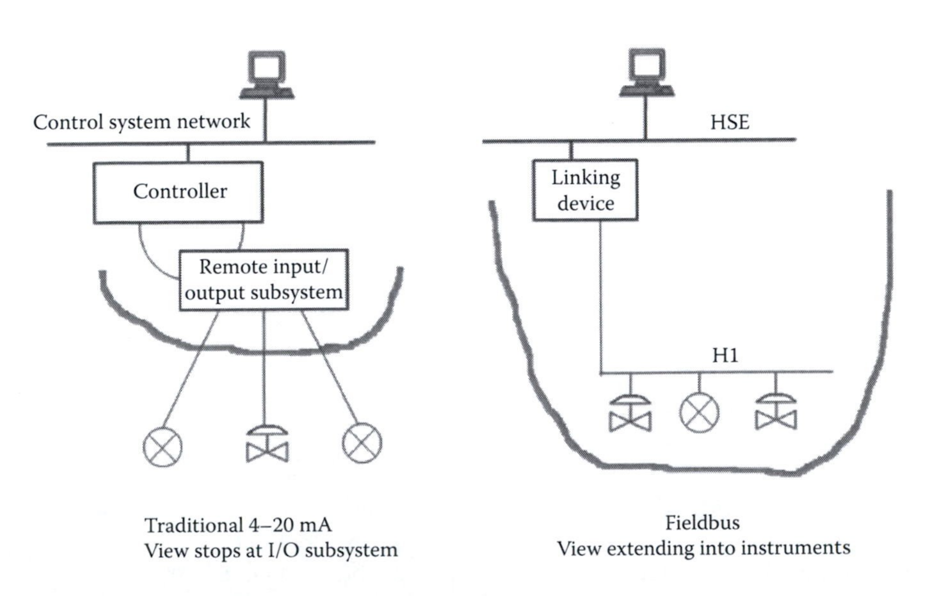

FIGURE 5.6 Extending the reach of fieldbus.

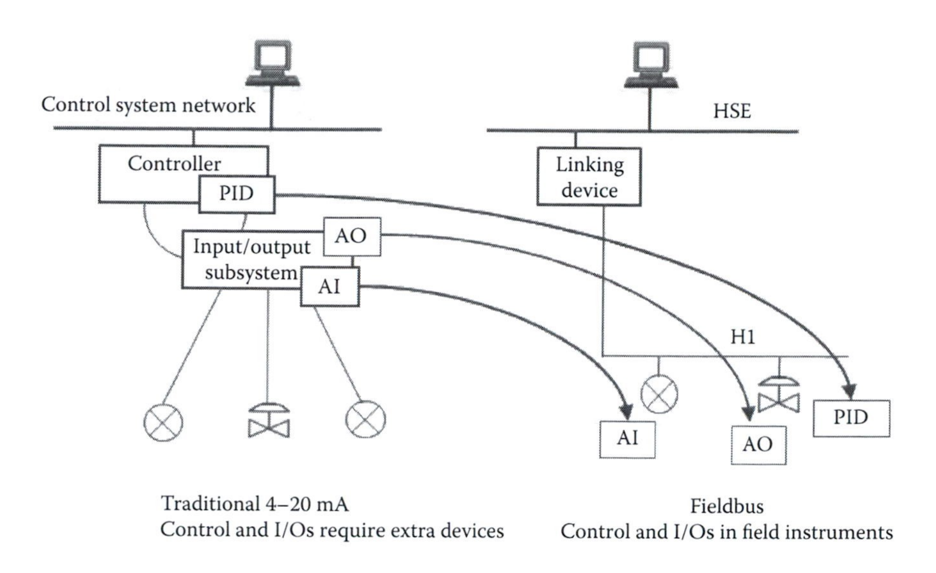

FIGURE 5.7 Control and I/Os in the field.

for the traditional system for which the controller is in the control room. Such a situation is shown in Figure 5.7.

Figure 5.8 shows a comparison between a conventional and a fieldbus-based system with regard to their applications in fire hazardous situations. For the former, for each device, an intrinsic safety (IS) barrier is needed, while for the latter, only one is employed for many field devices.

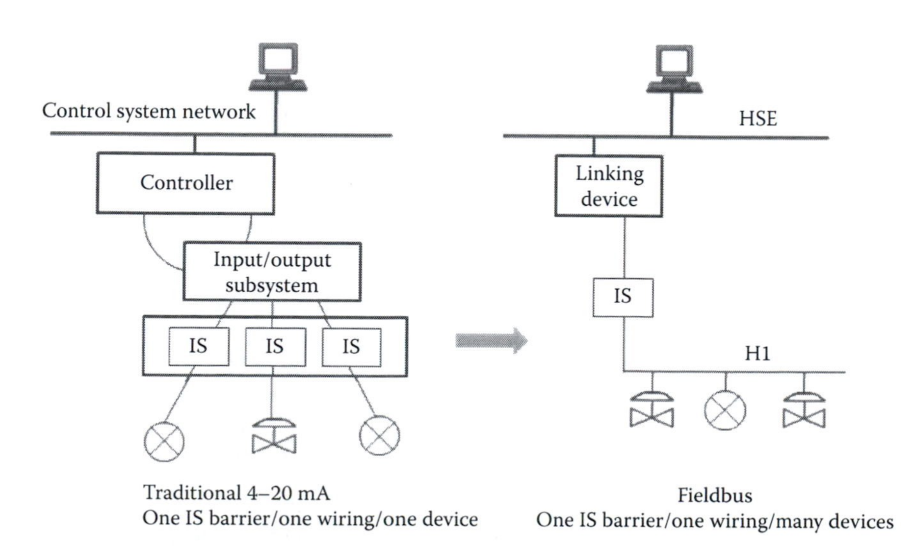

FIGURE 5.8 One IS barrier for many field devices.

5.2 TOPOLOGIES

Topology involves the manner in which the fieldbus devices are connected to the data highway. There are several possible topologies that are employed as per the needs of the plant geography. Several options employed are as follows: point-to-point, bus with spurs (multidrop), tree or chicken foot, daisy chain, and mixed. For clarity, power supply and terminators are not shown in the Figures 5.9 through 5.13 representing different fieldbus technologies.

5.2.1 Point-to-Point

Point-to-point topology is illustrated in Figure 5.9. In this topology, the segment consists of two devices. The two devices could be in the field, or else the device in the field and the host in the control room.

5.2.2 Bus with Spurs

A bus with spurs, also known as multidrop, is shown in Figure 5.10. The devices are connected to the segment via small cable lengths, called spurs. Spur length can vary between 1 and 120 m.

5.2.3 Tree (Chicken Foot)

In this topology, devices on a particular segment are connected via a junction box, marshaling panel, terminal, or I/O card (also called chicken foot). This scheme is suitable for devices situated in the same geographical place

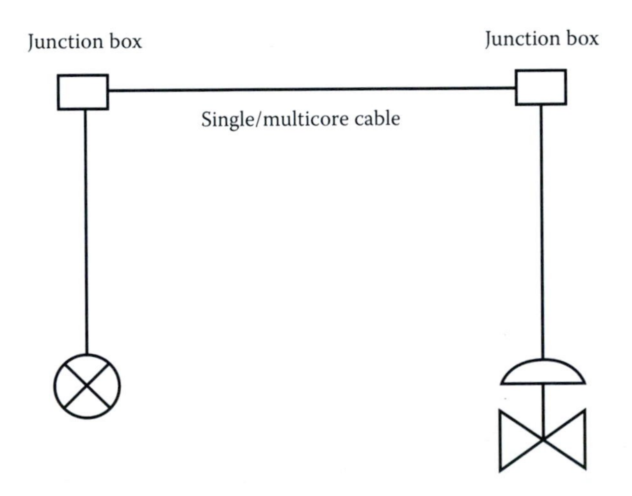

FIGURE 5.9 Point-to-point topology.

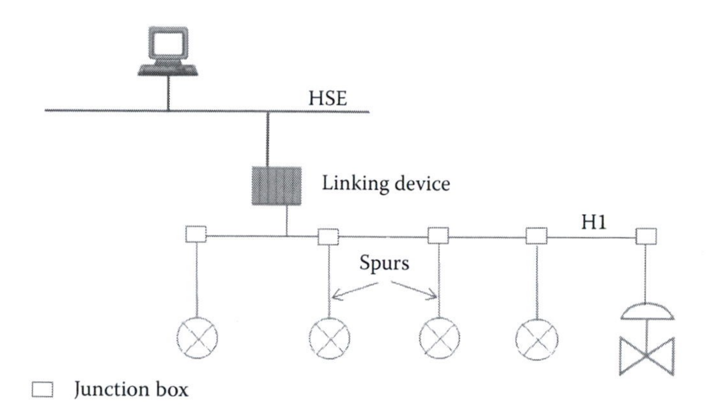

FIGURE 5.10 Bus with spurs or multidrop topology.

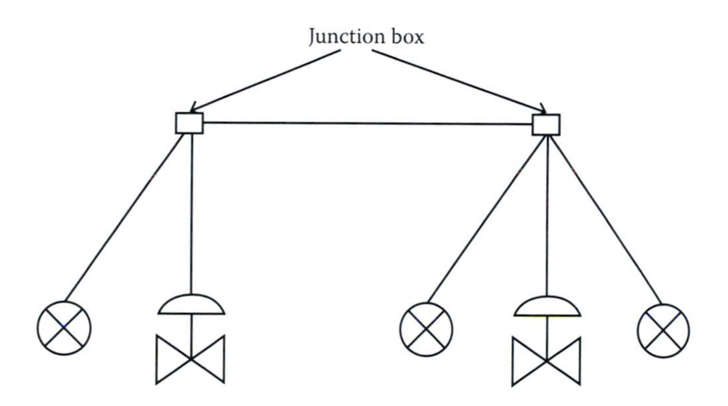

FIGURE 5.11 Tree or chicken-foot topology.

and can be connected to the same junction box and also obeying the rules of maximum spur length per segment. The scheme is shown in Figure 5.11.

5.2.4 Daisy Chain

This is shown in Figure 5.12 in which the devices are series connected in a particular segment. Device connections to the segment should be such that disconnection of a single segment would not lead to total isolation of that segment.

5.2.5 Mixed Topology

A mixed topology embraces more than one topology discussed above, and a possible combination is shown in Figure 5.13. Depending on the physical locations of the devices in the plant and the length restrictions for segment lengths, different topologies are employed to derive the advantages of individual topologies.

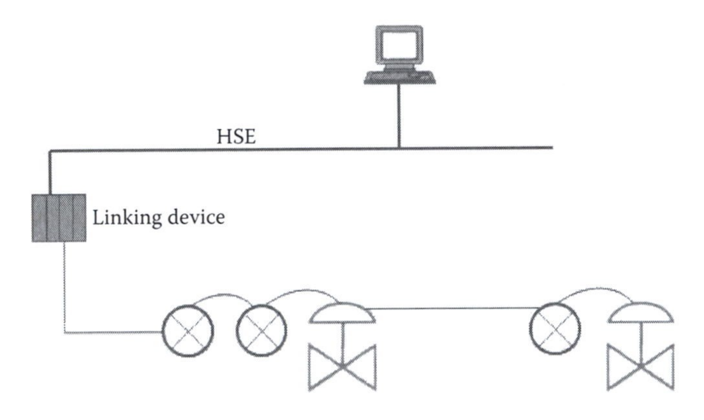

FIGURE 5.12 Daisy-chain topology.

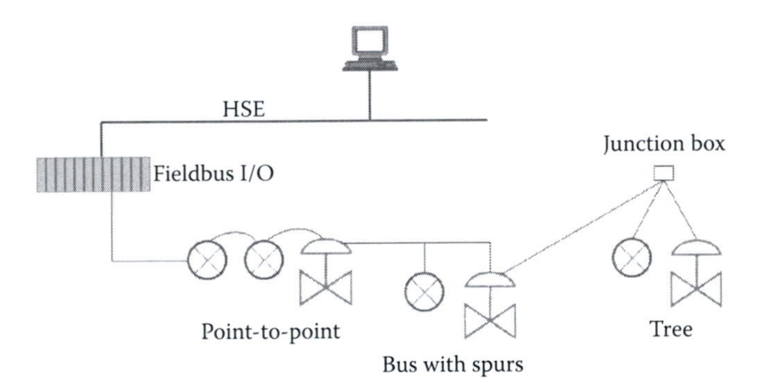

FIGURE 5.13 Mixed topology.

5.3 TERMINATORS

On either side of a segment, a terminator (*T*) is placed, which acts as an impedance module and has the same characteristic impedance of the line. Terminators do not communicate but they help in communication so that receiver error can be reduced to a minimum. A terminator prevents signal reflection and consequent communication errors. A terminator is a series combination of 100 Ω resistor and a 1 μF capacitor placed across the line. A fieldbus device transmits by changing its own current above a base level. The terminator converts this current change into an equivalent voltage change, which is then used by the processing circuitry.

A single terminator is sometimes used for robust fieldbuses, but again it is susceptible to sporadic noise. Deployment of more than two terminators per segment again may lead to reduction in signal strength and consequent failures. A network that cannot be accommodated in a single segment is realized by bridging the different segments by repeaters. Thus, on either side of a repeater, two terminators are placed—one each for adjacent sides of two segments.

5.4 FIELDBUS BENEFITS

There are different phases at which numerous benefits are derived by using fieldbus technology. The phases are planning, installation, operation, maintenance, and renovation. In the planning phase, total plant integration by digital communication employing fieldbus can be achieved. In the installation phase, a huge reduction in cost and time is achieved by replacing the traditional one-to-one wiring scheme with networking or multidrop communication. Optimum control due to enhanced functionalities in the field devices is derived in the operational phase by asserting control in devices. Self diagnostic/calibration can be done *in situ*—by software functionality in the maintenance phase. In the renovation phase, since network-based systems are modular, the upgrade cost and time would be minimum.

A single field device controller controls a single loop only, and its failure fails that particular loop. Thus, control action in fieldbus is distributed in the field devices. The field devices are smart and can carry out their own control/maintenance/diagnostics/calibration. The fieldbus system is intelligent enough to report its own failure and manual calibration can be done, if needed. By employing fieldbus-based control, efficiency in plant operations is enhanced and downtime is reduced. Fieldbus is not a product but in itself a technology and belongs to the lowest level in the hierarchy, i.e., field devices, and exchanges information with the higher levels.

In conventional systems, multiple analog-to-digital converters and digital-to-analog converters are used and each such conversion introduces error. Since fieldbus is all-digital in nature, it would then definitely introduce less error, and it would be more accurate and more precise. Since fieldbus devices are interoperable, interoperability does not become an issue and thus devices from different manufacturers can work together without any loss of functionality, which was not the case with proprietary protocols. This has become possible because of standardization of specifications of fieldbus instruments leading to interoperable field level networks. Fieldbus has eliminated the need for I/O subsystems and has the capability to detect/identify/assign addresses to devices.

Cost wise, in installation there is huge savings of around 80% in wiring cost over the conventional system. Device commissioning time is greatly reduced because of considerable configuration and diagnostic information available in fieldbus devices. A field device can be diagnosed for its problems in *in situ* condition, thereby reducing downtime. Lastly, since control has been shifted to the field devices, better and less complex central control system is required.

6 Highway Addressable Remote Transducer (HART)

6.1 INTRODUCTION

HART, an acronym for Highway Addressable Remote Transducer, is an open process control network protocol and was introduced in the late 1980s. It is a hybrid communication protocol that uses the Bell 202 Frequency Shift Keying (FSK) technique to superimpose digital communication signal on top of the analog 4–20 mA current loop signal. It is supported by the HART Communication Foundation (HCF). Unlike the other "open" digital communication technologies applied to process instrumentation, HART is compatible with existing systems.

For many years, the process automation and control industry is dominated by 4–20 mA analog signal to carry process variable (PV) and control signals to and from the control room. The HART protocol extends this analog communication with additional bidirectional digital communication, being carried by the same wiring at the same time. This digital signal, known as HART signal, carries device configuration, diagnostic information, calibration, and any additional process measurements. Some of the features of the HART protocol are as follows:

- Simultaneous analog and digital communication
- Compatible with conventional analog instrumentation schemes
- Supports multivariable field devices
- Flexible data access via up to two masters
- Open *de facto* standard
- Backward compatible
- Only protocol that supports both analog and digital, unlike other fieldbuses that are digital in nature
- Either point-to-point or multidrop operation
- Adequate response time of ~0.5 s

6.2 EVOLUTION AND ADAPTATION OF HART PROTOCOL

HART was introduced by Rosemount in the 1980s. Initially, the idea was to support the 4–20 mA loop current with digital PV signals. The diagnostic, calibration, troubleshooting capabilities, etc., of HART led to its increasing acceptance. This was even more so since the existing cables were utilized to carry the HART signals. The field personnel working at the sites did not have to learn much, and a small handheld master was all that was needed for *in situ* calibration, etc., of the instruments. With little training, the field personnel were able to adapt themselves with the working of HART signals. This led to the wide acceptance of HART technology. Fierce competition from other competing technologies led to its further development. Revision 4 was the first version that was introduced in the market having HART compatibility. Later, revision 5 was introduced in 1989 by Rosemount, which has a huge installed base. It supports the latter version of the HART protocol.

HART is the only protocol that supports both analog and digital (hybrid) communication at the same time over the same pair of wires. Because of its widespread acceptability, the HART User Group was formed in 1990 and HCF was formed in 1993. With the formation of HCF, Rosemount transferred ownership of all patents, trademarks, and technology to the Foundation.

6.3 HART AND SMART DEVICES

Developments in the fields of smart field devices and HART technology went on at around the same time. As more and more functionalities started emerging from the smart or intelligent field devices, more demands were placed from HART technology to complement them. This led to further developments in either of them. Smart instruments are connected to programmable logic controllers (PLCs), computers, etc., which collect sensor, actuator, or field instrument data and diagnostics by digital communication techniques.

Most smart instruments have the following capabilities: zero, span, and range adjustments; diagnostics to verify health; and memory to store status and configuration information. They support two-way communications with the controller, digitize the process signal, and digitally correct PVs. They contain advanced software that performs measurements, and control actions are taken accordingly. Smart instruments add more resolution, more accuracy, and above all more reliability to the process. At the site, various diagnostics can be verified in a few minutes by the handheld controller associated with HART technology.

Initially, HART was developed with the primary PV in mind. Later, PVs included upper and lower transducer limits, etc. Status information was later included to lend more and more functionalities.

Smart field devices of today are almost always multivariable in nature and supported by HART technology. For example, a flow meter includes a totalizer and a data logger. Initially, system manufacturers were slow to exploit the full capabilities associated with HART, but with time came the realizations and today, the huge installed base of HART-compatible multivariable field devices are fully utilized to support them.

6.4 HART ENCODING AND WAVEFORM

The HART protocol makes use of the FSK technique to superpose the digital communication on top of 4–20 mA signals. The digital FSK signal is phase continuous and has a 0.5 mA amplitude having two different frequencies, 1200 Hz and 2200 Hz, representing binary 1 and 0, respectively. Since the average DC value of a sine wave over a time period is zero, the FSK signal does not add any DC component on the 4–20 mA analog signal. Figure 6.1 shows the digital HART signal superimposed on the 4–20 mA analog signal.

The HART field devices and the central controller have FSK modems—the FSK-modulated HART signals are carried along with the 4–20 mA analog signal via the same wires. At the controller, the HART signal is demodulated and actions taken accordingly.

6.5 HART CHARACTER

HART uses an asynchronous mode for communication purpose. Thus, HART data are transmitted 1 byte at a time without any clock signal. A HART character is composed of 11 bits and shown in Figure 6.2.

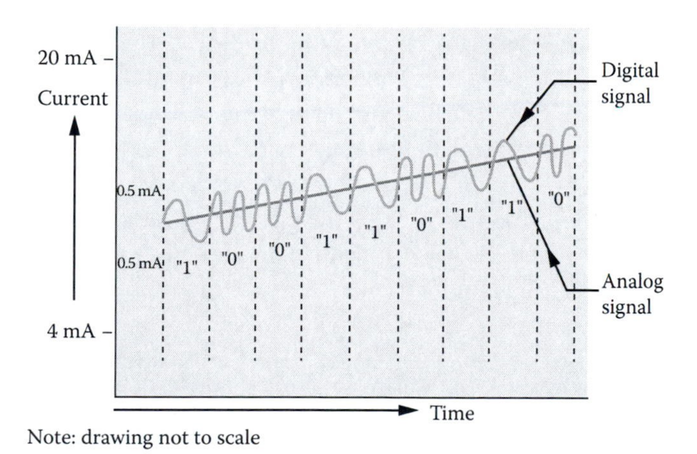

FIGURE 6.1 (See color insert.) HART digital signal superimposed on 4–20 mA analog signal. (HART Communication Foundation. *Application Guide HCF LIT 34. HART Field Communications Protocol.* Austin, TX, 1999. Available at http://www.pacontrol.com/download/hart-protocol.pdf.)

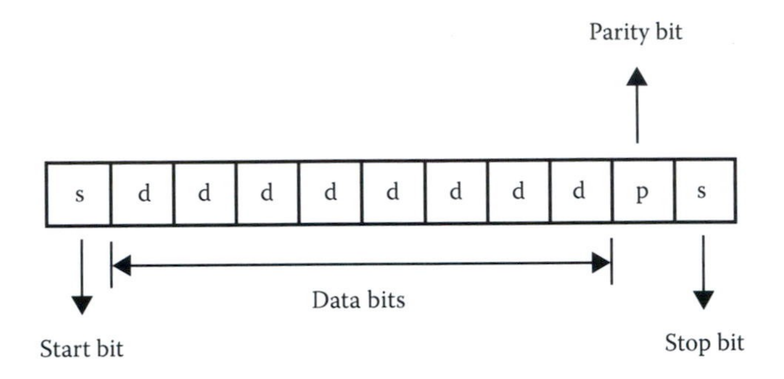

FIGURE 6.2 HART character.

The character starts with a start bit, followed by eight bits of data, one parity bit, and lastly one stop bit. The parity bit provides integrity to the data transmitted.

6.6 ADDRESSING

HART addressing is of two types: polling address and unique identifier. Polling address is single byte and is also known as "short address." Unique identifier is of 5 bytes and also called "long address."

The address field formats for the short and the long frames are shown in Figure 6.3. One bit of the short address distinguishes the two masters, while another bit indicates burst mode telegrams. The remaining four bits distinguish the field devices (from 0 to 15)—0 for single-unit mode and 1–15 corresponds to multidrop mode. The polling address format is used with old HART devices that do not support the long address format.

The 5-byte unique identifier is a hardware address that consists of 1-byte manufacturer code, 1-byte device type code, and a 3-byte sequential number. This 5-byte ID is unique for each device. HCF administers the manufacturer code, which eliminates the possibility of address duplication of any two HART devices. The master uses this unique long address to communicate with the slaves.

Master	Burst	0	0	Bit 3	Bits 2 to 0

(a) Short-frame format

Master	Burst	0	0	Bit 3	Bits 35 to 0

(b) Long-frame format

FIGURE 6.3 HART address field formats: (a) short frame; (b) long frame.

In single mode, the master polls address 0 to get the unique slave ID. In multidrop mode, the master checks all the polling addresses 0 to 15 to check device presence. The master then presents a list of live devices on the network. A user can alternatively enter the tag of the intended device and the master will broadcast the same. The slave with the unique ID and the tag responds against this query from the master. The polling address, in conjunction with the unique ID, indicates whether the message exchange is from a primary or secondary master and whether the slave is in the burst mode or not.

6.7 ARBITRATION

Arbitration ensures proper message transmission between master and the slave devices. There can be either master–slave or burst mode. In the former, it is the master that initiates message transmissions, by requesting the slave device. The slave device, in turn, responds only to the query from the master. A slave in the master–slave mode of operation can never initiate communication.

There can be two masters on the network—a primary master and a secondary master—which may typically be a handheld terminal. Arbitration between the two masters is based on timing.

The slave burst mode, like the master–slave mode, is initiated by a command from the master. In this, burst mode responses are generated by the requested slave without request frames from the master. Data is updated at a faster rate since the slave goes on transmitting without a request from the master.

A frame, either from a master or a slave, is transmitted only after ensuring that no transmission is taking place on the network at that point of time. It is the responsibility of the timer to allow access to the network to the primary masters, secondary masters, slaves, or slaves in burst mode. Both masters have equal priority in getting access to the bus. In case both masters have to repetitively access the bus, they would do so alternately. Burst mode slaves wait longer than the masters to transmit, allowing the masters to control such slaves—either to continue or abort the burst mode.

6.8 COMMUNICATION MODES

Communication employing HART protocol can either be master–slave or burst mode. In the former mode, communication is initiated by the master. A HART communication loop may have two masters: primary and secondary. The master is typically a system host—may be a distributed control system, a PLC, or a personal computer. The secondary master can be a handheld configuration tool (i.e., a handheld terminal) used for occasional configurations of different process parameters. A slave may be a transmitter or a valve positioner. Burst mode configuration provides the master with certain information on a continuous basis, until told to stop. It provides for a faster communication (about three to four data updates per second) than the master–slave mode and is used in single-slave configuration.

6.9 HART NETWORKS

HART networks can operate in two configurations: point-to-point and multidrop mode. In the point-to-point network, the traditional 4–20 mA current signal is used to control the process and remains unaffected by HART signal. The configuration parameters, etc., are transferred digitally over the HART protocol. The point-to-point scheme is shown in Figure 6.4. The digital HART signal is used for commissioning, maintenance, and diagnostic purposes.

HART multidrop communication networks are used when the devices are widely spaced. As shown in Figure 6.5, only two wires are required to communicate with the master. If required, intrinsic safety barriers and auxiliary power supply for up to 15 devices can be incorporated in this mode. All polling addresses of the devices are set at greater than zero and device current limited to a typical value of 4 mA.

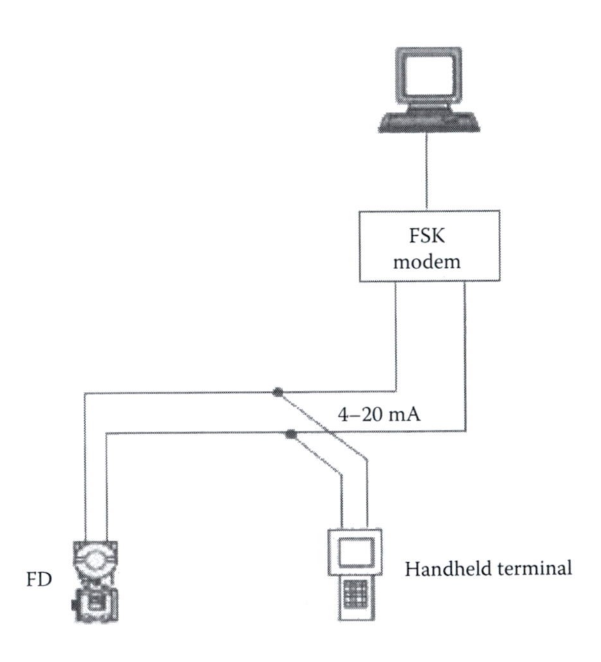

FIGURE 6.4 Point-to-point network configuration.

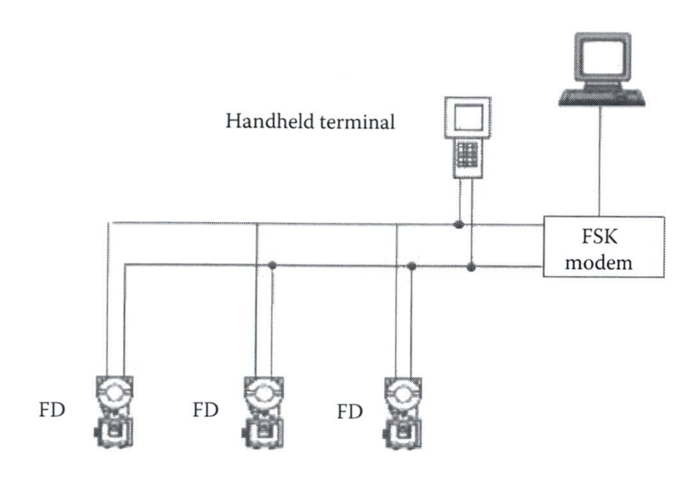

FIGURE 6.5 Multidrop network configuration.

Multidrop networks allow two wire devices to be connected in parallel. Information reading time from a single variable is typically 500 ms and with 15 devices connected to the network; approximately 7.5 s would be required to completely go through the network cycle once.

6.10 FIELD DEVICE CALIBRATION

HART devices need occasional calibration to ensure that the transducer output, via the different processing blocks, is representing the true process value. Any HART field device has a transducer block, a range block, and a data acquisition (DAQ) block for such calibration.

Calibration of such field devices include the calibration of the digital process value, its proper scaling, converting into some desired range, and finally sending the 4–20 mA current value. The process output successively goes to the transducer block, range or range conversion block, and finally to the DAQ block.

Figure 6.6 shows the calibration steps for a HART-enabled field device. The transducer block involves comparing a simulated transducer value with an internally generated traceable reference. This comparison determines whether calibration of the field device is required, which can be done by using the HART protocol. Calibration is usually performed by providing the field device with exact transducer values—one near the lower limit and the other near the upper limit. Using the HART protocol, the field device then performs the necessary adjustments, if needed.

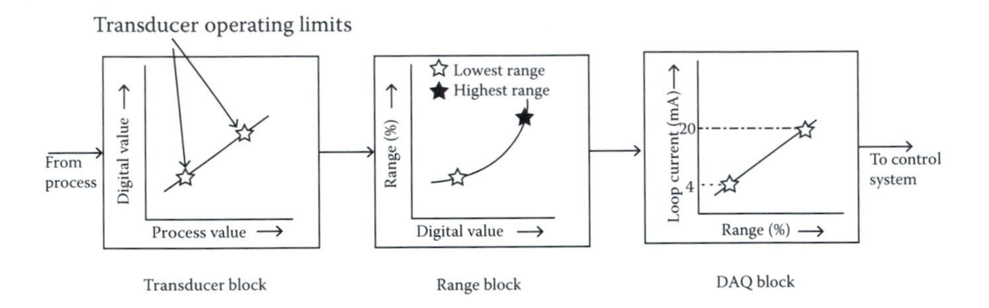

FIGURE 6.6 Calibration of HART field devices. (From B. G. Liptak. *Instrument Engineers' Handbook, Process Software Digital Network*, 3rd Edition. CRC Press, Boca Raton, FL, p. 554, 2002.)

The range block uses the lower and upper range values of 4 and 20 mA, respectively, to convert these transducer values into lower and upper range percent values—the former refers to 0% and the latter 100%. The range block output may be in linear or square root form—as in the case of flow-type PV.

The output of the range block is passed on to the DAQ block, which then converts the percent range values into loop current signals. Calibration of the current loop is not seriously needed because no moving parts are involved. Calibration at 4 mA corresponds to current zero trim and 20 mA is called the current span trim.

6.11 HART COMMUNICATION LAYERS

The HART protocol follows the seven-layer OSI (Open Systems Interconnection) protocol, although it uses only three layers: application, data link, and physical. The other four are not used, which is so for most of the field level protocols such as HART. The HART and the OSI protocol layers are shown in Table 6.1.

6.11.1 Physical Layer

HART uses an FSK physical layer that is based on the Bell 202 modem standard. It modulates digital "1" into 1200 Hz and "0" into 2200 Hz. This kind of modulation is robust and has very good noise immunity. A HART modem chip is used at both the sending and receiving ends for modulation and demodulation, respectively.

HART devices support both the conventional 4–20 mA current signal and modulated HART communications. These occupy different communication bands and are shown in Figure 6.7. Because of their noninterfering nature, as is apparent from Figure 6.7, both communications are possible

TABLE 6.1
HART Protocol Layers Compared with OSI Layers

OSI Layers	HART Layers
Application	HART commands
Presentation	
Session	
Transport	
Network	
Data link	HART protocol rules
Physical	Bell 202

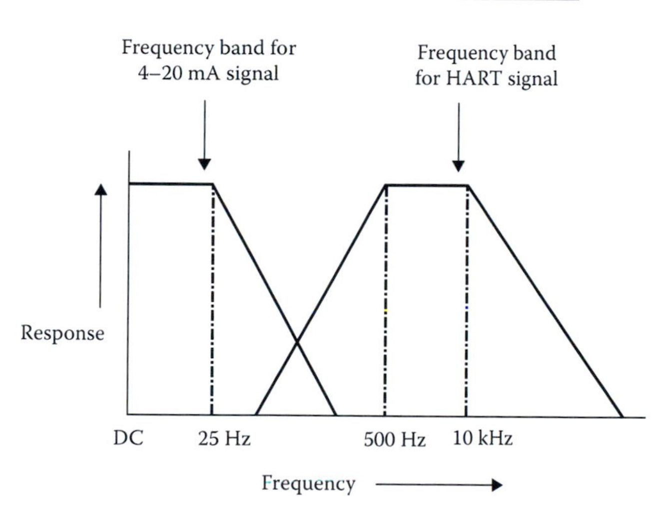

FIGURE 6.7 Frequency bands for 4–20 mA and HART signal. (From B. G. Liptak. *Instrument Engineers' Handbook, Process Software Digital Network*, 3rd Edition. CRC Press, Boca Raton, FL, p. 557, 2002.)

simultaneously. The HART communication signal is filtered out by the analog devices and as such they remain unaffected by the HART signal. Thus, devices with 4–20 mA input or output work nicely in control loops.

6.11.2 Data Link Layer

The HART message frame format, often called a HART telegram, is shown in Figure 6.8. It consists of nine fields.

Preamble is the first field in the message that is sent first, and it wakes up and synchronizes all the receivers of the connected devices in the network. The delimiter, a single-byte field, signifies the end of the preamble. The

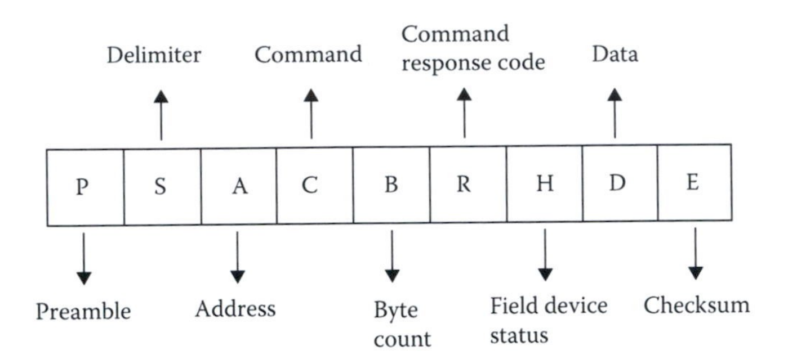

FIGURE 6.8 HART character.

delimiter is a start field and its content denotes if the frame is a request from a master, a response from a slave, or a request from a slave in the burst mode. It also indicates whether the address used is a polling address or a unique ID.

The address field that follows may be a single byte (short-frame format) for polling address or 5 bytes (long-frame format) for a unique ID address, already discussed in Section 6.6.

The fourth field is the command field and is of single byte. It represents the HART command associated with the message. The data link layer does not interpret the same but passes (accepts) the same to (from) the application layer. The byte count is of single byte length and indicates how many more bytes are still left for the message to be completed, excluding the checksum. The receiver can thus check for the end of the message from the information provided by this field.

The response code is also single byte and is included only in response messages. This response field from the slave would indicate the type of error that occurred in the received message, if it was received erroneously. Correct message reception would also be indicated by this field. The field device status, a single-byte field, is included in response messages to indicate the health of the device.

The data field may contain from 0 to 24 bytes. Data is not interpreted in the data link layer and is merely passed to the application layer. The checksum is of single byte and of longitudinal cyclic redundancy check type. If the data is received correctly, the transmitting end checksum value would be identical to the receiving end checksum.

6.11.3 Application Layer

HART commands are defined in the application layer of the HART protocol. The communication routines of the HART master devices and the programs are based on these commands. Commands from the master can seek for data, start-up service, or diagnostic information. The slave, in turn, responds by sending back the required information(s) to the master.

The HART command set includes three types of commands: universal commands, common practice commands, and device-specific commands. The host application may implement any of the command types for a particular application. A partial list of different types of HART commands is shown in Table 6.2.

TABLE 6.2
HART Commands

Universal Commands	Common Practice Commands	Device-Specific Commands
Read manufacturer and device type	Read selection of up to four dynamic variables	Read or write low flow cutoff
Read primary variable (PV) and units	Write damping time constant	Start, stop, or clear totalizer
Read current output and percent of range	Write device range values	Read or write density calibration factor
Read up to four predefined dynamic variables	Calibrate (set zero, set span)	Choose PV (mass, flow, or density)
Read or write 8-character tag, 16-character descriptor, date	Set fixed output current	Read or write materials or construction information
Read or write 32-character message	Perform self-test	Trim sensor calibration
Read device range values, units, and damping time constant	Perform master reset	PID enable
Read or write final assembly number	Trim PV zero	Write PID set point
Write polling address	Write PV unit	Valve characterization
	Trim DAC zero and gain	Valve set point
	Write transfer function (square root/linear)	Travel limits
	Write sensor serial number	User units
	Read or write dynamic variable assignments	Local display information

Source: HART Communication Foundation. *Application Guide HCF LIT 34. HART Field Communications Protocol.* Austin, TX, 1999. Available at http://www.pacontrol.com/download/hart-protocol.pdf.

6.12 INSTALLATION AND GUIDELINES FOR HART NETWORKS

HART devices use the conventional 4–20 mA current signal cable path to send the HART FSK signal. Thus, no extra communication paths are required to send HART signals that use existing telephone cables for communication. There are several ways to calculate HART cable lengths, which depend on the quality of cables used and consequently on the capacitance—the lower the cable capacitance, the longer would be the cable run. Capacitance limitation imposed by intrinsic safety considerations further puts a limit on the maximum cable run in such cases. It

TABLE 6.3
HART Communication Cable Length Limits

Pair	Shield	Twisted	Size	Length
Multi	Yes	Yes	0.2 mm^2 (AWG 24)	1500 m (5000 ft)
Single	Yes	Yes	0.5 mm^2 (AWG 20)	3000 m (10,000 ft)

Source: J. Berge. *Fieldbuses for Process Control: Engineering, Operation and Maintenance*. ISA, USA, p. 72, 2004.

TABLE 6.4
Relation between Number of Devices, Cable Capacitance, and Cable Length

Devices	65 nf/km (20 pf/ft)	95 nf/km (30 pf/ft)	160 nf/km (50 pf/ft)	225 nf/km (70 pf/ft)
1	2800 m (9000 ft)	2000 m (6500 ft)	1300 m (4200 ft)	1000 m (3200 ft)
5	2500 m (8000 ft)	1800 m (5900 ft)	1100 m (3700 ft)	900 m (2900 ft)
10	2200 m (7000 ft)	1600 m (5200 ft)	1000 m (3300 ft)	800 m (2500 ft)
15	1800 m (6000 ft)	1400 m (4600 ft)	900 m (2900 ft)	700 m (2300 ft)

Source: HART Communication Foundation. *Application Guide HCF LIT 34. HART Field Communications Protocol.* Austin, TX, 1999. Available at http://www.pacontrol.com/download/hart-protocol.pdf.

is seen that HART communication is not vulnerable even for degraded cable quality. Cable length is dependent on its size (cross section) and also whether it is single- or multicore cable (Table 6.3).

Table 6.4 shows the influence of cable capacitance and number of devices on cable length.

6.13 DEVICE DESCRIPTIONS

The universal and command practice commands ensure interchangeability and openness of the field devices irrespective of the manufacturer. This is so when the user needs only standard status and the fault messages. The universal and command practice commands become insufficient if some device-specific information or some special properties of a device are required. The software of the master device has to adapt itself for such occasional device-related information or a new device is included in the system. This is where the Device Description Language (DDL) was developed.

The DDL is not limited to applications related to HART only, but can also be applied to other fieldbuses. The DDL was developed and specified, in a general way for fieldbuses, by the "Human Interface" workshop of the International Fieldbus Group (IFG). The DDL is a language—like a programming language—that enables the device manufacturers to describe all communications in a complete and comprehensive manner. It is a powerful tool that provides significant benefits during configuration and device-specific features.

DDL supports all extensions. Thus, any device- or manufacturer-specific command can be executed by DDL, and the user is provided with a universally applicable and uniform user interface. This thus entails a clear, user-friendly, and safer operation and monitoring of the process.

A DDL encoder reads the device description as binary-encoded DD data that can be read by a master device. If a device has sufficient storage space, the said DD can be stored in its firmware. During the parameterization phase, it can be read by the master device.

6.14 APPLICATION IN CONTROL SYSTEMS

Because a combination of analog signal and digital data is available at the same time from HART-enabled multivariable field devices, a multitude of data—both the health of PVs and environmental conditions—can be monitored at the same time. HART field devices can update their data twice or thrice in a second. Although it is not very fast, this rate of data update is much faster than the time constraints associated with most of the processes. A HART multivariable pressure transmitter may have a temperature sensor that can monitor the process temperature. This sensor output would warn against any damaging temperature condition, thereby eliminating the need to employ separate temperature sensor for the above. This saves time, money, and space.

In the legacy 4–20 mA current transmission scheme, the field wiring is brought to a junction box. A multicore "home run" cable connects the junction box to the I/Os in the marshaling room. This is then connected to the controller via a backbone bus. The controller provides data to operator consoles, engineering workstations, etc. In this traditional scheme, only information is in the form of an analog 4–20 mA signal.

Later architectures employ remote I/Os, operator consoles, controllers, and engineering workstations, all connected by an "open" communication protocol. The remote I/Os are located close to the processes in which the existing cables are connected to HART field devices. The multicore cable is thereby dispensed with network cables providing two-way HART communication, which enhances both faster data availability and integrity.

Another major area in which HART-enabled systems score over the traditional ones is the availability of status information. This status information can be sent via the network cables to the control room. Thus, an *a priori* knowledge about plant's possible breakdown is available and preventive maintenance can be undertaken—reducing outages and downtime. This obviously leads to higher plant output and a more comprehensive maintenance schedule.

6.15 APPLICATION IN SCADA

The technology used in HART is very useful for SCADA applications. Since HART supports digital process values, it can be applied in SCADA. Various status parameters of PVs obtained from HART measurements can be used to schedule field trips for preventive maintenance. This would result in less downtime and less employment of manpower.

SCADA applications require updating field data in minutes. Since data updating using HART requires only a fraction of a second, it can very safely be applied in SCADA applications. Other application areas where HART can be applied in SCADA are inventory management, automated meter reading, pipeline monitoring situated in far off places, and remote monitoring of remotely located petroleum production units.

6.16 BENEFITS

The HART protocol is a unique communication facility that is backward compatible, ensuring that existing investments in plant cabling and power needs remain secure in the future. It is the only communication facility that supports both analog and digital communication (hybrid) at the same time. By preserving the traditional 4–20 mA signal, it extends the system capability for two-way digital communication with multivariable smart field instruments. Any process application can be addressed by the services offered by the HART protocol. It enjoys a phenomenal support base and offers the best solution for intelligent field devices.

The simplicity of the HART protocol and the very low cost of HART-compliant field devices is the basis for its widespread acceptability in process instrumentation systems. Some of the other advantages that accrue by employing HART protocols in process industries are as follows: fully open *de facto* standard, multiple smart devices along the same highway, multiple masters can control the same smart device, long-distance communication via telephone lines can be implemented by using a HART–CCITT converter, up to 256 PVs belonging to the same device can be taken care of, addition of new features without much difficulty, access to device parameters and diagnostics, common command and data structure, a large and still growing selection of available products.

7 Foundation Fieldbus

7.1 INTRODUCTION

The Foundation Fieldbus has its origin in the Fieldbus Foundation—a nonprofit consortium that recommends and develops the technical specifications for Foundation Fieldbus. These are based on ISA/ANSI S50.02 and IEC 61158. The standard specifies the communication requirements and methodology for the fieldbus. Manufacturers who follow the standards have to develop their products following the specifications recommended in the standard so that instruments from different manufacturers can be interoperable. The Fieldbus Foundation was formed in 1994 with the merging of Interoperable Systems Project and WorldFip North America.

The traditional analog 4–20 mA current signal is based on ISA S50.1. The ISA standard committee SP 50 met in 1985 to develop a standard to replace the analog current transmission with a digital communication link for the instrumentation and automation sector. It was the need of the hour to develop a common standard—from the economic and technical aspects—so that the best instruments from different manufacturers can be put together to run a plant without facing any interoperability problem.

The standards committee recommended and defined two different network requirements: a fairly low-speed H1 was recommended to be installed at the plant floor level (sensor level) to replace the 4–20 mA current transmission retaining the existing plant wiring. Second, a higher-speed H2 was recommended, which acts as a backbone to the H1 segments. Conformance testing for H1 was completed in 1998 by the Foundation Steering Committee. The same committee, in 1999, recommended the implementation specifications for H2 based on commercial off-the-shelf (COTS) high-speed Ethernet operating at 100 Mbps. It was completed in 2000, and thus H2 was replaced by High-Speed Ethernet (HSE).

7.2 DEFINITION AND FEATURES

The Fieldbus Foundation defines fieldbus as "a digital, two-way, multidrop communication link among intelligent and control devices." It is one of several local area networks dedicated to the industrial network.

The features associated with Foundation Fieldbus are as follows:

- It is a bidirectional, half-duplex, digital process control and automation protocol.
- When compared with the Open Systems Interconnection (OSI) reference model, the Foundation Fieldbus has three layers: physical layer (PHL), data link layer (DLL), and application layer (APL). It has an additional layer—layer 8, called the "user layer."
- It has a two-level architecture consisting of the lower level H1 and the upper layer H2. H1 operates at 31.25 kbps, while H2 operates at 1 or 2.5 Mbps.
- It supports both scheduled and unscheduled communications.
- It supports interoperability, i.e., devices from different manufacturers can be seamlessly connected.

7.3 FOUNDATION FIELDBUS DATA TYPES

Different data types are combined to form complex data objects. The different data types used in Foundation Fieldbus are Boolean, integer, floating point, unsigned, octet string, visible string, date, time of day, time difference, time value, bit string, null, and packed.

7.4 ARCHITECTURE

Foundation Fieldbus has a two-level architecture: H1 is the lower-level bus that connects the field devices together and H2 is the upper-level bus that interconnects the different H1 bus segments. The former operates at a speed of 31.25 kbps, while the latter at 100 Mbps. The two-level Foundation Fieldbus architecture is shown in Figure 7.1.

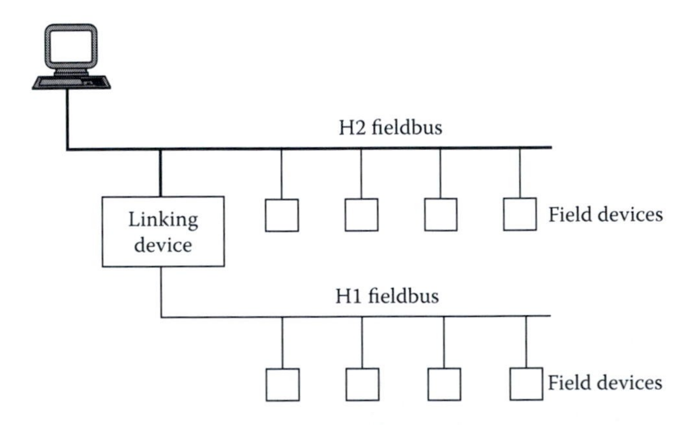

FIGURE 7.1 Two-level Foundation Fieldbus architecture.

The architecture supports control in the field devices such as control valve positioner and field instruments. It supports total control in the field (field devices) with no control room equipment. It also supports cascade control linking field devices and control loops in dedicated controllers.

Both H1 and H2 run on the same protocol and perform identical services. However, their physical and data link layers are quite different. The two-level architecture of Foundation Fieldbus is made possible with the implementation of a user layer—sometimes called "layer 8." The building block of the user layer lies in the "function block."

7.5 STANDARDS

The Instrumentation, Systems, and Automation Society (ISA) recommends standards that operate under ANSI rules. The latter, in turn, follows international standards as recommended by the United Nations. The Foundation Fieldbus has its standards based on ISA/ANSI S50.02 and IEC 61158. These two standards recommend communication specifications, requirements, and methods for Foundation Fieldbus. S50.02 came into being in 1993, while the standard IEC 61158 was introduced in 1999. The physical, data link, and application layers belonging to ISO 7498 of the seven-layer OSI are fully specified in this standard. The IEC standard was defeated in a ballot vote by a cartel of several manufacturers who thought their interests would be severely jeopardized in case such a standard was given the go ahead. A compromise was reached, which allowed several manufacturers to closely follow the-then existing digital communication buses in vogue. The fieldbus standard that was passed then became the IEC 61158 standard in the all-digital fieldbus field.

7.6 H1 BENEFITS

Application of H1 technology offers significant benefits in the control system life cycle. Numerous benefits accrue by employing H1 at the field level of Foundation Fieldbus technology including as follows: a huge reduction in the number of wires and marshaling panels, reduced number of I/Os, reduced number of IS barriers, reduced control room size, reduced number of equipments, reduced number of power supplies, and reduced downtime. Other benefits include ability to configure the devices remotely, increased accuracy in measurements, increased information availability with regard to health and status of equipment, *in situ* calibration, and increased plant safety.

Multivariables from each device can be brought to the control room for trend analysis, archival purposes, predictive maintenance, asset management and report generation, etc. An expanded view of the system from the control room level to the field level is possible by employing H1.

7.7 HSE BENEFITS

Different plant systems are integrated by HSE through the control backbone. Resource allocation for maintenance can be assigned via asset management with the help of HSE. The benefits of having HSE in the upper tier of the architecture are discussed in the following subsections.

7.7.1 Interoperability of Subsystems

A plant has different subsystems. A blast furnace, for instance, has a weighing system, excess gas bleeder station, charging of different inputs to the blast furnace, and tapping of liquid hot metal and slag intermittently. A furnace operator can integrate these operations because of open protocol. Using HSE, information for different plant operations can be accessed without customer programming. HSE's help is also taken for data integrity, device health checking, redundancy, etc.

7.7.2 Function Blocks

Function blocks are identical for both H1 and HSE devices. Foundation Fieldbus eliminates the need to have proprietary programming language. Instead, the same programming language is used for all plant operations. Function blocks are responsible for any loop shutdown, bumpless transfer, and the like.

7.7.3 Control Backbone

Linking devices bring data from individual segments (belonging to H1) to the HSE control backbone. Thus, different H1 networks are bridged by HSE. Thus, the control backbone has total control starting from individual processes up to the different plant areas spread across the total plant territory. HSE dispenses with the use of remote I/O networking level, enterprise and control level, thereby flattening the enterprise pyramid.

HSE supports peer-to-peer communication capability. A device can communicate with another one without having to go through the central computer—thus eliminating any risk in case of central computer failure.

7.7.4 Standard Ethernet

HSE uses standard Ethernet cables, interface cards, and networking hardware, which are available very cheap. Ethernet communications can be wireless, fiber optics, or twisted pair. Networking hardware from different suppliers are available both in commercial and industrial grades. Ethernet components are available in standard COTS category.

7.8 COMMUNICATION PROCESS

The communication process involves transportation of data and messages from one fieldbus device to another. The devices may reside on the same link or other links joined by bridges. Devices must have a device tag for proper device identification and a device network address for properly delivering data at the designated device.

Operation of different function blocks existing in the user application layer (layer 8) must be properly coordinated and linked and data put on the bus to be used by other devices, whenever required. Knowledge of the underlying communication technologies is the minimum basic requirement for understanding the communication process in fieldbus systems that include smart field devices.

7.8.1 OSI Reference Model

Foundation Fieldbus H1 technology is based on the OSI reference model. Figure 7.2 shows a comparison between the OSI reference model and the Fieldbus model. The Fieldbus model consists of three layers: PHL, DLL, and APL. For the Fieldbus model, the APL is divided into two sublayers: Fieldbus Message Specification (FMS) and Fieldbus Access Sublayer (FAS). FAS maps FMS into the DLL. There is a layer above layer 7—called the "user application layer." This is layer 8 of the Fieldbus model, although it is absent in the OSI reference model.

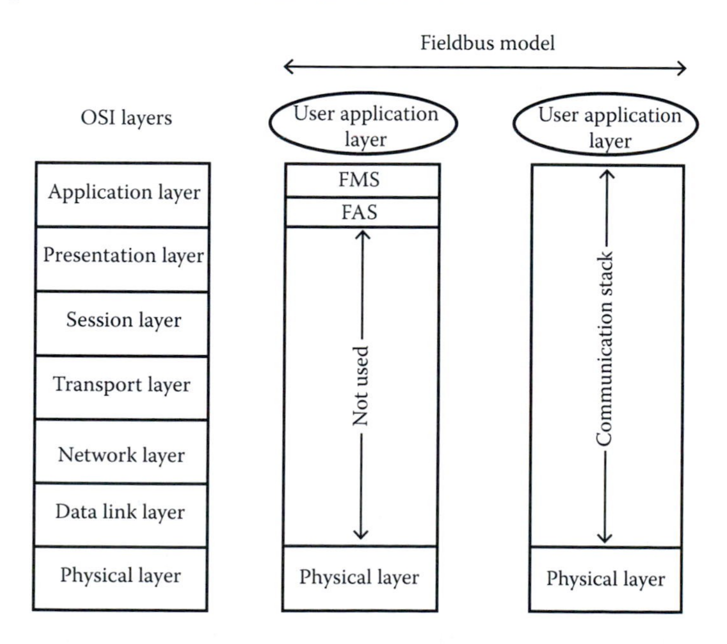

FIGURE 7.2 Fieldbus model compared with OSI reference model.

In the Fieldbus model, layers 3 to 6 of OSI are not used. Layers 2 to 7 are mostly implemented in software and termed as the "communication stack."

7.8.2 PDU

Figure 7.3 shows how user data is passed down the layers to reach the lowest level, i.e., physical layer. From the physical layer, it is then put on the fieldbus. It is seen that each layer appends Protocol Control Information (PCI) to the message that it receives from the higher layer. The total message that a layer shifts to its just lower layer is known as the protocol data unit (PDU) of the former. As an example, the FAS layer contains the FMS PDU (this is the total message received from the FMS layer) along with the PCI of FAS.

7.8.3 Physical Layer

As shown from Figure 7.3, the physical layer receives the DL PDU from the data link layer to which are appended the preamble, start delimiter, and end delimiter. This is then converted into physical signal and transmitted on the fieldbus medium by electrical or optical means. Physical layer interests are signal and waveform shapes, voltage levels, type of wires, etc.

On the physical wires, data is sent at the rate of 31.25 kbps. Although the speed of data transmission is not high enough, it serves the purpose of automation industries in the instrumentation sectors with the cabling remaining

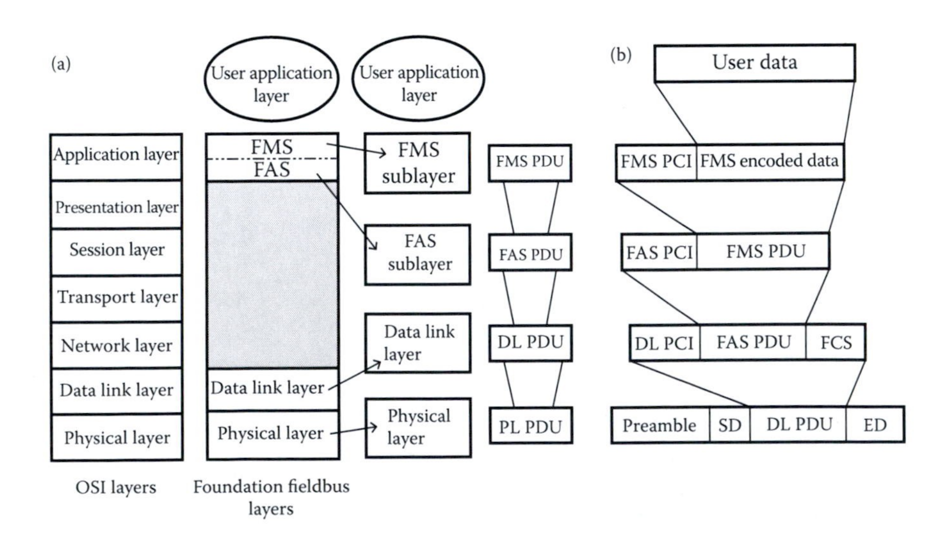

FIGURE 7.3 Concepts of PDUs in fieldbus model: (a) protocol layers and (b) data encapsulation.

intact. Explosive gases present in some areas in the plant prevent electronic devices to be placed there, and a 31.25 kbps transmission fits nicely, which demands devices with low power consumption.

7.8.3.1 Manchester Coding

Synchronous serial communication in half-duplex mode is used in Foundation Fieldbus. The Manchester Biphase-L coding scheme is used to code data generated from the field devices and transmitted along the bus. In this coding scheme, there is always a change in the coded pattern at the midpoint of each clocking, irrespective of data pattern. With clocking, a data of logic 1 state would be represented by logic 1 for the first 50% of clock duration and 0 state for the rest of the clock period. The reverse is true when a logical 0 is coded. The shape of the Manchester Biphase-L–coded data for an input data pattern 101100 is shown in Figure 7.4.

7.8.3.2 Signaling

A typical physical layer signal waveform is shown in Figure 7.5. The preamble is sent first, followed by the rest in the sequence shown. Preamble is an 8-bit sequence of alternating 1's and 0's and is used to synchronize the receivers on the bus. Start and end delimiters are also 8 bits in length. In both start and end delimiters, *N*+ (non-data positive) and *N*– (non-data negative) symbols, which are high and low level signals, span the whole clock duration. At the receiver, the start delimiter is used to identify the start of data bits and the end delimiter is used to signal end of data.

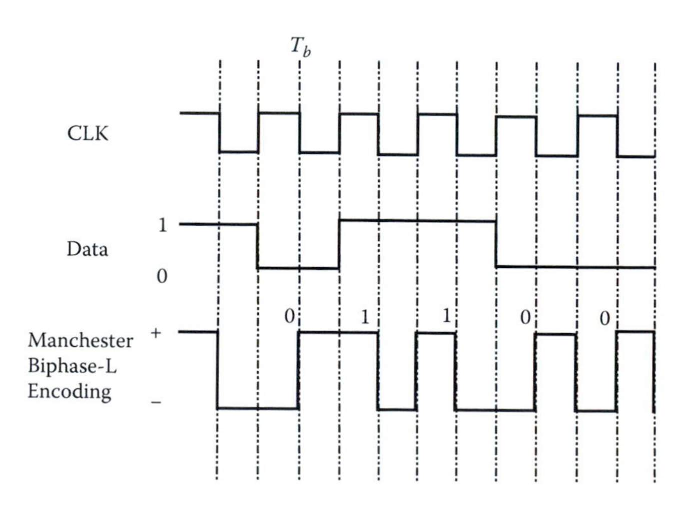

FIGURE 7.4 Manchester-coded data pattern. (From SMAR. *Fieldbus Tutorial—A Foundation Fieldbus Technology Overview.* USA, pp. 1–29. Available at http://www.smar.com/PDFs/catalogues/FBTUTCE.pdf.)

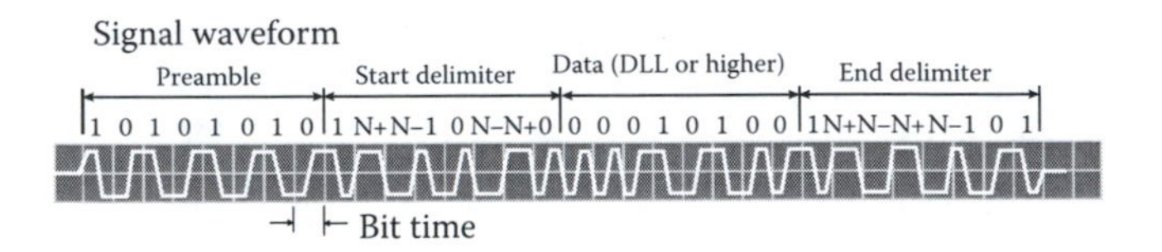

FIGURE 7.5 **(See color insert.)** Physical layer signal waveform. (From Yokogawa Electric Corporation. *Fieldbus Book—A Tutorial*, Technical Information TI 38K02A01-01E, pp. 1–33, 2000.)

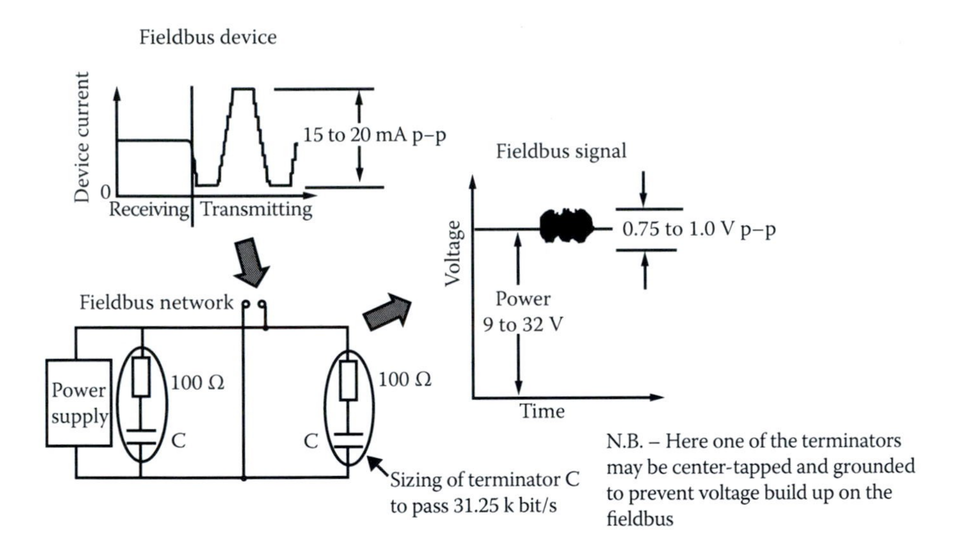

FIGURE 7.6 Voltage waveform on the bus for a transmitting fieldbus device. (From SMAR. *Fieldbus Tutorial—A Foundation Fieldbus Technology Overview*. USA, pp. 1–29. Available at http://www.smar.com/PDFs/catalogues/FBTUTCE.pdf.)

Fieldbus devices are supplied with a DC voltage in the range of 9–32 V. The device that is transmitting on the bus does so by delivering ±10 mA at 31.25 kbps into a 50 ohm equivalent resistor. This develops a 1.0 V peak-to-peak voltage modulated on top of the DC supply. Figure 7.6 shows the voltage wave shape on the bus when the device is transmitting.

7.8.4 Data Link Layer

The data link layer manages transfer of data from one node to another. It also manages the priority of data transfer among the field devices. Addresses, data transfer, their priority, and medium control are all managed by the data link control layer. It prevents access to the bus by two or more devices present in the segment at the same time.

The data link layer supports transmitting data as per urgency. There are three levels of urgency: URGENT, NORMAL, and TIME_AVAILABLE, in this order. An urgent message overrides all other message queues. Maximum data sizes belonging to the three data types are 64, 128, and 256 bytes, respectively.

7.8.4.1 Medium Access Control

The most important job of the data link layer is medium access control of the fieldbus devices. A link or segment is the particular part that shares the physical layer signal at the same time. A device, at a given instant of time, is allowed to use the link by proper software programming. A link active scheduler (LAS) controls this medium access in the data link layer.

7.8.4.2 Addresses

There are many devices in a link and there are many links that are connected by bridges. Devices on a fieldbus are identified by a DL-address. It consists of three fields: link, node, and selector, having their respective lengths—16, 8, and 8 bits. A link address identifies a link and if communication is between the same link, the link address is omitted. When a message goes from one link to another, only then the link address is inserted. The content of the node field (8 bits) gives the address of the node in a link. The selector field gives the device an internal 8-bit address. It identifies a virtual communication relationship (VCR). When a VCR is connected to another VCR, it is identified with Data Link Connection End Point. Data Link Service Access Point is used when a VCR is not connected to another VCR but can send/receive messages.

Foundation Fieldbus devices have node addresses in the range of 0×10 to 0×FF, which are classified into Link Master (LM) range, BASIC range, default range, and temporary range. LAS has a node address of 0×04, while a temporary device such as a handheld communicator has address in the temporary range. When a device loses its node address, it can communicate by using one address from the default range. The node address scheme is shown in Table 7.1.

7.8.4.3 LAS and Device Types

Foundation Fieldbus devices are classified into BASIC, LM, and Bridge types. BASIC class devices cannot become LAS, while LM class devices can. A Bridge class device can become LAS in addition to its original functionality to connect different links.

In a link, there can be more than one LM; however, only one can act as LAS at any given time. When there are no LAS in a link, LM devices try to acquire the LAS role, but the one with the least node address wins this contention. Figure 7.7 shows how an LM class device becomes a LAS.

TABLE 7.1
Node Address Scheme for Foundation Fieldbus Devices

0×10 ~ V(FUN)	Address for link master class devices
V(FUN) + V(NUN) ~ 0×F7	Address for basic class devices
0×F8 ~ 0×FC	Default address for devices with cleared address
0×FD ~ 0×FF	Address for temporary devices such as a handheld communicator

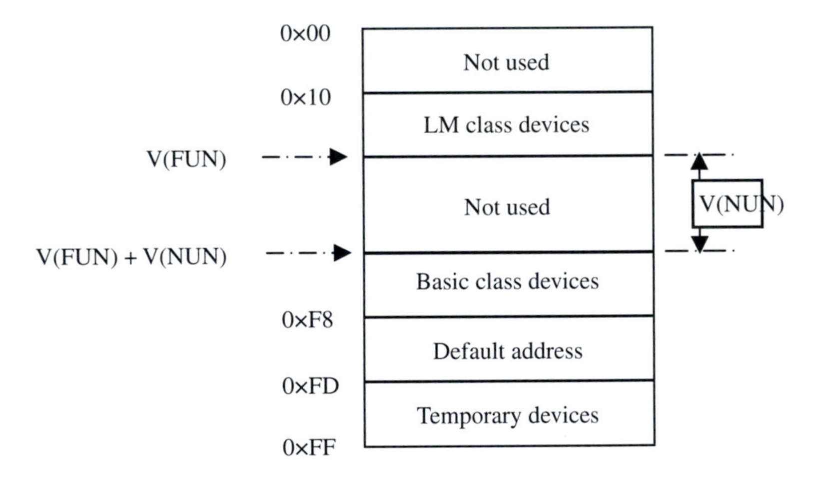

Source: From Yokogawa Electric Corporation. *Fieldbus Book—A Tutorial*, Technical Information TI 38K02A01-01E, pp. 1–33, 2000.

7.8.4.3.1 Scheduled Communication

Scheduled data transfers are used for regular cyclic transfer of control loop data between devices residing on the fieldbus. LAS controls the periodic data transfer from a publisher (source of data) to subscribers (data link) using network schedule. Thus, schedule transfers belong to a publisher–subscriber type of reporting for data transfer. Scheduled communication takes place in a synchronous manner.

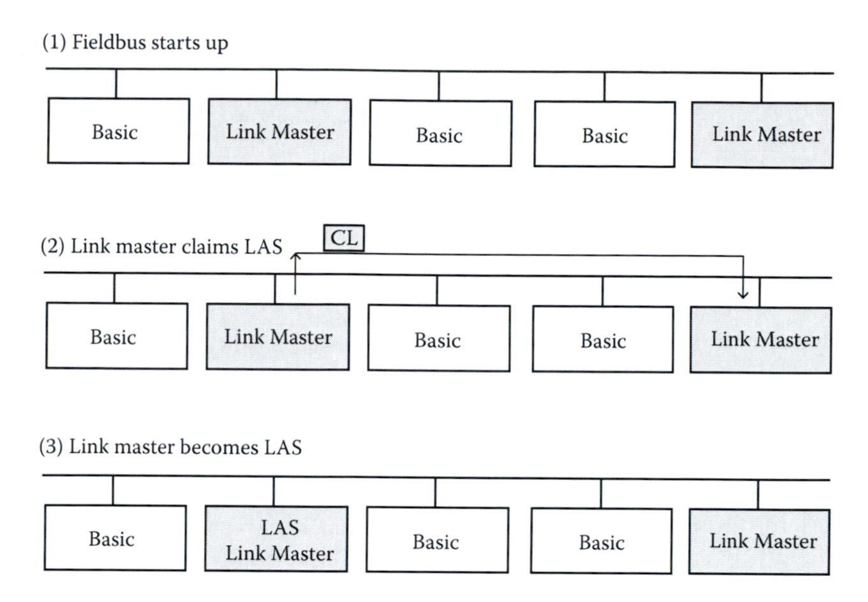

FIGURE 7.7 Process of an LM class device becoming a LAS. (From Yokogawa Electric Corporation. *Fieldbus Book—A Tutorial*, Technical Information TI 38K 02A01-01E, 2000.)

LAS issues a compel data (CD) to the device when its turn comes to upload data on the bus. The device then broadcasts or publishes its data in the buffer of all the devices connected to the bus. The device configured to receive data is called a subscriber. The scheduled data transfer scheme is shown in Figure 7.8.

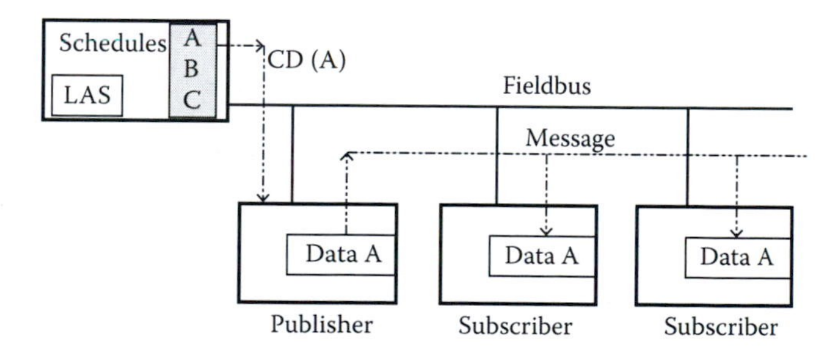

FIGURE 7.8 Scheduled data transfer technique. (From Glanzer D. A. *Technical Overview, Foundation Fieldbus*, FD-043, Rev 3.0, 1996 (Rev. 1998, 2003). Fieldbus Foundation, Austin, TX. Available at www.fieldbus.org/images/stories/technology/developmentresources/development_resources/documents/techoverview.pdf.)

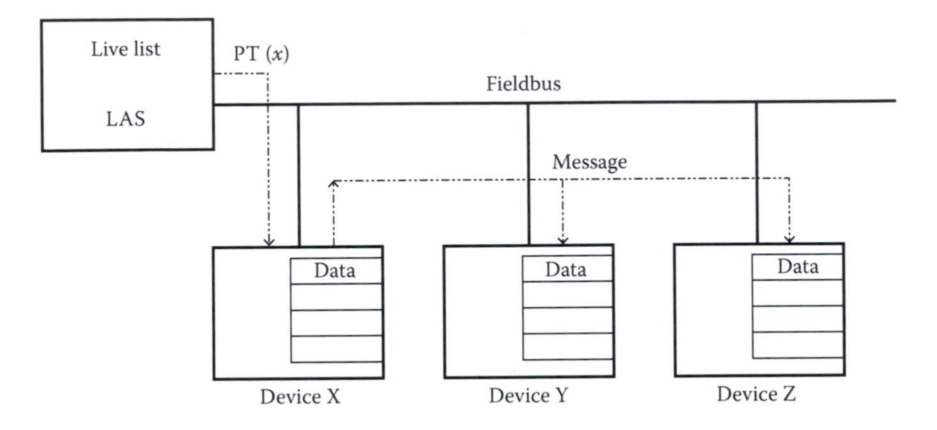

FIGURE 7.9 Unscheduled data transfer technique. (From Glanzer D. A. *Technical Overview, Foundation Fieldbus*, FD-043, Rev 3.0, 1996 (Rev. 1998, 2003). Fieldbus Foundation, Austin, TX. Available at www.fieldbus.org/images/stories/technology/developmentresources/development_resources/documents/techoverview.pdf.)

7.8.4.3.2 Unscheduled Communication

Unscheduled communication takes place in an asynchronous manner. LAS gives a chance to all the devices on the bus to send unscheduled messages between transmission of scheduled messages. LAS issues a pass token (PT) message to a device. This is shown in Figure 7.9. On receiving the PT, the device transmits messages until it is finished or until the "delegated token hold time" has expired—whichever is less. The message can be sent to a single destination or multiple destinations. In Figure 7.9, message from device x is sent to both devices y and z in a multicast manner.

Unscheduled data transfer uses either a client–server or a report distribution type of reporting for transferring data. Usually, such transfer scheme is undertaken for user-initiated changes such as set point change, tuning change, mode change, or upload/download.

7.8.4.3.3 LAS Operation

LAS operation can best be understood by the LAS algorithm shown in Figure 7.10. The flow chart shows the overall sequence of operations performed by LAS. If there is still some time left after completing the allotted job on issuance of CD, the algorithm may issue a probe node (PN), time distribution (TD), or PT as already programmed. Else, the system waits until it is time to issue the next CD. The system thus does the synchronous cyclic jobs interspersed with acyclic jobs assigned to it during programming.

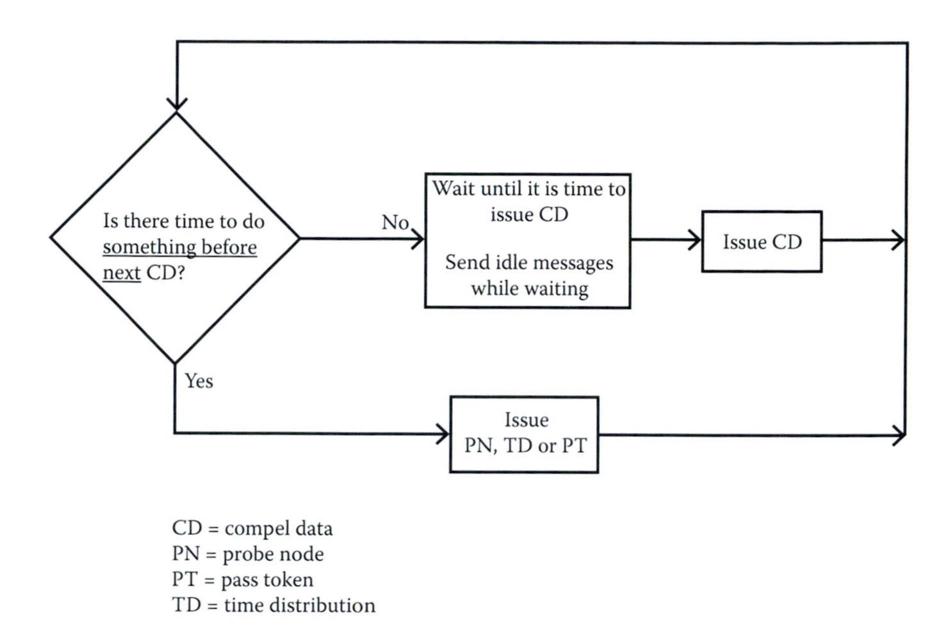

FIGURE 7.10 LAS algorithm. (From SMAR. *Fieldbus Tutorial—A Foundation Fieldbus Technology Overview*. USA, pp. 1–29. Available at http://www.smar.com/PDFs/catalogues/FBTUTCE.pdf.)

7.8.4.3.3.1 CD Schedule CD schedule contains a list of activities that are carried out by LAS in cyclic communication. The schedule is prefixed. Thus, at some precise point of time, LAS sends a CD message to a specific fieldbus device to put its data on the bus. The device then publishes its data on the bus and it goes to the subscribers. The CD schedule is the highest priority job of the LAS; the other jobs are carried out between the scheduled transfers.

7.8.4.3.3.2 Live List Maintenance "Live list" refers to the devices on the bus that responds to the PT issued by LAS. Devices may be added/deleted to/from the fieldbus as per the requirement. A device may also go off the list if it goes bad. It is the responsibility of LAS to maintain and update the list of live devices on the list.

LAS probes at least one node (i.e., one address) after it has completed one cycle of sending PTs to all the devices residing in the live list. A device may be added to the bus at any time. LAS periodically sends a PN message to the addresses not on the live list. If a device is detected at the node, it immediately sends back a probe response message and the device is added to the live list by the LAS. LAS also confirms the device by sending it a node activation message.

LAS removes a device from its live list if it does not respond to the PT or returns the same to the LAS consecutively for three successive tries.

Whenever a device is added or removed from the live list, LAS immediately broadcasts the changes in the live list to all the devices on the bus. Thus, each link master on the segment maintains a current and updated copy of the live list.

7.8.4.3.3.3 Data Link Time Synchronization A TD message is broadcast on the fieldbus by the LAS. This ensures that all the devices on the bus have exactly the same data link time. Scheduled cyclic communication on the bus and scheduled function block executions are based on such TD message information so that exact time synchronization is always maintained.

7.8.4.3.3.4 LAS Redundancy A segment or link has more than one link master. In case one LAS fails, another link master will take over. This will ensure that communication on the bus is not disturbed. Thus, the fieldbus is designed to be "fail operational."

7.8.5 Application Layer

The application layer consists of two sublayers: FAS and FMS. FAS manages data transfers, while FMS encodes and decodes user data.

7.8.5.1 FAS

The FAS provides services to the FMS with the help of scheduled and unscheduled communications of the DLL. The types of services provided by FAS are described by VCRs.

7.8.5.1.1 VCR

The VCR can be thought of as equivalent to the speed dial feature on a memory telephone. For an international call to be established, international access code, country code, exchange code, and finally the specific telephone number are required. These details are entered in the memory of the telephone and a "speed dial number" is assigned to it. Whenever a call is to be established with this particular telephone number, only this speed dial number is entered for dialing to take place. Similarly, after configuration, only the VCR number is needed to be communicated with another fieldbus device.

Just as there are different types of calls—such as calls to different individuals belonging to different countries, conference calls, and call collect—there are different types of VCRs to cater to various devices or applications at the same time. The VCR guarantees that message goes to the specific and correct partner.

When a message is transferred, it goes through the VCR by adding PCI before going to the physical wire. A VCR is identified by a device-local identifier called "index" specified in the application layer. A VCR can also be identified from other devices with a "DL-address" specified in the data link layer. A VCR has a memory to save messages.

Network management configures a particular VCR by providing it with correct index and DL-address. Thus, no two VCRs can have identical index and DL-address.

There are three different types of VCRs available: client–server VCR type, source–sink (report distribution) VCR type, and publisher–subscriber VCR type.

7.8.5.1.1.1 Client–Server VCR Type It is a user-initiated one-to-one, queued, unscheduled, prioritized communication between devices on the fieldbus. Queued means that messages are received as they are sent, with message priorities remaining intact. In this method, messages are not overwhelmed as in the publisher–subscriber type. The transfers in this VCR type is flow controlled and has a retransmission scheme in case of corrupted message transfers.

Operator-initiated messages—such as set point change, alarm acknowledge, tuning parameter change and access, and device upload/download—are done in client–server VCR type.

A device, when it receives a request from the LAS via PT, sends a request message to another device. The requester is the client, while device that received the request is called a server.

A typical example is a human–machine interface (HMI) wanting to read data from a function block. The former is the client and the latter is the server. When the request reaches the server from the client, the former sends the data to the client, which in this case is the HMI.

7.8.5.1.1.2 Publisher–Subscriber VCR Type It is used for buffered, one-to-many communications to transfer process critical data such as process variable. Buffering means the last published data on the network is maintained and held until a new data overwrites it. The data producer is called a publisher and the data receiver is called a subscriber.

In this VCR type, data can be transferred on a purely periodic basis by properly scheduling CD in the LAS. Unscheduled transmission of information is also possible. An attribute in the VCR would indicate which method is used.

An example of this VCR type is the output from an analog input (AI) linking process value input of the PID control block.

7.8.5.1.1.3 Source–Sink (Report Distribution) VCR Type It is a one-to-many, one-way communication without schedule. It is used to broadcast or

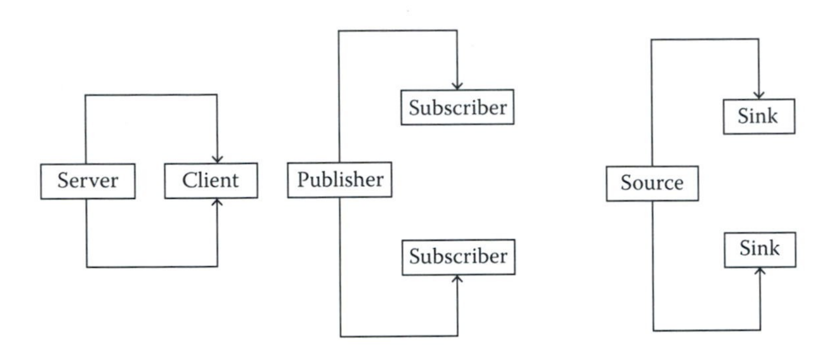

FIGURE 7.11 Three types of VCRs. (From N. Mathivanan. *PC-Based Instrumentation: Concepts and Practice*. Prentice Hall of India, p. 587, 2007.)

TABLE 7.2
Different VCR Types

Client–Server VCR Type	Source–Sink VCR Type	Publisher–Subscriber VCR Type
Bidirectional	Unidirectional	Unidirectional
Queued	Queued	Buffered
Scheduled by user	Scheduled by user	Scheduled by network
Used for operator messages	Used for event notification and trend reports	Used for publishing data

multicast event and trend reports. The destination address may be predefined or included separately with each report. It is sometimes called the report distribution VCR type.

Transfer in this mode takes place in a queued manner. Messages are delivered in the order in which they are transmitted. The transfers are unscheduled and take place in between the scheduled ones. This mode is typically used by fieldbus devices to send alarm reports to operator consoles.

The three types of VCRs, viz., client–server, publisher–subscriber, and source–sink (report distribution) types are shown in Figure 7.11.

7.8.5.1.1.4 Comparison A comparison between the different VCR types is shown in Table 7.2.

7.8.5.2 FMS

Applications use FMS services to send messages between themselves across the fieldbus using a standard set of message formats. FMS is a model for applications to interact over the fieldbus. Virtual field device (VFD) and object dictionary (OD) belong to this model. Message formats,

TABLE 7.3
Object Dictionary

Index	Object Description
0	–
1	
2	
–	–
n	*n*

communication services, and protocol behavior are the services that FMS provides for the user application layer.

7.8.5.2.1 Object Dictionary

OD can be thought of as a look-up table that gives information about a value, such as a data type, that can be read from or written into a device. Data communication over the fieldbus is described by an "object description." Such descriptions are structured together in OD. This is shown in Table 7.3.

As per the table, index 0 represents the object dictionary header. It provides a description of the dictionary itself.

7.8.5.2.2 Virtual Fieldbus Device

VFDs are used to remotely view local device data that are described in the object dictionary. An identifier, maintained in a VCR, identifies the VFD.

A Foundation Fieldbus device will have at least two VFDs—one is management VFD and the other is function block VFD. A field device may have two or more function block VFDs. Network and system management functions reside in management VFD. Such a VFD configures network parameters that include VCRs. It also manages fieldbus devices. A block schematic showing management and function block VFDs is shown in Figure 7.12.

As shown in Figure 7.12, a network management VFD contains objects that other layers of the communication stack use—these include FAS, FMS, and DLL. System management VFDs assign addresses and physical device tags. They also maintain function block schedules for the function blocks and distribute application time. System management VFDs also help in finding a device or a function block tag.

System management and network management VFDs can be read and written into by using FMS Read and FMS Write services.

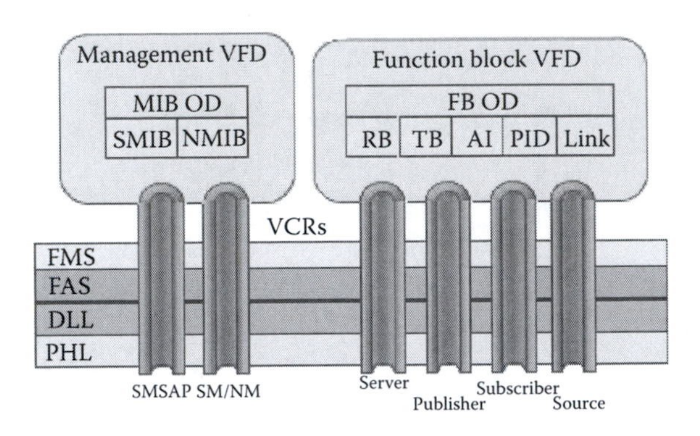

FIGURE 7.12 **(See color insert.)** Block schematic showing management and function block VFDs. (From Yokogawa Electric Corporation. *Fieldbus Book—A Tutorial*, Technical Information TI 38K02A01-01E, pp. 1–33, 2000.)

7.8.5.2.3 FMS Services

FMS provides different services to access different FMS objects. These are variable access, event management, context management, domain management, upload/download, and program invocation.

7.8.5.2.3.1 Variable Access Table 7.4 shows the following FMS services that allow the user application to access and change variables associated with an object description. Variable access is used for data publishing and reporting trends. When a variable is either a record or an array consisting of multiple variables, it is possible to transfer it as a whole or only one component assigned with a "subindex."

TABLE 7.4
Variable Access Services

Read	Read value of a variable
Write	Write value of a variable
Information Report	Send value as publisher or source
Define Variable List	Define a list of variables to send
Delete Variable List	Delete a list of variables

Source: From Glanzer D. A. *Technical Overview, Foundation Fieldbus*, FD-043, Rev 3.0, 1996 (Rev. 1998, 2003). Fieldbus Foundation, Austin, TX. Available at www.fieldbus.org/images/stories/technology/developmentresources/development_resources/documents/techoverview.pdf.

7.8.5.2.3.2 Event Management When an application detects something very important, this scheme is used. Failure, data update, and alarm are such examples. When such an event occurs, the source–sink model is used repeatedly until it is acknowledged by the client–server model. Such an event can either be disabled or enabled through another event-related service. Table 7.5 shows the different event management services.

7.8.5.2.3.3 Context Management Table 7.6 shows the context management services available that are used to establish and release VCRs with, and determine the status of, a VFD. Context management services help in managing the connection between applications. The first two services in the table (i.e., *initiate* and *abort*) are a must irrespective of the type of application.

TABLE 7.5
Event Management Services

Event Notification	Report an event as source
Acknowledge Read Value of a Variable	Acknowledge an event
Alter Event Condition Monitoring	Disable or enable an event

Source: From Glanzer D. A. *Technical Overview, Foundation Fieldbus*, FD-043, Rev 3.0, 1996 (Rev. 1998, 2003). Fieldbus Foundation, Austin, TX. Available at www.fieldbus.org/images/stories/technology/developmentresources/development_resources/documents/techoverview.pdf.

TABLE 7.6
FMS Context Management Services

Initiate	Establish communications
Abort	Release communications
Reject	Reject improper service
Status	Read a device status
Unsolicited Status	Send unsolicited status
Identify	Read vendor, type, and version

Source: From Glanzer D. A. *Technical Overview, Foundation Fieldbus*, FD-043, Rev 3.0, 1996 (Rev. 1998, 2003). Fieldbus Foundation, Austin, TX. Available at www.fieldbus.org/images/stories/technology/developmentresources/development_resources/documents/techoverview.pdf.

TABLE 7.7
FMS Upload/Download Services

Request Domain Upload	Request upload
Initiate Upload Sequence	Open upload
Upload Segment	Read data from device
Terminate Upload Sequence	Stop upload
Request Domain Download	Request download
Initiate Download Sequence	Open download
Download Segment	Send data to device
Terminate Download Sequence	Stop download
Generic Initiate Download Sequence	Open download
Generic Download Segment	Send data to device
Generic Terminate Download Sequence	Stop download

Source: From Glanzer D. A. *Technical Overview, Foundation Fieldbus*, FD-043, Rev 3.0, 1996 (Rev. 1998, 2003). Fieldbus Foundation, Austin, TX. Available at www.fieldbus.org/images/stories/technology/developmentresources/development_resources/documents/techoverview.pdf.

7.8.5.2.3.4 Domain Management Domain refers to a continuous memory area, which can either be program area or data area. Data can be downloaded in the domain for future use or uploaded from the domain with the help of FMS services. Table 7.7 shows the different FMS services that allow the user application layer to upload and download a domain in a remote device. This service is required for complex devices such as PLCs. For using such a service, a "domain" is used, which represents a memory space in a device.

7.8.5.2.3.5 Program Invocation Program execution in a device can be controlled remotely with the help of the Program Invocation (PI) service, shown in a tabular form in Table 7.8.

By using the download service, a device can download a program into a domain of another device and then remotely operate the program by issuing PI service requests.

7.8.5.2.3.6 OD Services The "Get OD" service provides an object description. Downloading an OD can be stopped by the "Terminate Put OD" service. Table 7.9 shows the different services available under OD services.

TABLE 7.8
FMS Program Invocation Services

Create Program Invocation	Create a program object
Delete Program Invocation	Delete a program object
Start	Start a program
Stop	Stop a program
Resume	Resume a program execution
Reset	Reset the program
Kill	Remove the program

TABLE 7.9
FMS Object Dictionary Management Services

Get OD	Read an OD
Initiate Put OD	Start downloading OD
Put OD	Download an OD
Terminate Put OD	Stop downloading OD

7.9 TECHNOLOGY OF FOUNDATION FIELDBUS

Communication in Foundation Fieldbus is based on OSI. It uses only three layers, viz., physical, data link, and application layer of the OSI. The data link and the application layers together are called the *communication stack*. For proper operation of the fieldbus, it defines another layer called layer 8 and also termed the user application layer. This layer is standardized by the Fieldbus Foundation based on blocks. The different blocks in the user layer are resource block, function block, and transducer block. Devices on the fieldbus are configured by resource and transducer blocks, while control strategy is built using function block.

Data transmission and reception on the fieldbus take place either by electrical wires or optical means at a rate of 31.25 kbps via the physical layer. Although a speed of 31.25 kbps may seem to be slow when compared with other high-speed transmissions, it is sufficient to cater to the needs of instrumentation and control and their corresponding automation aspects.

Signals on the fieldbus are encoded using the Manchester-coded Biphase-L technique. The voltage supplied to the system can vary from 9 to 32 V. When a device transmits, it delivers ±10 mA at a data rate of 31.25 kbps into a 50 Ω resistor generating a 1 V peak-to-peak voltage. The DC supply voltage is thus modulated by this 1 V signal.

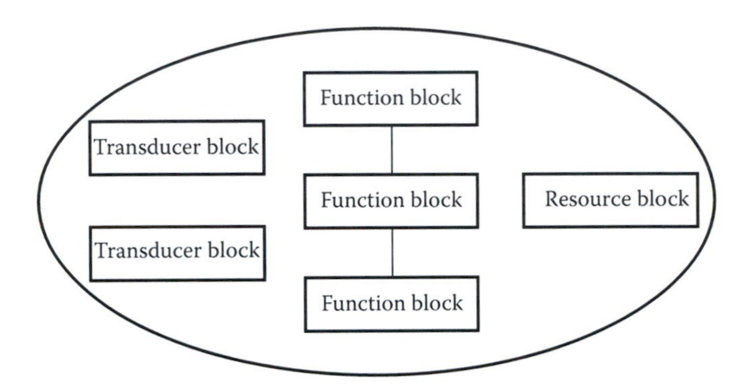

FIGURE 7.13 Blocks in user layer.

7.9.1 User Application Blocks

Layer 8, the user layer, consists of several blocks: resource block, function block, and transducer block. This is shown in Figure 7.13. The user layer defines blocks and objects that represent the functions and data available in a device. For Foundation Fieldbus, HMI is done through these blocks and objects—rather than through a set of commands as is followed in most communication protocols.

The blocks can be thought of as processing units. They accept inputs; their settings can be changed as per system requirements. They also have an algorithm, which when run, can produce outputs. The blocks also communicate with each other.

7.9.2 Resource Block

A resource block indicates the resources available with a device. This may include device type and revision, serial number, manufacturer ID, and resource state. A device can have only a resource block. The resource block controls the overall hardware of the device and function blocks within the VFD.

The resource block contains its hardware-specific characteristics. It does not have any input or output parameter. The algorithm within the resource block monitors and checks the device hardware performance. It must always be in automatic mode. With its help, the cause of any problem related to the device can be fixed and appropriate actions taken.

7.9.3 Function Block

Function block(s) in a fieldbus device performs the different functions required in a process control operation. A fieldbus device may contain one or more function blocks. In process control systems, operational requirements normally vary—depending on the particular process and the end-product quality. Thus, the Fieldbus Foundation has designed a standard function block as normally applicable in process industries. In addition, some other function blocks have been recommended. Most importantly, with the help of models and parameters existing in the function block, the design engineer can configure, maintain, and customize the applications.

The system architecture document of Foundation Fieldbus says

> One of these models, the function block model, has been specified within the architecture to support low level functions found in manufacturing and process control. Function blocks model elementary field device functions, such as analog input functions and proportional integral derivative (PID) functions. The function block model has been supplemented by the transducer block model to decouple function blocks from sensor and actuator specifics. Additional models, such as the "exchange block" model, are defined for remote input/output and programmable devices.
>
> The function block model provides a common structure for defining function block inputs, outputs, algorithms and control parameters and combining them into an Application Process that can be implemented within a single device. This structure simplifies the identification and standardization of characteristics that are common to function blocks.

A function block has input, output, and contained parameters. Data generated in a block is available at its output and acts as the input to another

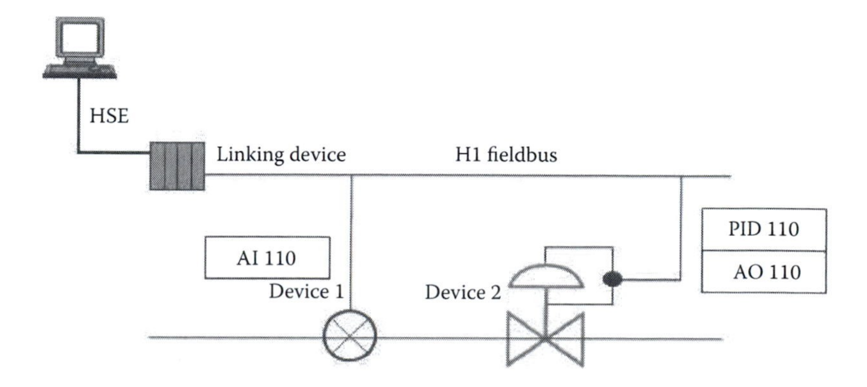

FIGURE 7.14 Control loop using function blocks. (From SMAR. *Fieldbus Tutorial—A Foundation Fieldbus Technology Overview*. USA, pp. 1–29. Available at http://www.smar.com/PDFs/catalogues/FBTUTCE.pdf.)

block. Thus an AI block's output acts as the input to the PID block. Its output is the input to the AO block, whose output may be fed to the PID block for process control purpose.

Function blocks are classified as (a) a standard block as specified by Fieldbus Foundation (b), an enhanced block for including additional parameters, and (c) an open block or vendor-specific block meeting the needs of specific process requirements.

Figure 7.14 shows a control loop that uses function blocks in fieldbus devices.

TABLE 7.10
Different Function Blocks

Part 2 Blocks	**Major in Control and Measurement**
AI	Analog Input Block
DI	Discrete Input Block
ML	Manual Loader Block
BG	Bias/Gain Station Block
CS	Control Selector Block
PD	P, PD Controller Block
PID	PID, PI, I Controller Block
RA	Ratio Station Block
AO	Analog Output Block
DO	Discrete Output Block
Part 3 Blocks	**Enhanced Blocks**
DC	Device Control Block
OS	Output Splitter Block
SC	Signal Characterizer Block
LL	Lead Lag Block
DT	Dead Time Block
IT	Integrator (Totalizer) Block
	(More blocks are under development)
Part 4 Blocks	**Multiple I/O Blocks**
MAI	Multiple Discrete Input Block
MDI	Multiple Analog Input Block
MAO	Multiple Discrete Output Block
MDO	Multiple Analog Output Block
Part 5 Blocks	**IEC61131 Blocks**
	(Under development)

Source: From Yokogawa Electric Corporation. *Fieldbus Book—A Tutorial*, Technical Information TI 38K02A01-01E, pp. 1–33, 2000.

7.9.3.1 Function Block Library

Table 7.10 shows a comprehensive list of the library of function blocks as defined by the Fieldbus Foundation. The 10 basic function blocks shown in Part 2 are the fundamental ones supported and required in process control and measurement. Thus, the parameters of such blocks must be known to all systems. Part 3 blocks are used for advanced measurements, while Part 4 blocks provide I/O interface to the outside world.

Attributes of a function block, such as data type, its name, minimum values, and maximum values, wherever applicable are contained in the DD (device description) block. For every function block, there is a DD—be it basic, extended, or custom-type function block. The basic and extended function blocks are documented by the Fieldbus Foundation in their respective function block specification.

7.9.3.2 Function Block Scheduling

Figure 7.15 shows the scheduling of function blocks. A schedule building tool is used to generate the proper time schedules for operations of both LAS and different function blocks. The schedules that have been built on the control loop operation are shown in Figure 7.14. The schedule building

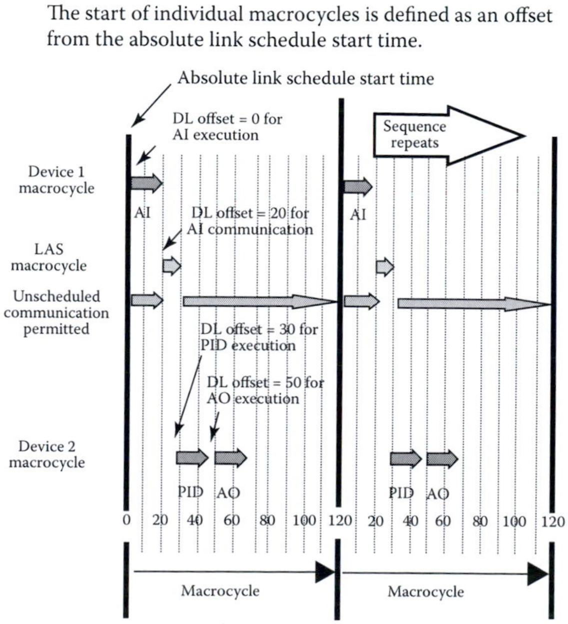

FIGURE 7.15 **(See color insert.)** Scheduling of function blocks. (From SMAR. *Fieldbus Tutorial—A Foundation Fieldbus Technology Overview.* USA, pp. 1–29. Available at http://www.smar.com/PDFs/catalogues/FBTUTCE.pdf.)

TABLE 7.11
Schedules of Different Function Blocks

Schedule of Different Blocks	Offset Values
AI execution	0
AI communication	20
PID execution	30
AO communication	50

Source: Yokogawa Electric Corporation. *Fieldbus Book—A Tutorial*, Technical Information TI 38K02A01-01E, pp. 1–33, 2000.

tool contains start time offsets of the different function blocks from the "absolute link schedule start time."

A "macrocycle" is defined as a single iteration of a schedule within a device. There can be device macrocycle or LAS macrocycle. Table 7.11 shows typical time offsets of AI, AO, and PID blocks from the absolute link schedule start time.

The AI block is executed at offset 0. At offset 20, LAS issues a CD to the AI function block buffer residing in the transmitter and then data in the buffer is published on the fieldbus. At offset 30, PID function block is executed, followed by execution AO function block at offset 50.

The different offsets represent the exact time at which a particular function is to start its operation with respect to absolute link schedule start time. The beginning of the macrocycle represents a common and unified start time for all concerned function block connected to the link and for the LAS link-wide schedule. Different function block execution and their corresponding data transfers on the bus are thus synchronized in time.

It is apparent from Figure 7.15 that from offset 20 to offset 30, the bus cannot be used for sending unscheduled messages. This is so because during this time, AI publishes (communicates) data on the bus. AI obtained this data during its execution from offset 0 to offset 20.

7.9.3.3 Application Clock Distribution

Foundation Fieldbus supports an application clock distribution function. It is usually set and adjusted to local time of the day or to some Universal Coordinated Time. System management contains a time publisher that periodically sends an application clock synchronization schedule message to all connected fieldbus devices. The data link scheduling time is sampled and sent along with the application clock message. This allows the receiving devices to adjust their local application time. In between the

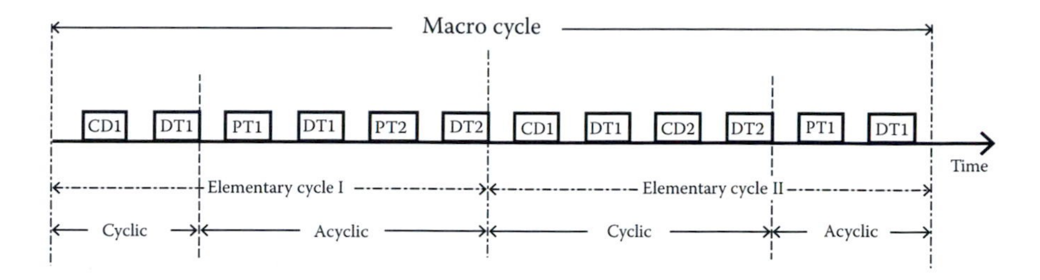

FIGURE 7.16 Macrocycle and elementary cycles.

synchronization time message, the application clock time is independently maintained in each device based on their internal clocks.

Application clock synchronization allows the fieldbus devices to time stamp data throughout the fieldbus network. If a backup application clock publisher is maintained in the network, the same would become active when the currently active time publisher should fail.

7.9.3.4 Macrocycle and Elementary Cycle

In process industries, continuous control of process variables is a basic requirement for maintaining the proper health of the system. A precise cyclic update of process variables is what is needed to ensure the above. Such demands and requirements from process variables are more or less fixed for a given plant. Also occasional and sporadic events (acyclic) such as alarm reporting and set point changes also need to be accommodated in the overall control scheme of things. LAS takes the overall responsibility of such cyclic and acyclic communications in macrocycles. A macrocycle is divided into elementary cycles, which consist of time needed for cyclic and acyclic communications. This is shown in Figure 7.16. It shows a system consisting of two devices requiring cyclic updates. An elementary cycle consists of cyclic and acyclic tasks. Scheduling of periodic tasks (cyclic) is so done that some time is assigned in the elementary cycle to address the need of aperiodic tasks (acyclic), should the need arises.

7.9.3.5 Device Address Assignment

For a device to work properly on a fieldbus, two requirements must be met: the device must have a unique network address and a physical device tag. Network address assignments are done by the configuration tool using system management services.

Assignments of network address follow these steps:

- Unconfigured (new to the network) devices join the network at one of the four special default addresses.

- A configuration tool chooses a used permanent address and assigns the same to the device using system management services.
- The same configuration tool will then assign a physical device tag to the device using system management services.
- The above procedure is followed for all new devices that want to have their entry into the network.
- These newly entered network devices store their physical device tag and node address in nonvolatile memory. This procedure ensures the devices retain their separate identifications even in case of power failure or system shutdown.

7.9.3.6 Tag Service

A device or a variable can be traced with the help of tag service. System management supports such a service. When a "find tag query" is broadcast to all the connected field devices, each of them starts searching its VFD for the requested tag and returns back all the required information. In case the tag is found (it includes network address, VFD number, VCR index, and OD index), the complete path is known and the host or maintenance device can then access data from the tag.

7.9.4 Transducer Block

A transducer block models sensors and actuators and connects function blocks to local input/output functions. Transducer blocks isolate the function blocks from the hardware details of a given device. Sensor outputs are accepted by transducer blocks and then write to actuators. Traditional sensors such as pressure and temperature transmitters can be mapped into a transducer block. It is connected to a function block through the channel parameter of the function block.

While a particular function block carries out the function that is assigned to it, a transducer block is dependent on the kind of measurement. That is, for temperature, pressure, or flow measurements, it employs differing measurement principles; however, in all the cases, it provides an analog value. A transducer block performs various functions such as digitizing, scaling, and filtering, which are needed to convert the sensor output to a value that can be accepted by the function block.

Generally there is one transducer block per device channel. Multiplexers allow multiple channels for a single transducer block. The different blocks, viz., resource, transducer, and function blocks, and their interconnections are shown in Figure 7.17.

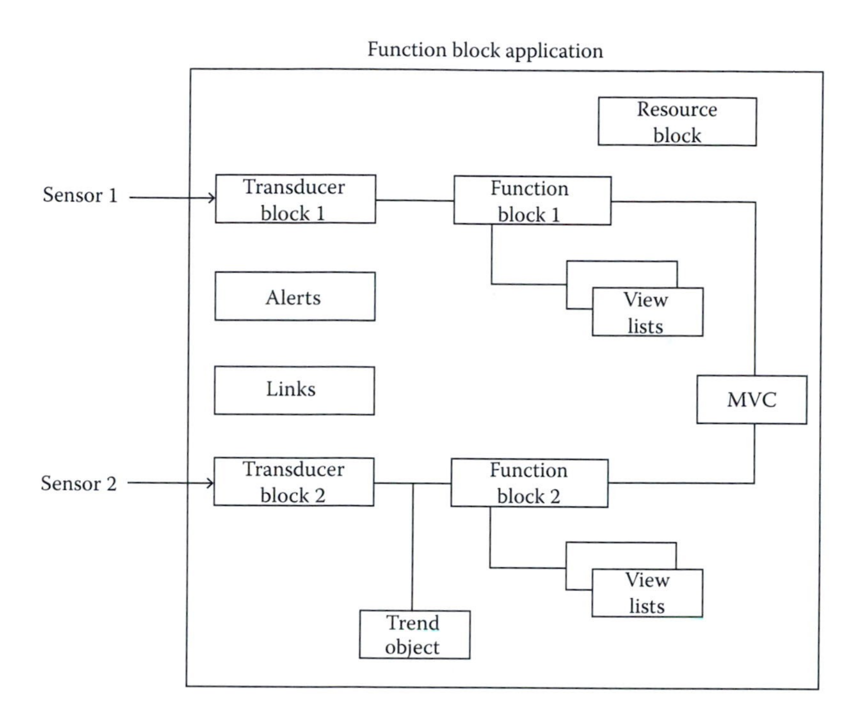

FIGURE 7.17 Different blocks and object representation in user layer. (From Glanzer D. A. *Technical Overview, Foundation Fieldbus*, FD-043, Rev 3.0, 1996 (Rev. 1998, 2003). Fieldbus Foundation, Austin, TX. Available at www.fieldbus.org/images/stories/technology/developmentresources/development_resources/documents/techoverview.pdf.)

7.9.5 Support Objects

Some additional objects are defined in the user application: link object, trend object, alert object, multivariable container (MVC) object, and view object. Link objects define the links between various function block inputs and outputs internal to the device and across the fieldbus network. Trend objects are used for local trending of function block parameters for host accessing or other devices. Alert objects allow alarm reporting and events on the fieldbus. MVC objects serve to encapsulate multiple function block parameters to optimize communications for report distribution and publisher–subscriber transactions. View objects refer to some predefined groupings of block parameter sets that can be displayed by HMI.

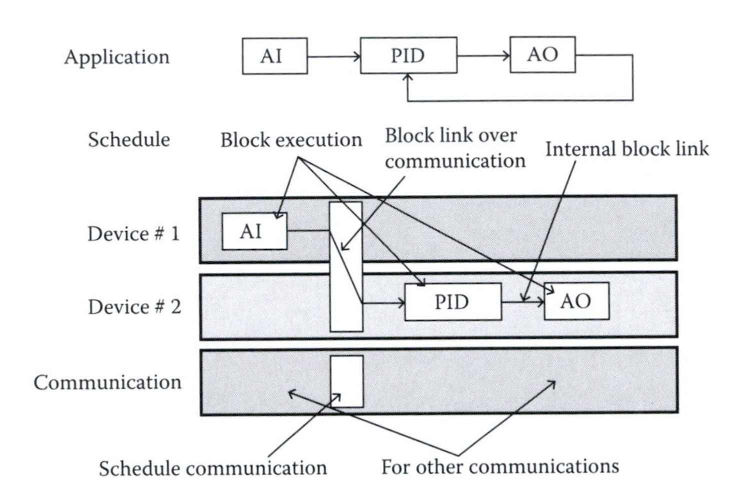

FIGURE 7.18 Linking and scheduling of blocks. (From Yokogawa Electric Corporation. *Fieldbus Book—A Tutorial*, Technical Information TI 38K02A01-01E, pp. 1–33, 2000.)

7.10 LINKING AND SCHEDULING OF BLOCKS

Figure 7.18 shows how the function blocks in a device or a function block and a VCR are connected via link objects. It shows the PID control of a process consisting of AI, PID, and AO function blocks.

A function block must get its input parameters before execution of is algorithm. The executed algorithm is then published over the fieldbus. The system management in a field device starts the function blocks according to the function lock schedule, which is preprogrammed. The publishing of the executed algorithm is done by LAS, which sends a CD schedule to the device. The LAS schedule and the function block schedule are defined as offsets in their respective macrocycles. They must be properly configured so that their actions—execution of actions by the function blocks and publication of data by the LAS on the bus—must not overlap. Device macrocycle and LAS macrocycle are shown in Figure 7.15.

7.11 DEVICE INFORMATION

Devices used on the fieldbus need to have HMI, fieldbus configuration, and maintenance procedure for their proper operation. These need more information about the devices. Foundation Fieldbus has standardized several files to make devices interoperable and help engineers' handling of the system much easier.

7.11.1 Device Description

In Foundation Fieldbus technology, a device is provided with three device support files to ensure interoperability: two DD files and one capability file. A DD provides information about blocks. A function block parameter can be read and displayed properly by its data type and display specifications. A device, when joined to the system, needs to install its DD to use its full functionality without any need to update the host software. DDs are platform and system independent.

A DD can be thought of as a "driver" of a device—like the drivers that a PC has to operate the printer, scanner, and other devices connected to it. A control system or host can operate with the device only if it has the device's DD.

7.11.2 Device Description Language

DDs are written in a standardized programming language called Device Description Language (DDL). Device functionality and data semantics can be described by DDL. This is then compiled with "tokenizer" software to generate "DD binary" files.

A DD binary has two files: a DD binary with extension ".ffo" and the second one is a DD symbol list with extension ".sym". Once these two files are installed, full access to the device would be ensured. Instead of writing the full functionalities of each device individually, the common portion of the devices' functionalities is included in DD library, which also includes a dictionary. The special features of each device are included in DDL.

7.11.3 DD Tokenizer

A DD tokenizer is a software tool that converts the DD source input files into DD output files by replacing keywords and standard strings in the source file with fixed tokens. The standard features associated with resource, function, and transducer blocks are included in DD library with the help of tokenizers.

7.11.4 DD Services

Device Description Services (DDS) is a software for HMI. It is a library function that helps in reading the device descriptions. A device can be added to the system, and it would work nicely if the DD of this device is added to the host or control system.

With the help of DDS, device descriptions only are made available to the system and not their physical values. The latter can be read from the FMS communication services. DDS helps in connecting devices from different manufacturers on the same fieldbus system.

7.11.5 DD Hierarchy

The hierarchical structure of DDs is shown in Figure 7.19. The Fieldbus Foundation has defined such a structure so that configuration of devices becomes very easy. There are four levels in the hierarchy and are referred to as universal parameters, function block parameters, transducer block parameters, and manufacturer-specific parameters.

Universal parameters refer to common attributes such as revision, mode, and tag. All blocks must include universal parameters. Parameters for standard function blocks are defined in function block parameters. In transducer function blocks, parameters for processes are defined, such as pressure, flow, and temperature. In the manufacturer-specific parameters, a manufacturer may add additional parameters to the function block parameters and transducer block parameters.

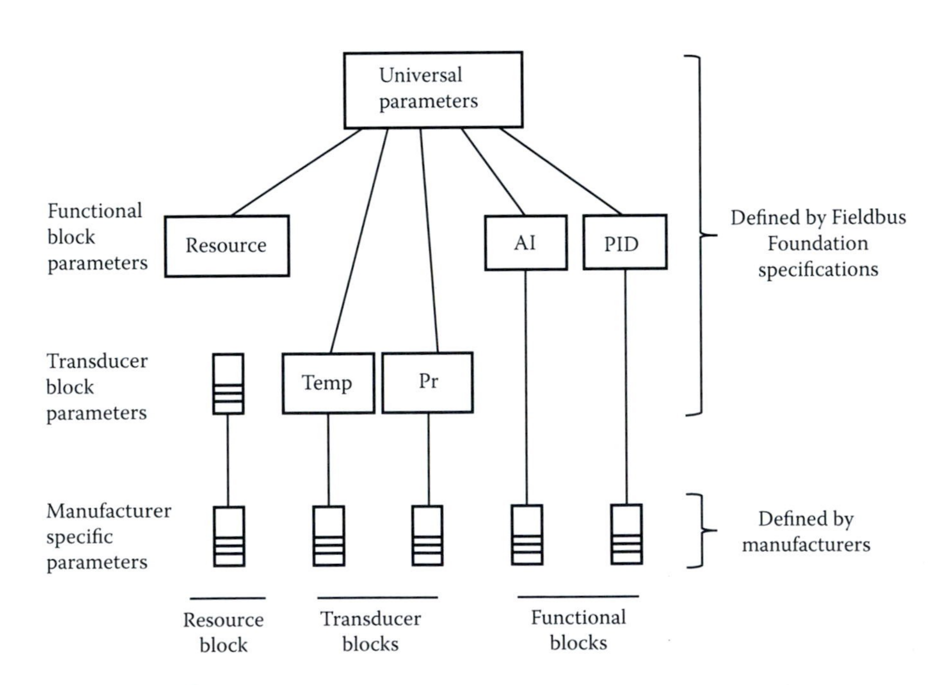

FIGURE 7.19 Hierarchical structure of DD. (From SMAR. *Fieldbus Tutorial—A Foundation Fieldbus Technology Overview*. USA, pp. 1–29. Available at http://www.smar.com/PDFs/catalogues/FBTUTCE.pdf.)

7.11.6 Capabilities File

The capabilities file tells the host about the resources a device has in terms of function blocks and VCRs. This enables the host to configure a device offline. It is the responsibility of the host to ensure that only functions supported by the device are allocated to it. A capabilities file has an extension ".cff". A capabilities file is often called "CFF," which stands for Common File Format.

7.11.7 Device Identification

A device on the fieldbus can be identified in one of the three following ways: device identifier (ID), physical device (PD) tag, and node (physical) address.

A device ID is a 32-byte unique number, and no two devices can have the same device ID. The device ID is burnt into the device by the manufacturer of the device and can never be changed. The PD tag is a unique name assigned to the device in the plant by the user. It is used to identify a device for the specific purpose (use) in the plant. It is also 32 bytes in length. When a device is replaced, the new device is assigned the old device's PD tag number. The physical or node address is a unique 1-byte number in a fieldbus segment and is assigned by the user at the time of configuring the network. Since the node address length is very small compared with either the device ID or physical device tag byte length, it is commonly used when communication is needed in the network.

7.12 REDUNDANCY

The Foundation Fieldbus is a two-level architecture—H1 at the device level and HSE at the host level of the control network. HSE is Ethernet based and is central to all control operations in the system. HSE ties up all the subsystems, and the visibility and activity of the devices at the field level and the control loops therein are all dependent on HSE. Availability of host-level network is therefore a must for proper system operation. Redundancy is included in the system at different levels to ensure availability of resources in times of breakdown. Redundancy can be added at host level, device level, media level, and network level. HSE has built-in robustness to ensure fault tolerance. Decentralization of operations is another measure of fault tolerance. When many controls are included in a single controller, failure of the same would mean a considerable part of the plant to be out of order and it may have some catastrophic influence on the safety of the plant. Decentralization of operations of the loops would ensure a small part of the plant to be out of order in case of single controller failure. Another major advantage of decentralization is localizing the fault.

7.12.1 Host-Level Redundancy

Redundancy at the host or central level is used to ensure high availability and thereby prevent disruption in production and consequent downtime and heavy losses. Host level ties the network and the field level devices together, and any breakdown here may be life threatening for the entire plant. There can be various methods to achieve host-level redundancy: media redundancy, network redundancy, and network and media redundancy.

7.12.1.1 Media Redundancy

Media redundancy is independent on the protocol used. Figure 7.20 shows a media redundancy scheme that uses star topology having redundant port receivers (dual port transreceivers). Such a transreceiver has three ports: one for the device itself and the rest two are for primary and secondary communication paths. The transreceiver switches from the primary to the secondary path when the former loses its link. When the transreceiver fails, both the links are lost and consequently both the paths fail.

Figure 7.21 uses a single-ring topology to achieve media redundancy. The linking devices are connected to the hubs. The topology provides a dual communication path—both clockwise and anticlockwise. When one path fails, communication can still take place through the other.

It is advisable to use several smaller hubs instead of a central hub connecting many devices. Failure of a smaller hub influences the system in a lesser way than the failure of a bigger hub would on the communication.

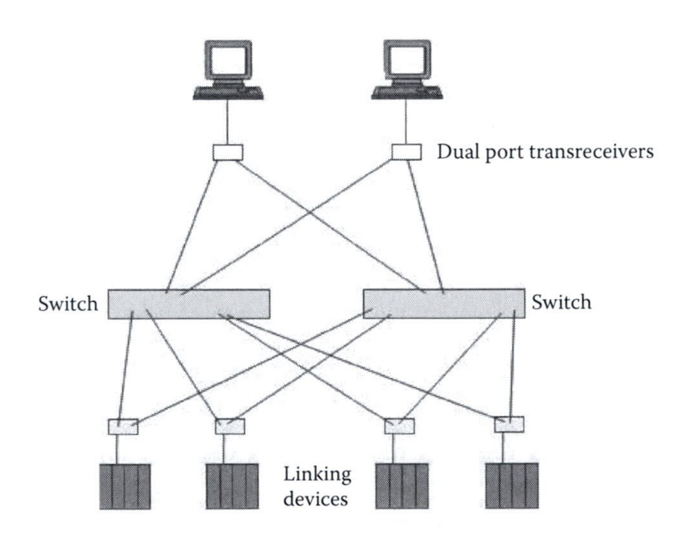

FIGURE 7.20 Media redundancy using dual-port transreceivers. (From J. Berge. *Fieldbuses for Process Control: Engineering, Operation, and Maintenance*. ISA, USA, p. 143, 2004.)

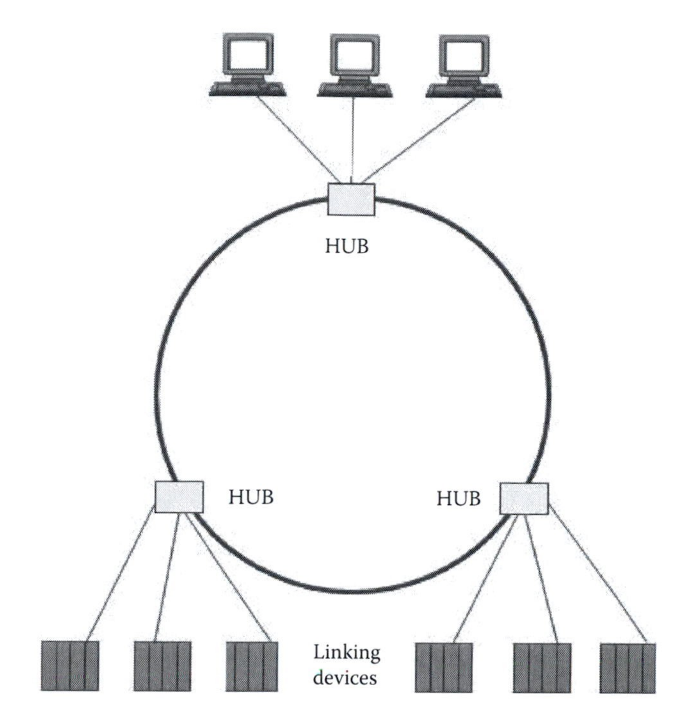

FIGURE 7.21 Media redundancy using single-ring topology. (From J. Berge. *Fieldbuses for Process Control: Engineering, Operation, and Maintenance*. ISA, USA, p. 144, 2004.)

Again, the distance from a hub to a device should be kept small to reduce the chances of damaging a nonredundant cable segment.

7.12.1.2 Network Redundancy

A complete host-level network redundancy scheme is shown in Figure 7.22. For its implementation, it needs to have two separate communication paths at each stage. It is used when very high availability is desired. The primary and its corresponding secondary (redundant) device must be kept on separate networks.

The scheme can be implemented in two ways; the primary and secondary devices of the redundant device may have a single port. Alternatively, the primary and secondary devices may have two ports. Both the ports have self-diagnostics, whereas communication takes place via a port at any given time. In this scheme, workstations are provided with dual network interface cards.

Alternate port and segment are automatically used in case of failure of primary network path. Such a changeover is totally bumpless and transparent to the operator. Each and every port on the network has its own separate IP address, making diagnostics very easy.

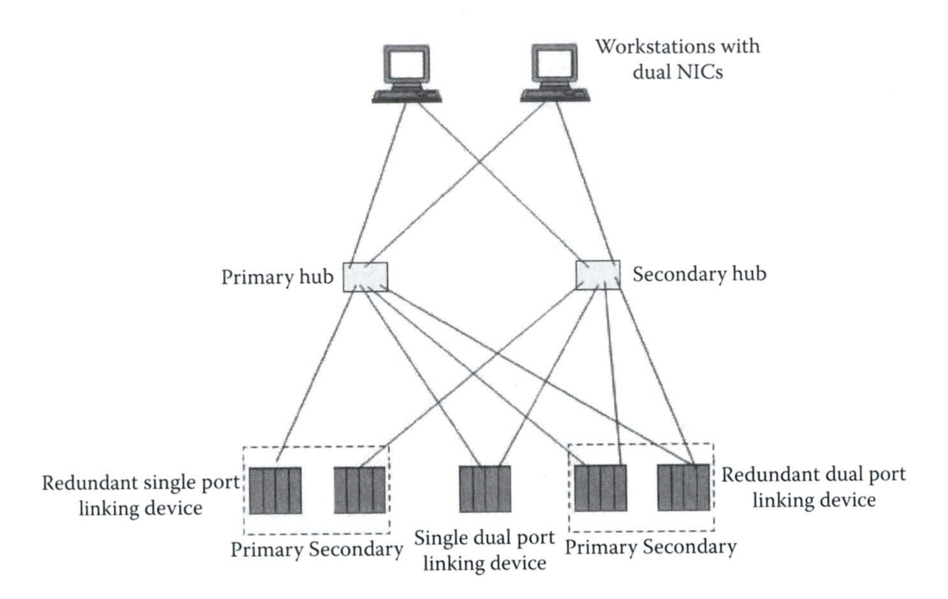

FIGURE 7.22 Network redundancy scheme. (From J. Berge. *Fieldbuses for Process Control: Engineering, Operation, and Maintenance*. ISA, USA, p. 385, 2004.)

7.12.1.3 Media and Network Redundancy

Such a redundancy scheme is undertaken when extremely reliable fault tolerance is needed. This is shown in Figure 7.23. It uses a dual-ring topology scheme creating four paths across the network.

7.12.2 Sensor Redundancy

Sensor redundancy is achieved by employing at least two sensors to measure the value of a process variable at a single point. The two sensor outputs are connected to the two input points of the transmitter. With the help of the standard selector block in the Foundation Fieldbus programming language, the better of the two sensor outputs is selected. In case of failure of the "good" sensor, the other sensor takes over, thereby obviating the need of shutdown. Thus, redundant sensors result in less numbers of shutdowns. Only if both the sensors fail would it result in shutdown. This is an example of two-out-of-two (2oo2) configuration.

7.12.3 Transmitter Redundancy

Transmitter redundancy involves employing several transmitters for accessing the same process point. Foundation Fieldbus uses status bytes for each process variable to select the "good" transmitter output and reject the others. Foundation Fieldbus takes the help of standard selector block

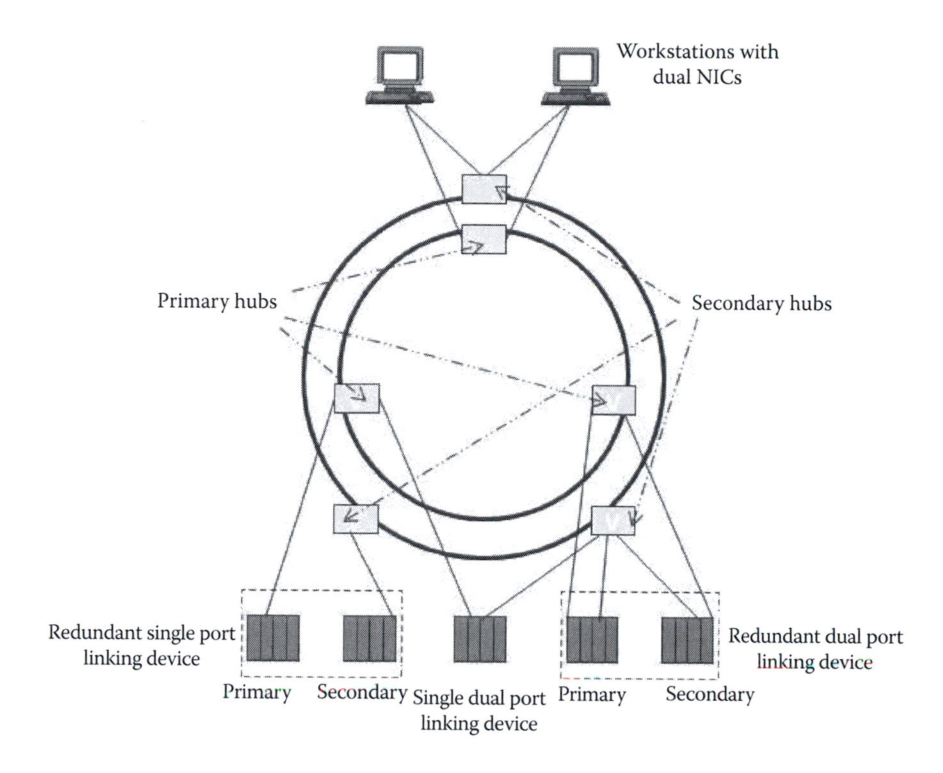

FIGURE 7.23 Media and network redundancy using dual-ring topology. (From J. Berge. *Fieldbuses for Process Control: Engineering, Operation, and Maintenance*. ISA, USA, p. 386, 2004.)

for the above. Such a selector block can handle three or even more inputs. Initially, the configuration strategy is so designed that in case of failure of a transmitter, the other automatically takes over. This eliminates any possibility of the loop shutdown of which these transmitters are a part.

When four such transmitters are so used for accessing and connecting the same process point, this is called four-out-of-four (4oo4). Transmitters from different manufacturers are used to eliminate identical types of transmitter failures, thereby increasing transmitter redundancy. Foundation Fieldbus technology ensures that transmitters from different manufacturers are interoperable.

7.13 HSE DEVICE TYPES

HSE devices are of four types: host device, linking device, gateway device, and Ethernet device. A combination of these four device types is sometimes used to form a single device type.

Host devices are non-HSE devices that can communicate with HSE devices. Examples include operator workstations, OPC server, and configurators. A linking device is one that connects H1 networks to the HSE network. It provides gateway services that map FDA system management and FMS messages to their counterparts. A gateway device interfaces other network protocols (e.g., PROFIBUS, MODBUS) so that they can work seamlessly with Foundation Fieldbus protocols. The Ethernet device provides services for communication over Ethernet media, Ethernet stack initialization, and time synchronization. Function blocks may be executed by an Ethernet device.

7.14 SYSTEM CONFIGURATION

There are two phases for configuring a system: system design and device configuration. The system design has to be completed first as per the requirements of the system under consideration and then only the device configuration is done. The system becomes operational only after the devices have been configured in the right manner.

7.14.1 System Design

For Foundation Fieldbus systems (or for that matter for any fieldbus system), the system design is akin to Distributed Control System (DCS). However, a fieldbus system design differs with DCS-based design on two counts. First, the conventional point-to-point 4–20 mA current loop is replaced by a digital bus where many smart field devices are connected to the same bus. Second, the control and I/O functions are brought into the fold of fieldbus devices. The ability of fieldbus devices to distribute the functions in these devices drastically reduces the number of remote-mounted I/O equipments as also the number of rack-mounted controllers. The devices residing on the fieldbus must have their individual unique physical device tag and a corresponding network address.

7.14.2 Device Configuration

Following a control strategy, devices are configured by connecting the function block inputs and outputs together for each device. Once the above are completed, each such fieldbus device would generate its own information. The system becomes operational once all the fieldbus devices are configured and their individual input and output blocks are interconnected as per the control strategy.

8 PROFIBUS

8.1 INTRODUCTION

PROFIBUS (PROcess FIeldBUS) is an open fieldbus standard developed in 1989 catering to the needs of both process automation and factory manufacturing automation. Initially developed by Siemens, it is particularly suitable for fast, time-critical applications and also involves complex communications. It is an open, vendor-independent (thus providing interoperability) fieldbus standard, adhering to the Open Systems Interconnection/International Standards Organization (OSI/ISO) model for communication. It supports single cable wiring having multi-input sensor blocks, intelligent devices operator interfaces, and smaller subnetworks such as AS-i. It is based on the German National Standard DIN 19 245 Parts 1 and 2 and also has been ratified by the European National Standard EN 50170 Vol. 2. Three PROFIBUS versions are available: PROFIBUS-FMS, PROFIBUS-PA, and PROFIBUS-DP. The general PROFIBUS features are tabulated in Table 8.1.

PROFIBUS supports two types of devices: master device and slave device. The former is called an "active station," while the latter is called a "passive station." A master device has the right to control the bus when it has bus access. Then it can transmit messages without any remote request. Transmitters, sensors, and actuators are examples of slave devices. A slave device acknowledges any received message and on receiving a request from a master, can send messages to that master.

8.2 PROFIBUS FAMILY

There are several PROFIBUS standards: PROFIBUS-DP (master–slave), PROFIBUS-FMS (multimaster/peer-to-peer), and PROFIBUS-PA (intrinsically safe). DP, Fieldbus Message Specification (FMS), and PA stand for Decentralized Periphery, Fieldbus Message Specification, and Process Automation, respectively.

PROFIBUS-DP handles fast communication processes such as drives, remote inputs/outputs (I/Os) normally encountered in factory automation. In this mode, multimasters are also used in which case a slave is assigned to one master only. It means that multiple masters can read inputs from a specific device but only one master can write outputs to that device.

TABLE 8.1
Salient PROFIBUS Features

Communication methods	Master–slave, Multimaster slave Publisher–subscriber	
Network speed	9.6 kbps–12 Mbps	
Data transfer rate	Up to 244 bytes	
Transmission media/ technologies, max. no. nodes	RS-485 STP copper: 126 Fiber optic: 126 IR: 126 RF: 126 Slip ring: 126 MBP-IS: depends on power budget	
Max. distance	RS-485 STP copper segment/with nine repeaters	9.6 kbps: 1000 m/10,000 m 12 Mbps: 100 m/1000 m
	Fiber optic (between fiber optic repeaters)	Plastic: 50 m Multimode glass: 400 m Single-mode glass: 15 km
	IR and RF	Varies with vendor product
	MBP-IS	31.25 kbps: 1.9 km max depending on cable type
Diagnostics	Standard: 6 bytes	
	Detailed: Up to 238 bytes total	Device related Module related Channel related

Source: www.isa.org/Template.cfm?Section=Books1&template=Ecommerce/FileDisplay.cfm&ProductID=6959&file=Chapter1_PROFIBUS.pdf.

Field devices are generally connected to PROFIBUS-PA. It differs with PROFIBUS-DP on three major counts: the devices can be powered on the bus cable; it supports devices in explosion hazardous areas; and data transfers on the physical layer take place vide IEC 61158-2. It allows higher freedom in the selection of bus topology and longer bus segments.

PROFIBUS-FMS is a peer-to-peer messaging format. This enables masters to communicate with each other. Up to 126 nodes are available, like PROFIBUS-DP, and all can be masters. FMS messages have more overhead than PROFIBUS-DP messages.

Apart from the above three types, a "combi mode" is sometimes used that uses both FMS and DP simultaneously in the same network. This is used when a programmable logic controller (PLC) is used along with a personal computer (PC), in which case the primary master communicates with the secondary master via FMS.

8.3 TRANSMISSION TECHNOLOGY

Different transmission technologies are used for PROFIBUS, which can be RS-485, RS-485-IS, MBP, and Fiber Optic. Table 8.2 summarizes the different transmission technologies with regard to their data rate, maximum cable length, protection, cable types, safety issues, data security, topology, etc., at the physical level.

The most commonly used is RS-485, which uses a shielded twisted pair cable with transmission rates up to 12 Mbps. The bus structure used allows the addition and deletion of a station without affecting other stations. In a particular segment, there are up to 32 devices (including the total of the master and slave devices). Either side of a segment is terminated with an active bus terminator. When more than 32 devices are used or there is a need to expand the network, repeaters are used.

RS-485-IS is used in potentially explosive areas (type EEx-i) that use a four-wire medium. When using this technology, maximum current and voltage levels must be adhered to avoid any potential explosive situation.

MBP stands for Manchester coding with bus powered. It is a synchronous transmission scheme with a 31.25 kbps rate of transmission. It is mainly used for process automation systems, with special emphasis on the chemical and petrochemical industries.

Industries having high electromagnetic disturbances or devices spaced at considerable distances apart employ fiber optic transmission schemes. Obviously, devices in the network must be able to integrate with the fiber optic transmission technology in the physical layer.

8.4 COMMUNICATION PROTOCOLS

The three PROFIBUS versions—FMS, DP, and PA—all use a standard bus access protocol. This protocol is implemented by layer 2 of OSI, which in the case of PROFIBUS is termed as Fieldbus Data Link (FDL) layer. Along with handling transmission protocols, FDL handles both data security and error detection. The protocol is so designed that the three variants work seamlessly together by offering high-speed, high-deterministic operation at the field level; reduced costing by employing two wire connections for PA; and an extended capability at the control level for FMS.

Of the three protocols, FMS is the first communications protocol and is designed for operations at the cell level, where normally PLCs and PCs communicate with each other. A wide range of functions are offered by FMS, making it quite complex to implement at an average transmission speed.

DP has three variants: DP-V0 to DP-V2. The original one, i.e., DP-V0, offers the basic functionalities that include cyclic I/O communication and

TABLE 8.2
Physical Layer Transmission Technology

	MBP	RS-485	RS-485-IS	Fiber Optic
Data transmission	Digital, bit synchronous, Manchester encoding	Digital, differential signals according to RS-485, NRZ	Digital, differential signals according to RS-485, NRZ	Optical, digital, NRZ
Transmission rate	31.25 kbps	9.6–12,000 kbps	9.6–1500 kbps	9.6–12,000 kbps
Data security	Preamble, error protected, start/end delimiter	HD = 4, Parity bit, start/end delimiter	HD = 4, Parity bit, start/end delimiter	HD = 4, Parity bit, start/end delimiter
Cable	Shielded, twisted pair, copper	Shielded, twisted pair, copper, cable type A	Shielded, twisted four-wire, cable type A	Multimode glass fiber, single-mode glass fiber, PCF, plastic
Remote feeding	Optional available over single wire	Available over additional wire	Available over additional wire	Available over hybrid line
Protection type	Intrinsic safety (EEx ia/ib)	None	Intrinsic safety (EEx ib)	None
Topology	Line and tree topology with termination; also in combination	Line topology with termination	Line topology with termination	Star and ring topology typical; line topology possible
No. of stations	Up to 32 stations per segment; total sum of max. 126 per network	Up to 32 stations per segment without repeater; up to 126 stations with repeater	Up to 32 stations per segment, up to 126 stations with repeater	Up to 126 stations per network
No. of repeaters	Maximum 4 repeaters	Maximum 9 repeaters with signal refreshing	Maximum 9 repeaters with signal refreshing	Unlimited with signal refreshing (time delay of signal)

Source: PROFIBUS, Technology and Application: System Description, PROFIBUS International Support Center, Karlsruhe, Germany, Copyright by PNO 10/02, October 2002. Available at www.pacontrol.com/download/profibus-overview.pdf.)

diagnostic reporting. DP-V1 offers both cyclic and acyclic communication and alarm services, diagnostics, parameterization, and field device controls, while DP-V2 supports some services that are particularly needed in the field of drive control. These include functions for producer–consumer communication between slave devices, isochronous slave mode, time synchronization, and time stamping. PROFIBUS-DP can address extremely time-critical communication tasks. The last one, PROFIBUS-PA, is especially oriented toward meeting the requirements of process automation communication.

Layer 2 (the data link layer [DLL] or FDL as it is termed here) corresponds to the bus access protocol that manages the communication procedure between the master–slave and the token-passing method for the multimaster system. FDL also handles data security and data frames. Layer 7, or the application layer, acts as an interface between the application programs and the different profiles—FMS, DP, or PA existing in the user layer.

8.5 DEVICE CLASSES

PROFIBUS devices can be classified into three device types: class 1 PROFIBUS-DP master (DPM1), class 2 PROFIBUS-DP master (DPM2), and PROFIBUS slaves.

Class 1 masters are usually used for cyclic data exchange with the slaves connected to it. They are normally PLCs or PCs programmed for data exchange with the slaves on a precise time-sharing basis. Class 1 masters have the following characteristics: tokens are passed between the masters; can write data into the slaves assigned to it and can read data from a slave in the network; sets the data rate; and the connected slaves detect the same data automatically.

Class 2 masters are used as a tool for device and system commissioning. In DP-V1 and DP-V2, this master class is used for setting/altering device parameter values acyclically. DPM2s are not required to remain permanently connected to the system and are used only for initial device configuration. Normally, PROFIBUS master devices support the functionalities of both DPM1 and DPM2. Class 2 masters have the following characteristics: act as supervisory masters, used for diagnostic purposes and slave commissioning, control slaves at any given point of time, and can only read slaves but do not have write access.

PROFIBUS slave devices respond to master polling by sending device data. The slaves are field devices such as transducers, valves, and remote I/Os. Slaves can be of two types: compact devices and modular devices. Compact devices have a fixed I/O configuration, unlike the case of a modular device in which it may vary. A modular device contains a station to which the fieldbus interfaces are connected through individual slots specifically meant for them. Multivariable devices can be thought of as modular devices. By combining and modifying the different modules, if the need so arises, modular slaves can be configured to satisfy specific I/O needs. Slave devices as required in process automation may either be of discrete or word I/O type. Slaves have the following characteristics: they do not possess bus control, can only respond to a master's request, can acknowledge messages, and General Station Description (GSD) files define a slave for a master.

Finally, in process automation, there are some devices that can act as both master and slave device.

8.6 PROFIBUS IN AUTOMATION

Figure 8.1 shows how PROFIBUS technology is applied at different levels in the automation hierarchy. The three different versions of PROFIBUS—DP, PA, and FMS—all share the same transmission medium, RS-485. Figure 8.1 shows a typical network having both process automation and factory automation tasks shown together in the same network.

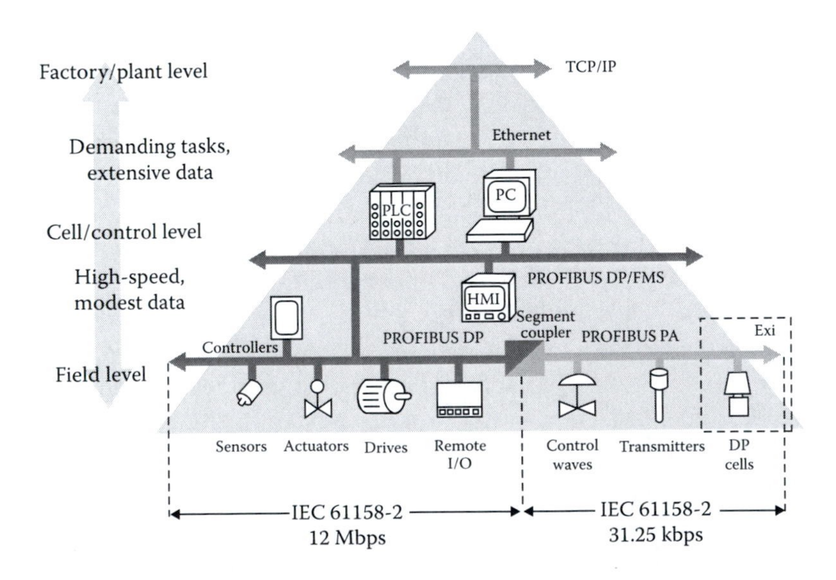

FIGURE 8.1 **(See color insert.)** Typical PROFIBUS network includes both process and factory automation. (From A. A. Verwer. *Introduction to PROFIBUS*. Manchester Metropolitan University, Department of Engineering, Automation Systems Centre, PROFIBUS Competence Centre, 2005.)

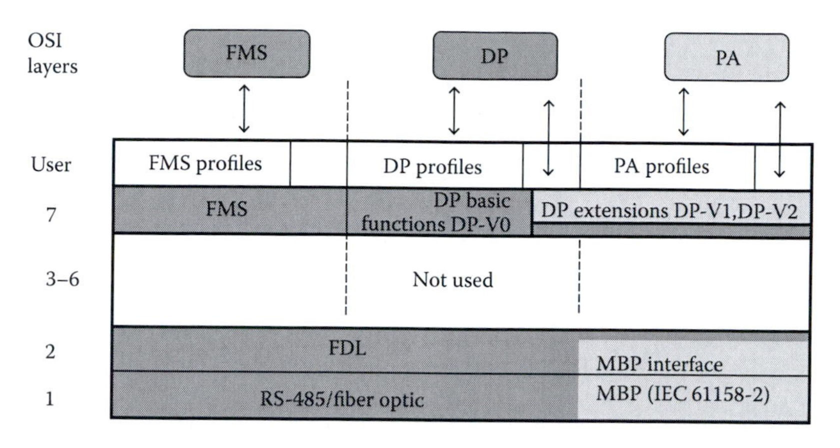

FIGURE 8.2 PROFIBUS protocol stack related to ISO/OSI model. (From A. A. Verwer. *Introduction to PROFIBUS*. Manchester Metropolitan University, Department of Engineering, Automation Systems Centre, PROFIBUS Competence Centre, 2005.)

8.7 OSI MODEL OF PROFIBUS PROTOCOL STACK

A uniform bus access mechanism is followed by all the three versions (DP, PA, and FMS) of a PROFIBUS system. This protocol is implemented by layer 2 of the OSI model. This layer in PROFIBUS is known as FDL. FDL also includes data security and error detection mechanism. The PROFIBUS protocol stack is shown in Figure 8.2.

The PROFIBUS protocol ensures that all the three variants work seamlessly and provide high-speed, deterministic operation and communication at the field level (DP), cost-effective two-wire connection (PA), and the extended connection at the cell level (FMS).

8.8 PROFIBUS-DP CHARACTERISTICS

PROFIBUS is used for fast communications at the device level of process automation systems such as chemical, paper, food, and automobile industries. It is included in EN 50254 and IEC 61158 standards.

PROFIBUS-DP is the high-speed solution for PROFIBUS systems. It is largely used in time-critical solutions. Originally, it was developed for communication between automation systems and decentralized equipment.

Figure 8.3 shows the bus cycle times for a DP monomaster system assuming that each slave has 2 bytes of input and output data. From Figure 8.3, it is seen that approximately 1 ms time would be required at a bit rate of 12 Mbits/s for transmission of 512 bits of input and 512 bits of output data distributed over 32 stations. Bus cycle time would depend on both the number of stations and the transmission rate.

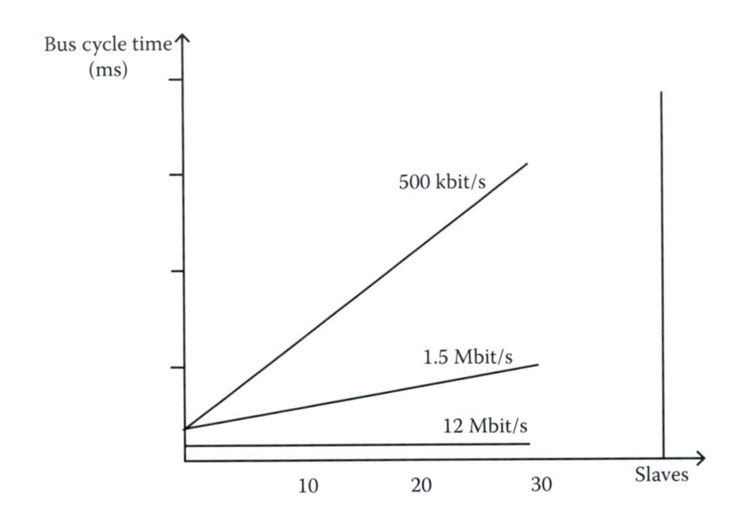

FIGURE 8.3 Bus cycle time for DP monomaster system. (From PROFIBUS, Technology and Application: System Description, PROFIBUS International Support Center, Karlsruhe, Germany, Copyright by PNO 10/02, October 2002. Available at www.pacontrol.com/download/profibus-overview.pdf.)

PROFIBUS-DP is used in most cases involving slaves. It is available in three variants: DP-V0 (1993), DP-V1 (1997), and DP-V2 (2002). The variants continued to be developed because of technical developments and increasing demands for more stringent automation systems.

8.8.1 Version DP-V0

The basic functionalities of DP are provided in DP-V0. These include cyclic data exchange between master and slaves, module diagnosis, and channel-specific diagnosis and station diagnosis.

The master reads the input information from the slaves in a cyclic manner and writes output information into the respective slaves cyclically.

8.8.1.1 Diagnostic Functions

Diagnostic functions are used for fast location of faults. The diagnostic message is sent over the bus and ultimately received by the bus. Such functions can be of three types: device-specific diagnosis, module-related diagnosis, and channel-related diagnosis.

Any device-specific parameter—such as overvoltage, undervoltage, or overheating—is taken care of by device-specific diagnosis services. Module-related diagnosis relates to any pending diagnosis in a specific I/O subdomain of a station. Any fault related to an individual I/O bit (channel) is manifested in channel-related diagnosis.

8.8.1.2 Synchronization and Freeze Mode

DPM1 handles cyclic data transfers automatically. In addition to this, it can send control commands to a slave or a group of slaves simultaneously. These are called multicast commands. A master can send a "sync" command to a slave or a group of slaves. The outputs of the addressed slaves are then frozen in their current states. During subsequent data transmissions, the output data is stored in the slave and the output state remains unaltered. The stored output data is not sent to the outputs until the next sync command is received by the slave. A sync command can be terminated by issuing an "unsync" command.

When a master issues a "freeze" command to a slave, the corresponding slave enters into the freeze mode. The states of the slave inputs are then frozen at their current values. This remains so until the slave receives another freeze command, when the input data is updated. A freeze command can be terminated by an unfreeze command.

8.8.1.3 System Configuration

PROFIBUS systems support monomaster and multimaster systems. A maximum of 126 devices, which include both master and slave, can be connected to the bus. System configuration involves the following considerations: number of stations, bus parameters, diagnosis message formats, and assigning station addresses to the I/O addresses.

8.8.1.4 Time Monitors

A time monitor is a very efficient and effective protective mechanism to combat incorrect parameterization or failure of transmission-related functions. Time-monitoring mechanisms are fitted both at the master and the slaves for the above purpose. During configuration, the time interval of a time monitor is specified.

At the master level, DPM1 uses a Data_Control_Timer function to monitor data communication with the slaves. Each slave has its own time monitor. The timer is tripped if no correct user data transfer takes place within the specified time interval, and the user is notified accordingly. In case the automatic error handling is exerted, i.e., Auto_Clear = True option is enabled, DPM1 comes out of the normal state of data communication mode if error occurs, outputs are put into fail-safe mode, and then DPM1 moves to the clear mode.

The slaves use their respective watchdog timers to detect errors of the master or transmission error. The outputs of the slaves are put into fail-safe mode if no data communication with the master occurs within the watchdog control interval, which is preset.

In a multimaster system, it must be ensured that only the authorized master gets the access to the corresponding slave.

8.8.1.5 Token-Passing Characteristics

The characteristics associated with the token-passing mechanism are discussed below:

- The token-passing technique involves more than one master.
- The token is passed from one master to another in ascending order.
- Each master is responsible for the addition or removal of stations in its address range.
- The system does not need initialization in case of loss of token. The master with the lowest address creates a new token after its token timer is timed out.
- After a power up, the master station with the lowest station address commences initialization. It claims the token if it sees no bus activity after waiting for a predefined period. It informs other master stations on the network about its own activity and transmits a "request field data link status" to each station in an increasing order. The station that responds to this request is passed on the token.

8.8.2 Version DP-V1

DP-V1 is an enhancement to DP-V0 in process automation and control field. Both cyclic and acyclic communications are implemented in this version. This permits online access to stations using engineering tools. In addition, parameter assignment, alarm handling of intelligent devices, calibration of field devices, etc., are also handled by the version.

8.8.2.1 Cyclic and Acyclic Communication

Both cyclic and acyclic communications are features of PROFIBUS-DP V2. DPM1 devices such as PLCs exchange information with slaves cyclically at predefined time slots. The master has the token and sends messages to or retrieves them from slave 1 and then from slave 2, etc., in that sequence until it reaches the last slave in the list. The master can then utilize the remaining available time of the program cycle to set up an acyclic communication with any slave. The cyclic and acyclic communication services are shown in Figures 8.4 and 8.5, respectively.

Cyclic communication is mostly carried out in a transparent manner. The master has its own memory in which it stores the incoming data from the slaves. Any data that is coming out from the master is first written into the master's memory before being sent out to the concerned slave. Figure 8.6 shows the cyclic data transfer scheme for a PROFIBUS system.

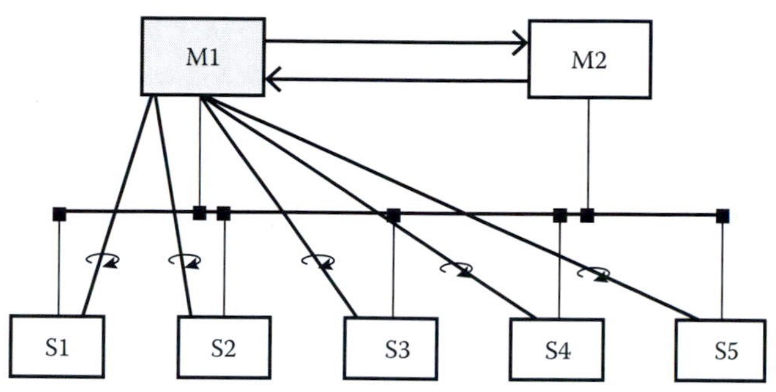

M1: master class 1

M2: master class 2

FIGURE 8.4 PROFIBUS-DP bus access by cyclic communication. (From B. G. Liptak. *Instrument Engineers' Handbook, Process Software Digital Network*, 3rd Edition. CRC Press, Boca Raton, FL, p. 580, 2002.)

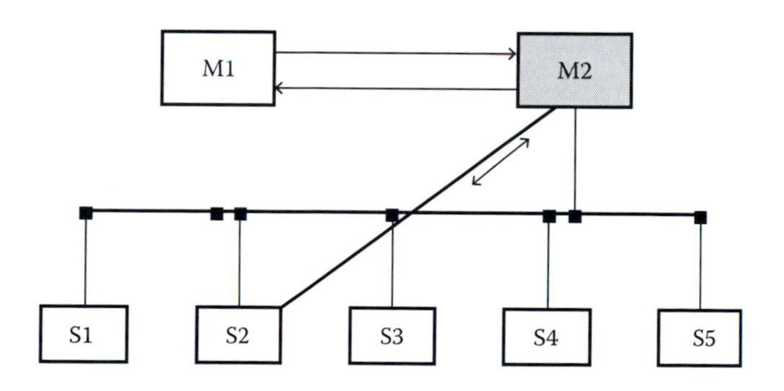

M1: master class 1

M2: master class 2

FIGURE 8.5 PROFIBUS-DP bus access by acyclic communication. (From B. G. Liptak. *Instrument Engineers' Handbook, Process Software Digital Network*, 3rd Edition. CRC Press, Boca Raton, FL, p. 580, 2002.)

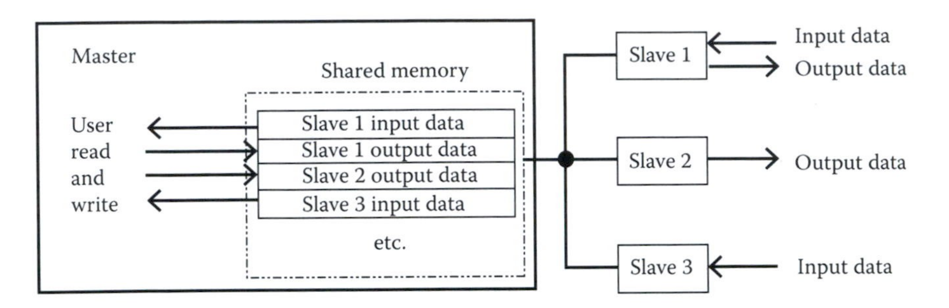

FIGURE 8.6 Transparent nature of cyclic data transfer. (From A. A. Verwer. *Introduction to PROFIBUS*. Manchester Metropolitan University, Automation Systems Centre, PROFIBUS Competence Centre, UK, 2005.)

8.8.3 Version DP-V2

Version DP-V2 is still another improvement over DP-V1. Additional functionalities of DP-V2 include the isochronous mode, direct slave-to-slave communication without any help from the coordinating master. This publisher–subscriber model helps reduce bus response times by up to 90%. DP-V2 can also be implemented as a drive bus to control fast movement sequences in drive axes.

8.8.3.1 Slave-to-Slave Communication

It is a direct communication between slaves without going through the master. It saves time considerably, to the extent of up to 90%. One slave acts as the publisher and the other ones as the subscribers. This occurs without the master. Thus, this process enables a slave(s) (subscriber(s)) to read data from another slave (publisher). Figure 8.7 explains the above process.

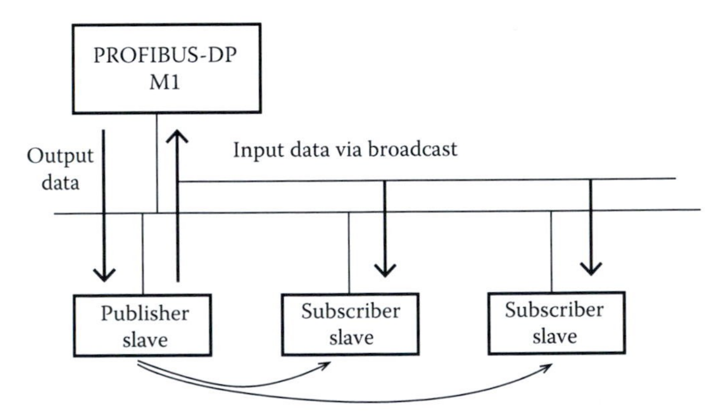

FIGURE 8.7 Slave-to-slave communication. (From PROFIBUS, Technology and Application: System Description, PROFIBUS International Support Center, Karlsruhe, Germany, Copyright by PNO 10/02, October 2002. Available at www.pacontrol.com/download/profibus-overview.pdf.)

8.8.3.2 Isochronous Mode

This facilitates clock synchronization between masters and slaves irrespective of bus load. It provides highly precise positioning systems with clocks that are precise within 1 μs. All the devices connected to the network are synchronized with the master through a "global control broadcast message." A "special sign of life" enables monitoring of synchronization.

8.8.3.3 Clock Control

All slaves connected to a master are synchronized by *clock control.* That is, all slaves are time stamped by the master to within less than a millisecond. This time stamping of the slaves are done by the new connectionless MS3 channel. Clock control is a major facility that allows the network to track occurrence of events very precisely. This is particularly useful for a multimaster system. Clock control helps in diagnosing and identification of faults as well as chronological planning of events.

8.8.3.4 Upload and Download

This is a special function that enables loading of data of any size in a device with very few commands. It allows programs to be updated or devices replaced without any need for manual loading processes.

8.8.3.5 HART on DP

A large installation base of HART devices has led many users to integrate PROFIBUS systems on them. An immediate advantage of the HART devices is that they can benefit from the PROFIBUS communication techniques.

For such integration of HART on DP, PROFIBUS has implemented a specification profile above layer 7 in both master and slave devices. This ensures the mapping of HART client–master–server model on the PROFIBUS. Figure 8.8 shows such an integration in which the HART–client application is integrated in a PROFIBUS master and the HART master in a PROFIBUS slave.

8.8.3.6 Comparison between DP-V0, DP-V1, and DP-V2

The three different variations of PROFIBUS-DP, viz., DP-V0, DP-V1, and DP-V2, have various differences among them and are shown in Figure 8.9.

8.8.4 Communication Profile

The communication profile of PROFIBUS-DP is shown in Figure 8.10. It shows layer 1, or the physical layer, which uses different physical mediums such as fiber optics, infrared, or other wireless communication systems. Layer 2, the data link layer, is called the FDL here. Layers 3 to 7 are empty

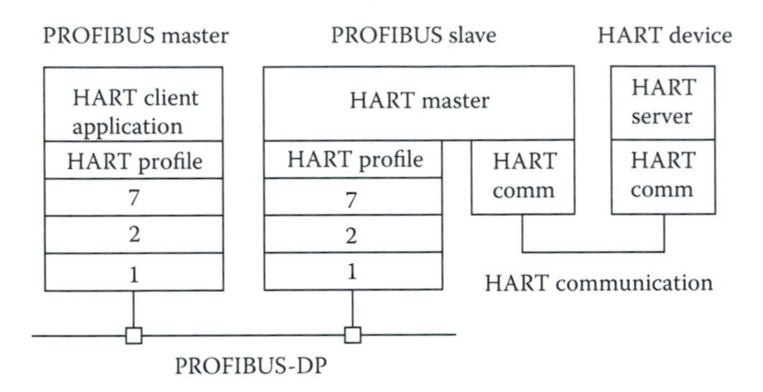

FIGURE 8.8 Integrating HART into PROFIBUS-DP. (From PROFIBUS, Technology and Application: System Description, PROFIBUS International Support Center, Karlsruhe, Germany, Copyright by PNO 10/02, October 2002. Available at www.pacontrol.com/download/profibus-overview.pdf.)

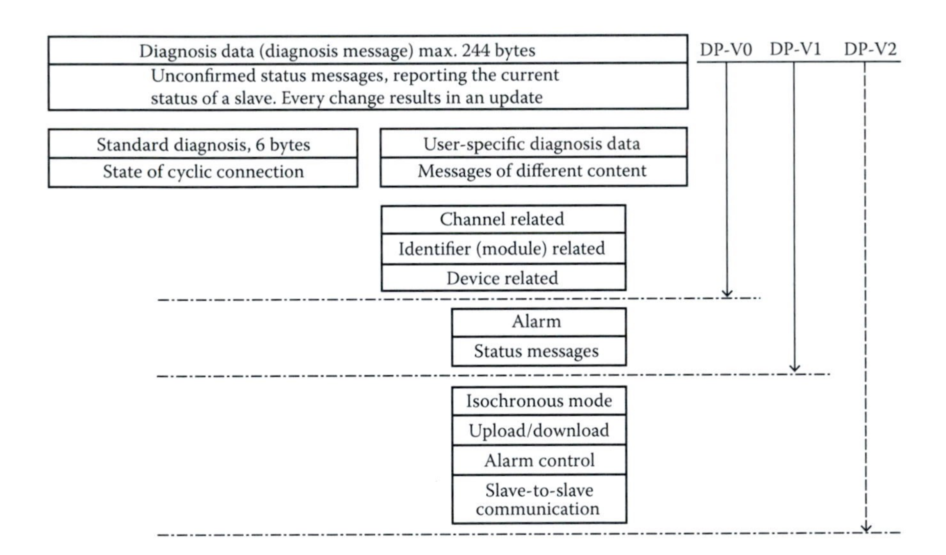

FIGURE 8.9 Comparison between DP-V0, DP-V1, and DP-V2.

in PROFIBUS-DP configurations. Other functions required in a fieldbus are implemented by two upper layers called Direct Data Link Mapper (DDLM) and user interface.

8.8.5 Physical Layer

Mostly, the physical layer of PROFIBUS-DP is specified by the RS-485 standard. Some other configurations involve using fiber optics, infrared,

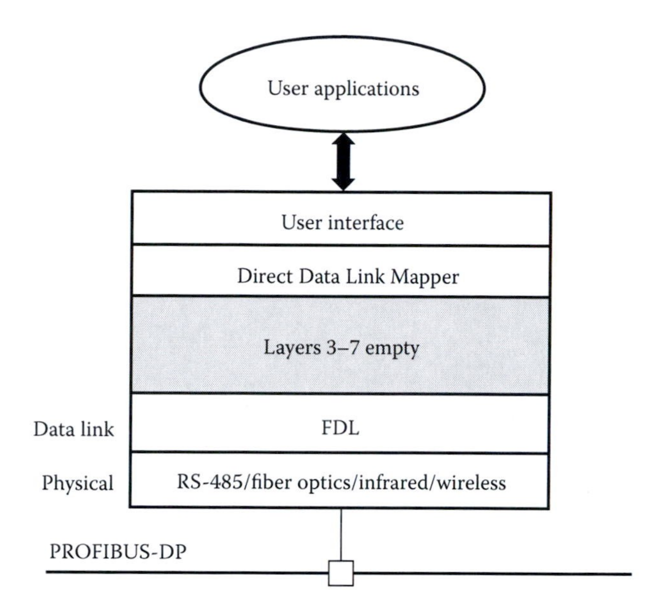

FIGURE 8.10 Communication profile of PROFIBUS-DP. (From B. G. Liptak. *Instrument Engineers' Handbook, Process Software Digital Network*, 3rd Edition. CRC Press, Boca Raton, FL, p. 623, 2002.)

or other wireless transmission medium. For RS-485, the physical medium used is a twisted pair cable that may be shielded, if required. A maximum of 126 stations can be connected.

The physical layer of PROFIBUS-DP, based on RS-485, has the following features:

- The topology is a linear bus and terminated at both ends.
- Transmission is via a twisted pair cable, with optional shielding. For transmission rates greater than 500 kbaud, type A cable is preferred, while type B cable is used for short-distance communication and having a baud rate lesser than the above.
- Stubs are possible along the line.
- Different physical media may be used, such as fiber optic, infrared, and wireless transmission.
- PROFIBUS-DP, PROFIBUS-FMS, and PROFIBUS-PA can together be used on a common bus line.

8.8.5.1 Transmission Speed vs. Segment Length

Transmission speed varies from 9.6 kbps to 12 Mbps, and the maximum length of a segment is related to transmission speed, as shown in Table 8.3.

TABLE 8.3
Transmission Speed vs. Maximum Segment Length for PROFIBUS-DP

Transmission speed (kbps)	9.6	19.2	93.75	187.5	500	1500	6000	12,000
Maximum segment length (m)	1200	1200	1200	1000	400	200	100	100

8.8.6 Data Link Layer

The data link layer in PROFIBUS-DP is known as FDL. It performs the following tasks:

- Bus access control or medium access control
- Data security
- Telegram structure
- Availability of data transmission services
  - SDN (serial data with no acknowledge)
  - SRD (send and request data with reply)

Data communication between master and slave takes place via the logical token-passing method. Every station knows the address of the station from which it received the token and also the address of the station to which the token is to be handed over.

Data transmission with SRD is a confirmed one, and the source station can send a maximum of 246 bytes to the selected destination station. The latter responds with a maximum of 246 response bytes. It is a very reliable and confirmed way of data exchange method between stations. With SRD, the master either issues a command or sends data to the concerned slave and receives a reply (either an acknowledgment or data from the slave) in a defined time span. It is shown in Figure 8.11.

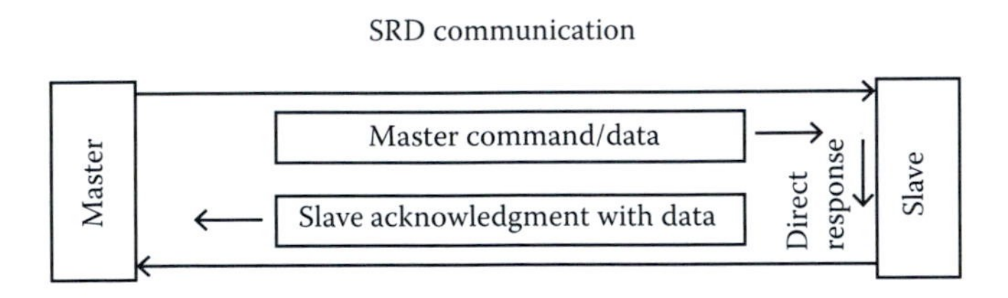

FIGURE 8.11 SRD method of data exchange between master and slave. (From A. G. Samson. PROFIBUS PA Technical Information, Part 4 Communications, L453EN, Samson A. G. Mess und Regeltechnik – Weismullerstrabe 3-D60314 Frankfurt am Main, http://www.samson.de.)

For SDN services, no confirmation is forthcoming from the destination station. The source station can send a maximum of 246 bytes—as is the case with SRD. SDN is mainly used for broadcast or multicast messages.

8.8.7 DDLM and User Interface

The user interface and the DDLM reside at the top of layer 7 of the OSI protocol. These two together form the interface between the application program and layer 2, i.e., DLL or FDL, and are responsible for the correct execution of all operations specified by the standard. Figure 8.12 shows the interface between the user interface, DDLM, and FDL. DDLM does the job of mapping the requests coming from the user interface layer into the FDL. DDLM provides several asynchronous service functions in PROFIBUS-DP V1, e.g., DDLM_Read, DDLM_Write, DDLM_Intiate, DDLM_Abort, and DDLM_Alarm_Ack. The user interface uses these DDLM functions for communication purposes such as start-up, maintenance, alarm messages, and diagnostics. For the purpose of interfacing, PROFIBUS-PA utilizes several descriptions and specifications such as device database files (GSD), device profiles, Electronic Device Description (EDD), and Field Device Tool (FDT) description.

PROFIBUS-DP is based on the master–slave concept and also has mono- or multimode master options. In the latter case, a token-passing method is used to determine which master is active at any given instant of time. The token-passing mechanism is precisely timed. Since a part of the transmission bandwidth is utilized for token transmissions, use of several masters has a negative effect on efficiency. Thus number of masters that

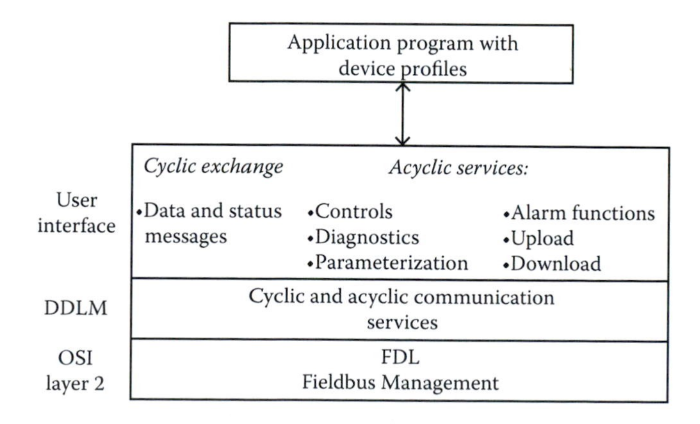

FIGURE 8.12 User interface, DDLM, and FDL. (From A. G. Samson. PROFIBUS PA Technical Information, Part 4 Communications, L453EN, Samson A. G. Mess und Regeltechnik – Weismullerstrabe 3-D60314 Frankfurt am Main, http://www.samson.de.)

should be employed in a multimaster system should always be judiciously chosen.

A master in a PROFIBUS-DP system carries out several functions on its slave with the help of the user layer and DDLM. Some of these are reading diagnostic data, setting parameter data, exchanging cyclic data, checking configuration data, and emission of global control commands.

A slave reports to the master, when requested, about its own status in diagnostic message while the parameter message sets the limits of the slave operations as dictated by the master. Cyclic data exchange is undertaken in cyclic messages between the master and the slave on a precise time basis. In parameter message, the master checks the parameter values that were set during initialization and in global control command message, a master uses it to synchronize the input and output operations such as freeze and sync commands. The clear command forces all the addressed slaves to put their outputs in safe state.

8.8.8 State Diagram of Slave

The state diagram of PROFIBUS-DP slave is shown in Figure 8.13. A slave looks for its address immediately after power on. If the address is already set in hardware, it goes into the wait_prm state; otherwise, it waits for a Set_slave_address message from a master. Once in the wait_prm state, the slave, i.e., the device, waits for the parameter message. The slave decides

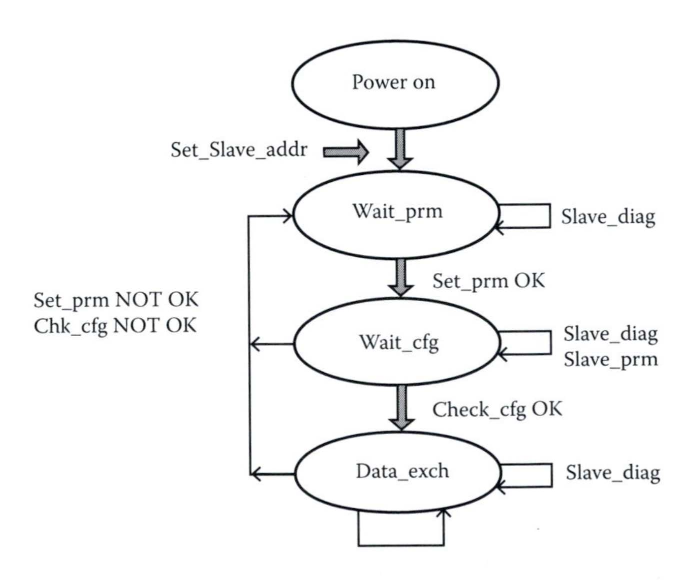

FIGURE 8.13 State diagram of slave. (From B. G. Liptak. *Instrument Engineers' Handbook, Process Software Digital Network*, 3rd Edition. CRC Press, Boca Raton, FL, p. 625, 2002.)

whether to accept the received parameters or not. If yes, it enters into wait_cfg state; otherwise, it remains in the wait_prm state.

In the wait_cfg state, the slave compares its own configuration with the configuration it has received from the master. If the two match, it enters into the Data_exch state. In this, the master polls the slave cyclically and exchanges data as specified by configuration message.

8.8.9 Addressing with Slot and Index

PROFIBUS assumes devices to be modular in nature—the individual modules are either physical or logical subdivisions of the device. This model, based on modules, is used by basic DP functions for cyclic data communications. A simple device may consist of a single module and the rather complex ones would consist of several modules. A module can be an input, an output, or a combination of both. The addressing of the modules is based on identifiers. Thus, a slave is configured on the basis of identifiers—each identifier represents a single module within the slave.

A *slot number* addresses a module and the *index* addresses the data blocks assigned to a module. The scheme is shown in Figure 8.14. A data block can have a maximum of 244 data bytes. The modules begin at slot number 1 and numbered in increasing order. Slot number 0 identifies the device itself.

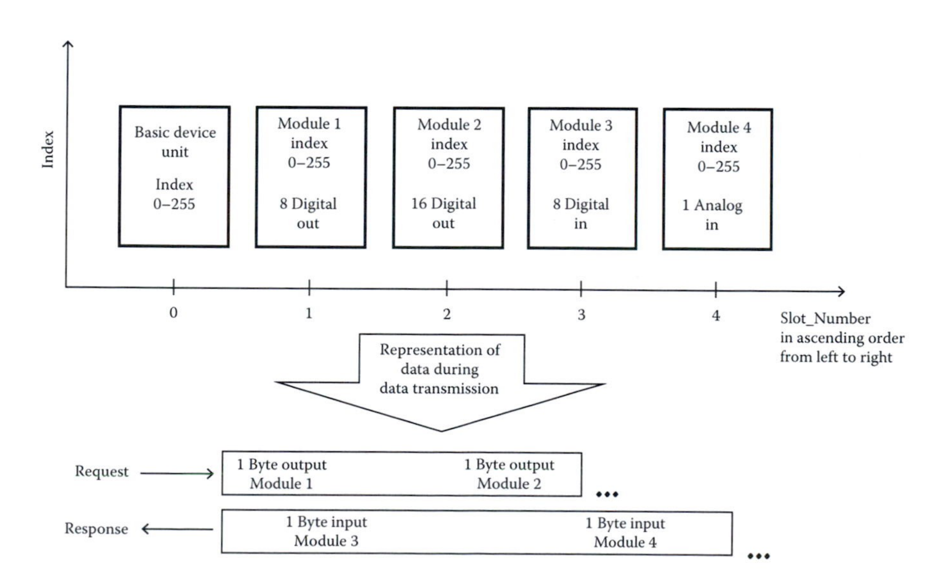

FIGURE 8.14 Addressing with slot and index. (From PROFIBUS, Technology and Application: System Description, PROFIBUS International Support Center, Karlsruhe, Germany, Copyright by PNO 10/02, October 2002. Available at www.pacontrol.com/download/profibus-overview.pdf.)

8.9 PROFIBUS-PA CHARACTERISTICS

PROFIBUS-PA caters to the needs of process automation and control systems. The field instruments, viz., transmitters, positioners, valve actuators, etc., are all connected by a PROFIBUS-PA segment. The connection to PROFIBUS-DP is made by using a DP–PA coupler.

Potential advantages accrue by using such PA technology as failure safety systems, autodiagnosis, reliable information transfer, equipment rangeability, high-resolution measurement, and integration to higher-speed discrete control. It results in significant cost reduction in installation, less downtime, higher reliability of operations, easy integrability for future expansion, more functionality and safety, less start-up time, etc.

The PROFIBUS-PA communication protocol uses identical protocol with that of PROFIBUS-DP. The twisted pair cable is used for both data communication as well as power supply to the devices. Three major differences exist with PROFIBUS-DP:

1. Devices are powered on the bus cable.
2. Devices can be used in explosion hazardous areas.
3. Data are transmitted via the IEC 61158-2 physical layer.

In PROFIBUS-PA, the IEC transmission specifications define a digital bit synchronous data transmission at a rate of 31.25 kbits/s. The Manchester bus-powered current signal waveform is shown in Figure 8.15. The topology can be line or tree with 126 addressable devices and up to 1900 m line length.

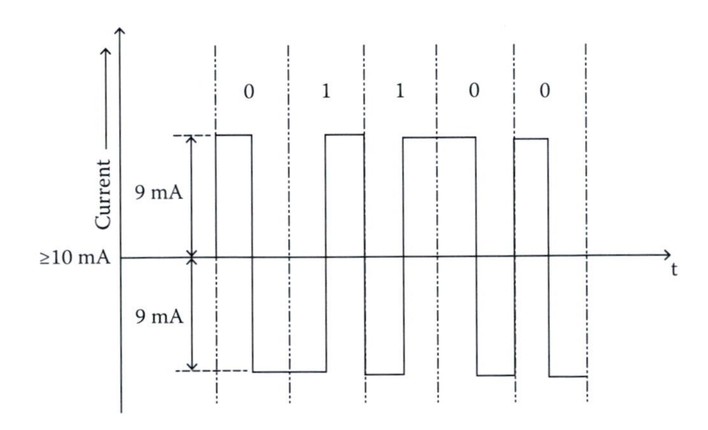

FIGURE 8.15 Manchester-coded bus-powered current signal waveform. (From A. A. Verwer. *Introduction to PROFIBUS*. Manchester Metropolitan University, Automation Systems Centre, PROFIBUS Competence Centre, UK, 2005.)

8.9.1 Bus Access Method

The master–slave configuration is used in PROFIBUS-PA systems to regulate bus access. In a multimaster system, the token-passing method is used so that at any instant, only one master is active and regulates the traffic in DP–PA environment. Each master has bus control under it for a precisely defined time. The devices on the PROFIBUS-PA segment are accessed by a segment coupler or link. Figure 8.16 shows the diagram of how the PROFIBUS-PA accesses the bus via a segment coupler.

Segment couplers are transparent to the PROFIBUS-DP master and as such are not engineered in the PLC. The couplers only route the signals to the proper devices in the PROFIBUS-PA environment. The couplers are not assigned any PROFIBUS-DP address. Instead, each field device is given a PROFIBUS-DP address through which they communicate with the masters. As far as the PROFIBUS-DP is concerned, the devices merely act as slaves.

A link, on the other hand, is recognized by the PROFIBUS-DP master and must be engineered in the PLC. Since the link is opaque in nature, the PROFIBUS-DP master cannot see the PROFIBUS-PA devices. The link acts as a bus master on the PROFIBUS-PA side, while it acts as a slave on the PROFIBUS-DP side. Figure 8.17 shows how the PROFIBUS-PA bus access is effected via a link.

Each field device is allocated a PROFIBUS-PA address. This address is unique for that link only and is not valid for other PROFIBUS-PA segments. The master polls the devices cyclically and stores the respective data values in a buffer.

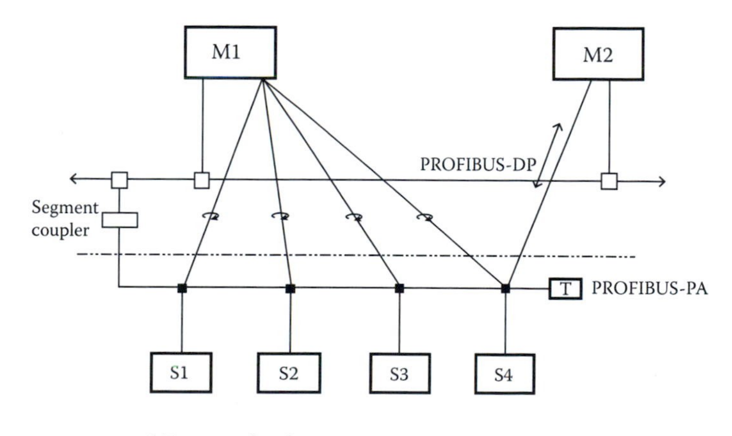

FIGURE 8.16 Bus access in PROFIBUS-PA system via segment coupler. (From B. G. Liptak. *Instrument Engineers' Handbook, Process Software Digital Network*, 3rd Edition. CRC Press, Boca Raton, FL, p. 581, 2002.)

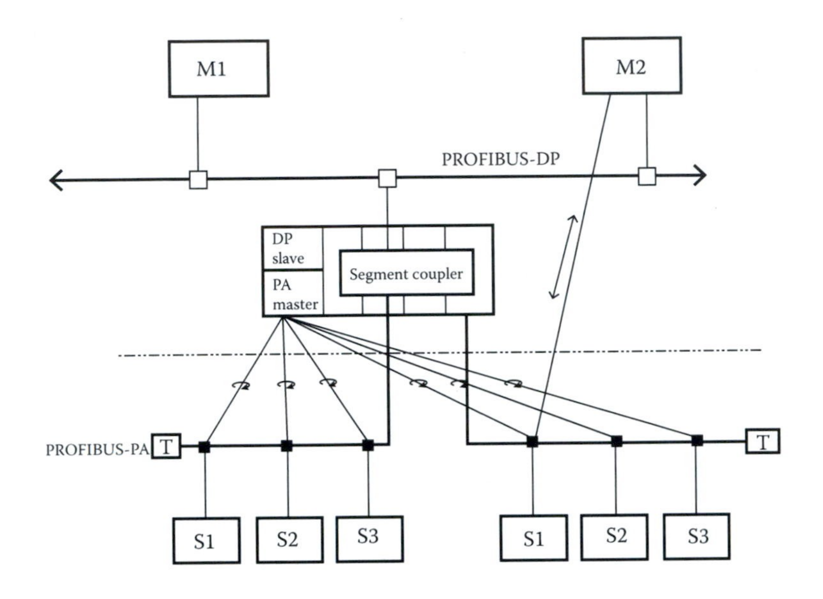

FIGURE 8.17 Bus access in PROFIBUS-PA system via a link. (From B. G. Liptak. *Instrument Engineers' Handbook, Process Software Digital Network*, 3rd Edition. CRC Press, Boca Raton, FL, p. 582, 2002.)

On the PROFIBUS-DP side, the link that acts as a slave is assigned a PROFIBUS-DP address. The master polls the devices cyclically, and each device data is packed in respective telegrams by the link.

8.9.2 Data Telegram

The PROFIBUS Data telegrams of the IEC-61158-2 transmission are to a large extent identical to the FDL telegrams of the asynchronous RS-485 transmission. FDL data telegrams have different versions:

- Telegrams that do not have any data field and having 6 control bytes
- Telegrams having one data field of fixed length (8 data and 6 control bytes)
- Telegrams having a variable data field (0–244 bytes of data and 9–11 control bytes)
- Very brief acknowledgment (1 byte)
- A token telegram for bus access control (3 bytes)

Figure 8.18 shows the FDL telegram (upper) and bit synchronous transmission of the IEC telegram.

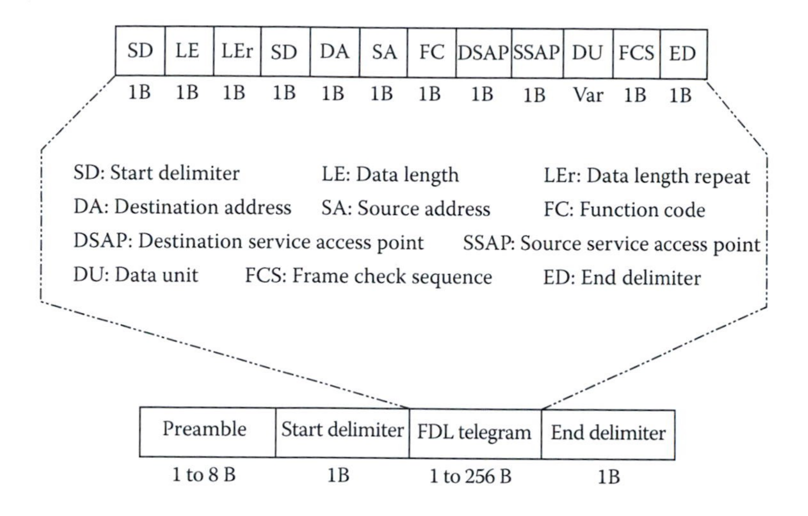

FIGURE 8.18 FDL telegram (upper) and bit synchronous transmission of IEC telegram. (From A. G. Samson. PROFIBUS PA Technical Information, Part 4 Communications, L453EN, Samson A. G. Mess und Regeltechnik – Weismullerstrabe 3-D60314 Frankfurt am Main, http://www.samson.de.)

A Hamming distance of 4 is employed to detect three errors with certainty. It is to be noted that the FDL telegram shown in the upper part of Figure 8.18 is transmitted asynchronously in the form of UART characters over the RS-485 lines, but the transmission on the IEC segments is bit synchronous in nature (shown at the bottom of the same figure).

8.9.3 Device Profile

A process control system is best accepted by end users if operation, communication, and monitoring of device parameters and functions are totally standardized such that interoperability between devices from different manufacturers is ensured. Device profiles ensure that the properties and functions of field devices are predefined within suitable limits. Device profile includes parameterization of different variables such as measured value, alarm limits, status flag, and scaling factor.

There are two different profile classes for a PROFIBUS-PA system: class A and class B. Class A profiles include the basic parameters that are absolutely necessary for process automation systems. These include process variable, status of the measured value, physical unit, and the tag number.

Class B profiles include all the basic parameters of class A. In addition, class B includes several other functions. It differentiates between parameters that are mandatory and those that are optional in nature.

TABLE 8.4
Profile Definition of a Control Valve

	Profile Class A	Profile Class B
Physical Block	m	m
Analog Output Function Block	m	m
Transducer Block	–	m
Electropneumatic Transducer Block	–	s
Electric Transducer Block	–	s
Electrohydraulic Transducer Block	–	s
Additional Function Blocks	–	o
Additional Transducer Blocks	–	o

Source: A. G. Samson. PROFIBUS PA Technical Information, Part 4 Communications, L453EN, Samson A. G. Mess und Regeltechnik – Weismullerstrabe 3-D60314 Frankfurt am Main, http://www.samson.de.

Note: m, mandatory; s, selected; o, optional.

Table 8.4 shows how the function blocks are assigned to the profile classes A and B for a control valve. The table suggests that class A profile is a subset of class B.

8.9.4 PA Block Model

Blocks are commonly used to describe the characteristics and functions of a measuring point. Automation applications are generally represented through a combination of several such blocks. PROFIBUS-PA devices requiring PROFIBUS User Organization e.V. (PNO) certification, need to have a set of universal parameters that must be implemented. It would then ensure configuration compatibility between identical device types made from different manufacturers.

Figure 8.19 shows a PA block model. It presents a complete overview of the PA device in block form and shows the data flow path through the three data channels, viz., MS0, MS1, and MS2. It consists of four blocks: transducer block, physical block, function block, and device management block. The sensor output first goes to the transducer block from which it goes to the function block. The physical block is not a part of process signal flow but contains information about the device itself, such as the serial number of the device, manufacturer code, installation date, and diagnostic features.

Except the device management block, the beginning of every other block contains the standard parameters that are used to identify the block. The parameters assigned to an individual block use the data structure and

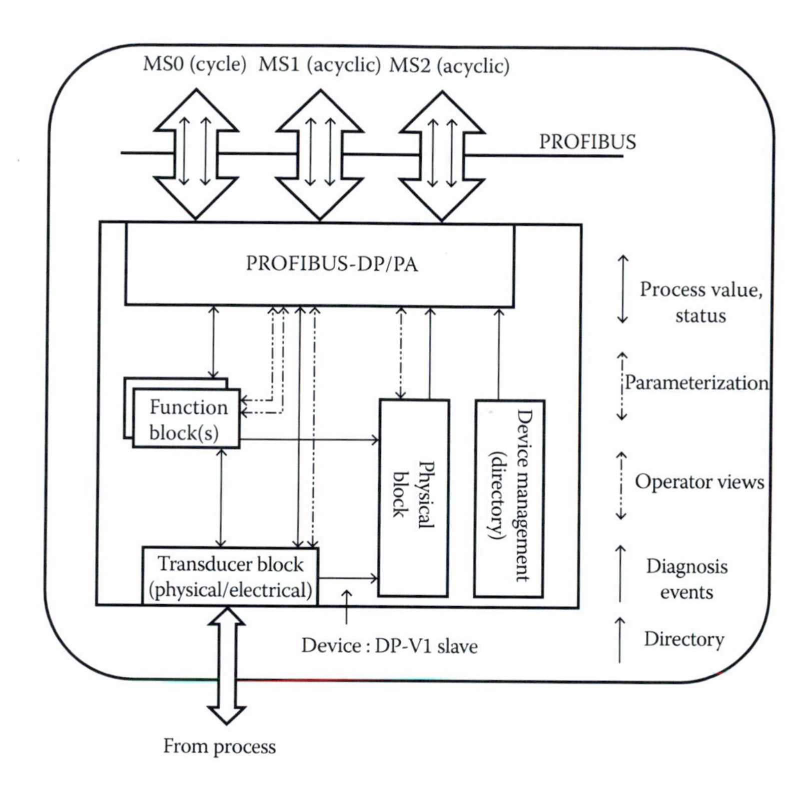

FIGURE 8.19 Block model of a PROFIBUS-PA device. (From PROFIBUS, PA System Description: Open solutions for the World of Automation, Karlsruhe, Germany, August, 2007. Available at www.profibus.jp/tech/document/PROFIBUS-PA-system-descr_e_Aug07.pdf.)

data formats as used in PROFIBUS standard. The data structure ensures that data stored and transmitted all follow an orderly manner.

8.9.4.1 Transducer Block

The transducer block accepts the sensor output. It processes the same and its output goes to the function block. The measurement principle of the transducer block is dependent on the basic process parameter that is to be measured and controlled. Currently available transducer blocks are temperature, pressure, level, flow, discrete input/output for switches, and electromagnetic and pneumatic transducer blocks for actuators. Multifunctional sensors having more than one sensor will have a corresponding number of transducer blocks.

8.9.4.2 Physical Block

The physical block contains the properties of the field device, i.e., device parameters and functions that are independent of the measurement method. It contains the serial number of the device by which it can be uniquely identified, the code of the particular manufacturer, data about the installation date, device operational data, and diagnostic features—both standard and manufacturer specific.

8.9.4.3 Function Block

A function block contains all the requisite functions so that final processing of the measured value is done in this block. It is then ready for transmission to the control system. The blocks are designed to be independent of the sensor and the fieldbus. There are several function blocks available:

1. Analog Input Block: It delivers the measured value from the sensor to the control system, i.e., class 1 master after processing, simulation and proper scaling.
2. Analog Output Block: It provides the device with a value as specified by the class 1 master.
3. Discrete Input Block: It provides a digital value from the device to the class 1 master.
4. Discrete Output Block: It provides the device with a value as determined by the class 1 master.
5. Totalizer Block: It is used when a process variable is to be integrated, i.e., to be summed over a period of time. An example is a flow totalizer, the output value being collected by the class 1 master.

8.9.4.4 Device Management Block

Today's devices contain a lot of information and can execute a number of functions hitherto done by PLCs and controllers. Such tasks can be executed correctly with the help of certain software tools needed for commissioning, maintenance, engineering, and parameterization of the devices.

PROFIBUS has developed the necessary software tools for device descriptions ultimately leading to device management blocks. A device management block provides the information about the following:

- Which blocks are present in a device
- Number of blocks a device has
- Where the starting address of a device is located

8.9.4.4.1 GSD

GSD is a general and device-specific specification for communication. It is supplied by the manufacturer of the PROFIBUS system. With the help of a keyword, the configuration tool can read the device identification,

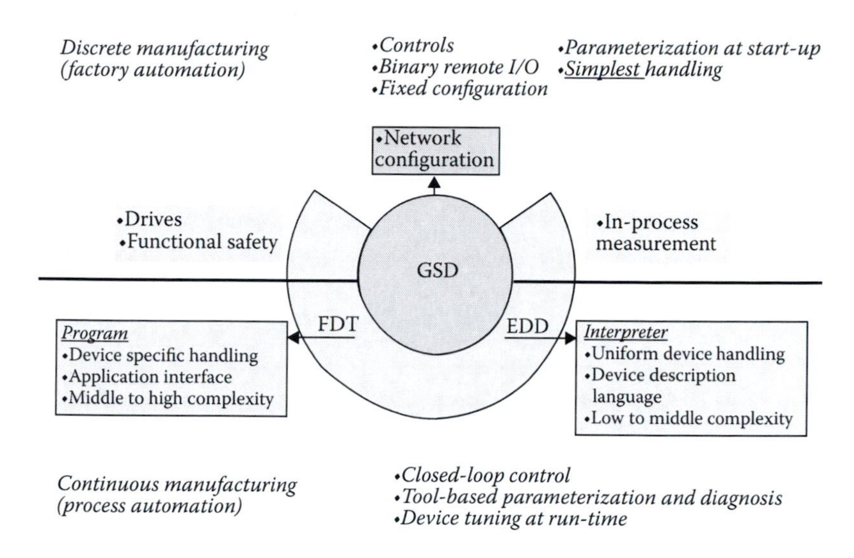

FIGURE 8.20 (**See color insert.**) Technology integration by GSD, FDT, and EDD. (From PROFIBUS, Technology and Application: System Description, PROFIBUS International Support Center, Karlsruhe, Germany, Copyright by PNO 10/02, October 2002. Available at www.pacontrol.com/download/profibus-overview.pdf.)

data type, and the permitted limit values from the GSD. A keyword like Vendor_Name is mandatory, while the keyword Sync_Mode_Supported is an optional one. A GSD can check for errors in input and data consistency. Technology integration by GSD along with FDT and EDD is shown in Figure 8.20.

A GSD file can identify a PROFIBUS-DP device (master–slave). It contains various information, such as vendor ID, baud rate supported, options/features supported, I/O signals, and the length and timing of I/O. It permits plug-and-play interoperability between devices from different manufacturers. The GSD file of each device is compiled into a master parameter record.

8.9.4.4.1.1 Specifications There are three GSD specifications: general specification, master specification, and slave specification.

The general specifications contain device and vendor names, hardware and software versions, transmission rate, signal assignment on the bus connector, time interval to monitor times, etc. The master specifications contain master-related information, such as the maximum number of connectable slaves and upload/download options. The slave specifications contain slave-related information such as the number and type of I/O channels, diagnosis text specifications, and available modules in case of modular devices.

8.9.4.4.1.2 Uses GSD can have two types of uses: GSD for compact devices and GSD for modular devices. The block configuration of the former is known on delivery, i.e., the device manufacturer completely designs the GSD. For modular devices, the block configuration is not fully specified by the manufacturer. The system user configures the GSD by using the configuration tool in line with the module configuration.

An optimum use of GSD can be done by reading the GSD into the configuration tool.

8.9.4.4.2 EDD

EDD is a very powerful software tool based on Electronic Device Description Language (EDDL). It can describe the application-related parameters and functions of a field device such as configuration parameters, range of values, measurement units, and default values.

Using EDDL, a device manufacturer can create the relevant EDD files. These files provide the information to the engineering tool and to the control system. Application areas of EDD include commissioning, runtime, engineering, asset management, documentation, and eCommerce. Using EDD has its own advantages, such as reducing training expenses, input data validation, and only one tool for all applications.

8.9.4.4.3 FDT/DTM

FDT is a manufacturer-independent, open interface specification that helps in integrating field devices with the operator program using Device Type Manager (DTM). FDT/DTM technology is subject to international standardization (IEC 62453).

Configuration and parameterization of field devices using existing languages have their own limitations:

- Intelligent field devices have their own complex, nonstandard diagnostics that cannot be properly utilized by the existing software.
- Both preventive maintenance and maintenance techniques are not properly included in the existing software.

Thus, these nonstandardized tasks must be included in some "auxiliary tool" that would enable device manufacturers of automation systems to integrate intelligent field device characteristics in the control system and at the same time allow the users with an expanded view of the field device characteristics through these special software auxiliary tools.

FDT provides an universal interface that includes the necessary software to address all the engineering and other automation requirements of the field devices. The specific functions of a field device, such as parameterization, configuration, diagnosis, and maintenance, along with user interface, are mapped in a software component called the DTM. A DTM is a device operator program that helps in implementing either device functionality or communication capabilities. The manufacturer programs the DTM in a device-specific way and contains a separate user interface for each device. DTM can be generated in any one of the following ways:

- Uses an existing DD with the help of a compiler or interpreter
- A DTM toolkit that uses MS Visual Basic
- May use some higher programming language

Some of the user benefits that accrue by using FDT/DTM are as follows:

- Provides an integrated solution for parameterization, configuration, diagnosis, and maintenance of the field devices.
- Provides ways for corporate management.
- Because of its protocol independent nature, FDT/DTM supports additional new options to the user by mapping numerous device functions in software tools.

8.9.4.4.4 ID

Every master and slave in a PROFIBUS system must have their individual IDs for proper system operation. On the basis of this ID, a master can identify the connected devices with the help of IDs without having an extensive overhead. A master compares the ID number of a device with the ID number provided in the configuration data. Data transfer does not take place until and unless the correct device type having the correct address is connected to the bus. This will eliminate any possibility of configuration error. PROFIBUS User Organization allocates the ID numbers of all devices—be it slave or master.

8.10 NETWORK CONFIGURATION

A manufacturer of a PROFIBUS controller adapter provides the GSD and network configuration program to set up the network configuration. A GSD file provides the following: vendor ID, features/options, baud rate supported, and the length and timing of I/O data. One major job of a GSD file is to support interoperability among different devices belonging to different manufacturers.

Each and every device belonging to the network has a GSD file of its own. These individual GSD files are loaded into the master parameter record. The master has an address allocation list of all the devices connected to the network.

During start-up operation, the master parameter record is used to set up communication with each assigned slave. When a new device is added to the network, it must be configured by the master so that it is inducted into the master parameter record. After this initialization, the system is ready to be used. A system reset is performed before using the network. The master will try to establish contact with all the slaves connected to the network, before initiating data exchange. The process begins with the lowest address of the slave and ending with the highest one. Each slave has a unique valid address in the range of 0 to 125.

8.11 BUS MONITOR

Bus monitor, also called a protocol analyzer, is a software tool for monitoring and troubleshooting network activity. With its help, timing and packet content verification can be ascertained. Bus monitors are typically PCs. These PCs must have a special PROFIBUS interface card and data capturing software.

All PROFIBUS activities can be monitored and captured by such a monitor. Each captured message is time stamped with very high time resolution. Thus, such a monitor can very precisely tell the occurrence of such captured data/event. The bus monitor is very helpful in indicating and diagnosing a problem occurring with an individual device.

Finally, such a monitor does not have a station address nor does it affect the speed or efficiency of the network.

8.12 TIME STAMP

Occurrence of certain important events and actions, such as diagnosis and fault location, must be precisely known. Thus, such events can very precisely be *time stamped*, enabling precise time assignment for such events.

PROFIBUS has a time stamp profile for this purpose. This needs a clock in the slave devices—realized by a master clock in the system. An event can be given a precise time stamp and can be read accordingly. Messages are graded as per their priority, with alarms falling under high priority and events under low priority. Figure 8.21 shows the technique of time stamping of alarms and events. The master reads acyclically such time-stamped alarms and events from the buffer of the device.

8.13 REDUNDANCY

Redundancy ensures increased system availability. Redundancy can be applied at different stages, such as master redundancy, media redundancy, segment coupler redundancy, ring redundancy, and slave redundancy.

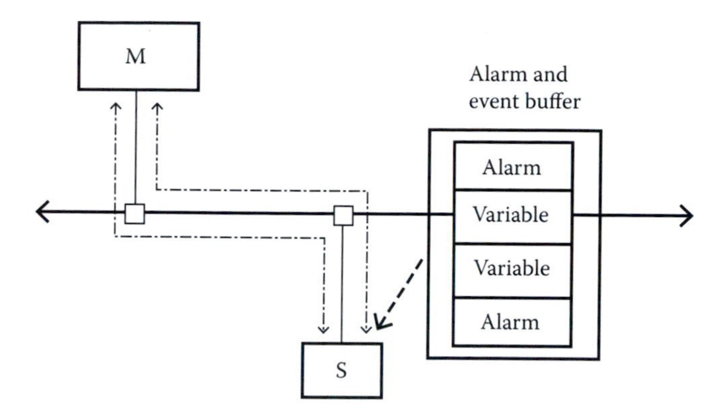

FIGURE 8.21 Time stamp and alarm messages. (From PROFIBUS, Technology and Application: System Description, PROFIBUS International Support Center, Karlsruhe, Germany, Copyright by PNO 10/02, October 2002. Available at www.pacontrol.com/download/profibus-overview.pdf.)

Master redundancy ensures the availability of a second master in case the primary one fails, i.e., the availability of the controller is ensured. Media redundancy ensures that the cabling is designed with redundancy. Segment coupler redundancy means that if a DP–PA gateway fails, the other will take over its function. Neither the master nor the slaves remain unaware of such a switchover. Ring redundancy ensures media redundancy on the PA side.

Slave redundancy ensures installation of field devices with redundant communication. Figure 8.22 shows a PROFIBUS system with redundant communication.

The devices must have the following characteristics for slave redundancy to be ensured:

- Slave devices must contain two independent PROFIBUS interfaces. One is called the *primary* and the other is the *backup* (also known as slave interface). There can be one or two devices to ensure the above.
- There must be two independent protocol stacks in the devices with a special redundancy expansion.
- Within such a device, a *redundancy communication* (*RedCom*) must run between the two protocol stacks. Under normal conditions, communications take place over the primary slave, which also sends the diagnostic data of the slave. When the primary slave fails, either the secondary slave takes over seamlessly or is requested by the master to do so.
- Slave redundancy can be realized on a single PROFIBUS line or on two PROFIBUS lines. The latter one ensures line redundancy. Slave redundancy provides high data availability, short reversing time, and no data loss and a fault-tolerant system.

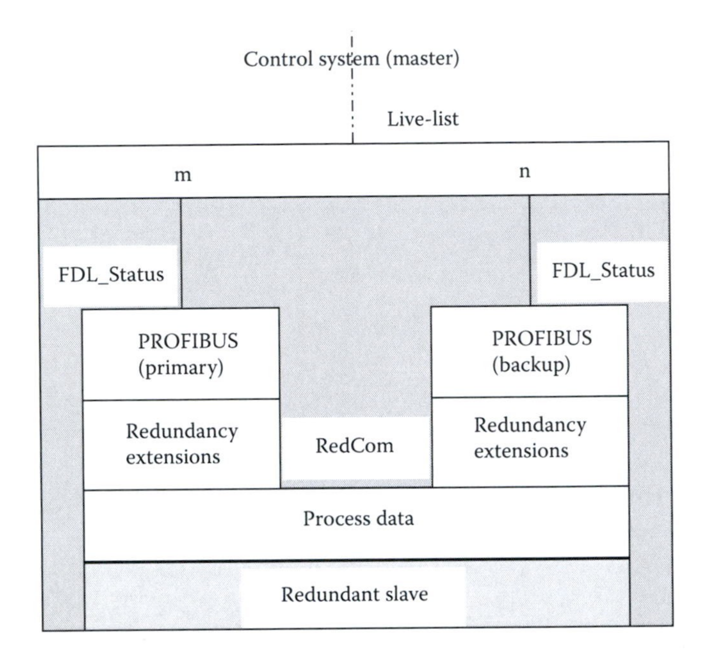

FIGURE 8.22 Slave redundancy in a PROFIBUS system. (From PROFIBUS, Technology and Application: System Description, PROFIBUS International Support Center, Karlsruhe, Germany, Copyright by PNO 10/02, October 2002. Available at www.pacontrol.com/download/profibus-overview.pdf.)

8.14 PROFIsafe

There are several important issues that need addressing in serial bus communication. Some of these are loss or repetition of data, some data packet reaching the destination with delay, corrupt messages, incorrect sequences of data reception, etc. PROFIBUS has designed PROFIsafe to counter these error possibilities.

PROFIsafe is a single-channel software solution implemented in the devices above layer 7 of OSI. Figure 8.23 shows the fail-safe mode with PROFIsafe.

PROFIsafe defines how fail-safe devices such as emergency stop push-buttons can communicate over PROFIBUS with the help of fail-safe controllers. This safety-related automation task can be used up to KAT4 compliant with EN954 or Safety Integrated Level (SIL3). This is implemented over a special format of user data and a special protocol. PROFIsafe increases the transmission safety of the PROFIBUS protocol.

To achieve this fail-safe position, PROFIsafe has taken into consideration the following:

- Numbering of consecutive safety telegrams
- An identifier between sender and receiver in the form of a password

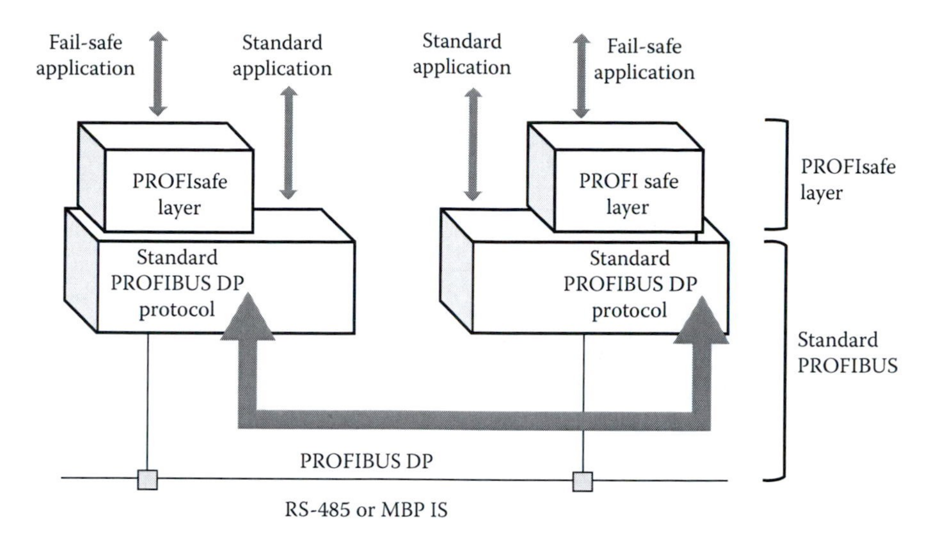

FIGURE 8.23 Fail-safe mode as used with PROFIsafe. (From PROFIBUS, Technology and Application: System Description, PROFIBUS International Support Center, Karlsruhe, Germany, Copyright by PNO 10/02, October 2002. Available at www.pacontrol.com/download/profibus-overview.pdf.)

- Additional check on data in the form of CRC
- Timeout of incoming message frames and their acknowledgment

By combining these remedial measures, PROFIsafe has achieved a safety level of SIL3 and beyond. Devices that have PROFIsafe profile incorporated in them operate in tandem with standard devices. Finally, PROFIsafe uses acyclic communication and RS-485, fiber optic, or MBP transmission technology.

The safety-related issues are incorporated in the software and superimposed on the PROFIBUS protocol. The safety layer checks the data during cyclic data exchange and indicates an error when detected. When data is sent, the safety layer incorporates the safety-related functions into the data.

8.15 PROFIdrive

PROFIdrive is used in the area of electric drives. Its application ranges from simple frequency converters to highly sophisticated servo controls. The protocol defines six different application classes, depending on the complexity of the application.

Standard drive (class 1) is controlled by a set point value whereby the rotational speed is controlled by a drive controller. *Standard drive with technological functions* (class 2) involves dividing the main automation process into several smaller ones and then shifting some of the automation functions from the main controller to the drive controller. An example of

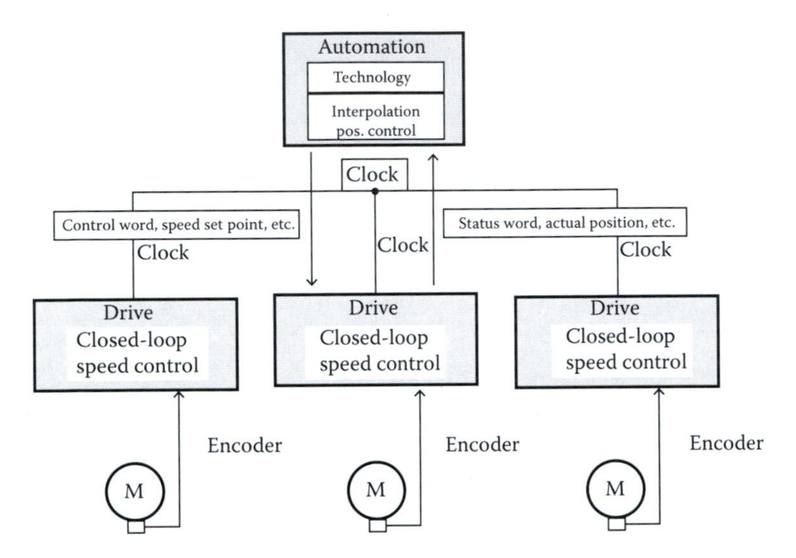

FIGURE 8.24 PROFIdrive position control loop and clock synchronism. (From PROFIBUS, Technology and Application: System Description, PROFIBUS International Support Center, Karlsruhe, Germany, Copyright by PNO 10/02, October 2002. Available at www.pacontrol.com/download/profibus-overview.pdf.)

class 2 type is slave-to-slave communication. With *positioning drive* (class 3), an additional position controller is incorporated in the drive. An example is the twisting on and off of bottle tops. The positioning is precisely done by the drive controller. In *central motion control* (classes 4 and 5), multiple drives coordinate the motion sequence and the motion is controlled by a computer controlled numerical machine. Other applications include motion control of linear motors. Figure 8.24 shows a class 4 PROFIdrive system. *Distributed automation* (class 6) has applications in slave-to-slave communication with the help of clocked processes and electronic shafts.

8.16 PROFInet

Process automation technology is increasingly moving toward modular designs with distributed intelligence. PROFInet uses Ethernet-based communication technology and has the following features fulfilling the needs of automation technology:

- It is open system with other ones.
- It can implement IT standards.
- It is a vendor-independent communication technology, leading to interoperable systems.
- It communicates from level 1 to level 5 consistently.
- It integrates PROFIBUS segments without any need to change them.

PROFInet is available as a specification- as well as operating system-independent source software. PROFInet has an engineering model and a communication model. The engineering model is a vendor-independent engineering concept that fits into a PROFInet system by using a user-friendly configuration tool. The model also supports manufacturer-customized functional expansions. The communication model defines a vendor-independent communication specification for data transfer over the Ethernet using conventional IT procedures (called runtime communications). This model uses TCP/IP and COM/DCOM protocols. Figure 8.25 shows the device structure of PROFInet that implements the above protocols.

Integration of PROFIBUS-PA into PROFInet ensures protection of current investments and at the same time enjoys the benefits of PROFInet technology. Mapping of PROFInet into PROFIBUS-PA ensures easy integration.

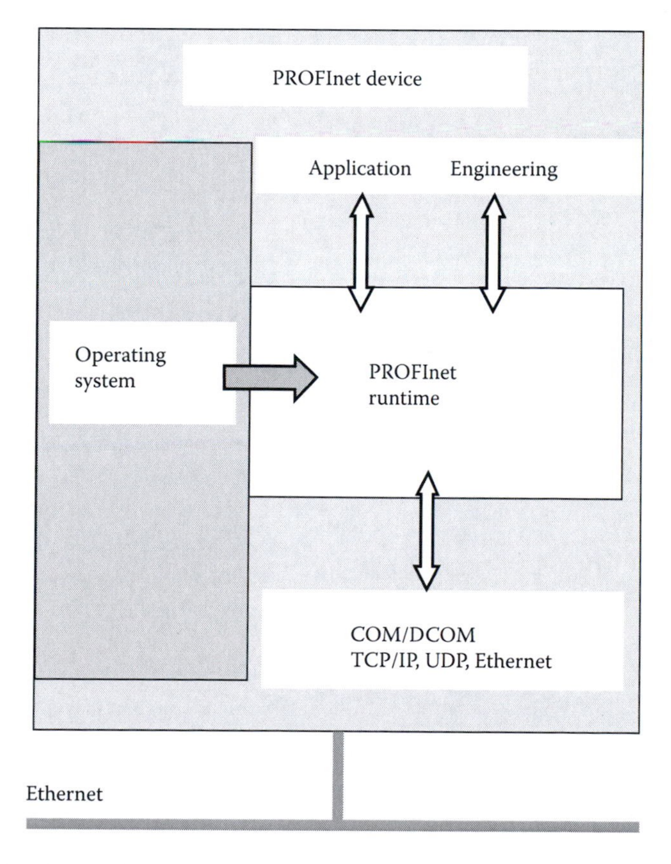

FIGURE 8.25 Device structure using PROFInet. (From PROFIBUS, Technology and Application: System Description, PROFIBUS International Support Center, Karlsruhe, Germany, Copyright by PNO 10/02, October 2002. Available at www.pacontrol.com/download/profibus-overview.pdf.)

8.17 PROFIBUS INTERNATIONAL

PNO was established in 1989 to promote PROFIBUS technology in manufacturing and process automation with a view to develop and maintain the market dominance of the protocol. PROFIBUS International (PI) was established in 1995—now having 23 members, of which PNO is one. PI has the following tasks to perform:

- Development and maintenance of PROFIBUS technology
- Enhancing acceptability of PROFIBUS technology worldwide
- Extending technical support for PROFIBUS technology through competence centers
- Quality assurance through device certification
- Enhancing members' interests through standardization committees and associations

8.18 FOUNDATION FIELDBUS AND PROFIBUS—A COMPARISON

A comparison between Foundation Fieldbus and PROFIBUS is shown in Table 8.5.

TABLE 8.5
Comparison between Foundation Fieldbus and PROFIBUS

Foundation Fieldbus	PROFIBUS
Is designed to address the needs of process automation.	Is designed to address the needs of discrete manufacturing and building automation.
Foundation Fieldbus H1 uses the physical layer as defined in IEC 1158-2.	PROFIBUS-PA uses the physical layer as defined in IEC 1158-2.
Supports the requirements of intrinsic safety, including the new FISCO model.	Supports the requirements of intrinsic safety, including the new FISCO model.
A peer-to-peer protocol. Devices can communicate with each other without a host and they can initiate communications without a specific host command.	PROFIBUS-PA is a master–slave protocol. A field device is a slave and it can only respond to a command from the master.
Control can be in the field device, or the host, or partially in both.	With PROFIBUS-PA, control resides only in the host.
When control is in the field devices, the host can be disconnected without halting the loop.	Host must be present for control to function.

(continued)

TABLE 8.5 (Continued)
Comparison between Foundation Fieldbus and PROFIBUS

Foundation Fieldbus	PROFIBUS
Addresses interoperability by a combination of device descriptions and function blocks. A single host application can configure and access all device information and functionality.	A PROFIBUS-PA host uses a standard profile for basic functionality. For additional vendor-specific functionality, the host must have the corresponding software.
It uses device description technology to make all information available to all devices, host systems, and applications.	The device configuration and management host use device descriptions to configure and interact with the device. The control host uses profiles to access device information.
The device address can be manually/automatically assigned. It uses a special message to detect and identify a new device. A device can be added or deleted when the segment is in operation.	Device addresses are set by setting DIP switches or user-entered software addresses. To add a device, the segment must be shut, and the device address and configuration parameters are configured in the host. The segment is then restarted.
It supports device and function block tags in the field devices. Thus, a device can be located by simply asking for it by its tag.	It supports tag in the host. Tag data base is manually entered in the host.
Provides a distributed real-time clock on the bus.	Does not provide a real-time clock on the bus.
It is appropriate for real-time control on the bus—with or without a host.	Appropriate for host-based control only.

9 MODBUS and MODBUS PLUS

9.1 INTRODUCTION

MODBUS is a serial communication protocol initially developed by AEG-Modicon. It was initially designed to operate with programmable logic controllers (PLCs). It is an application layer messaging protocol, operating at layer 7 of the Open Systems Interconnection (OSI) protocol, and provides client–server communication between devices connected on different types of networks. The MODBUS protocol layers are shown in Figure 9.1, with the OSI protocol layers shown beside it. It defines a method of accessing and controlling a device by another irrespective of the type of physical network involved.

For MODBUS, no interface is required, as is the case with many other buses. A user has the option of choosing between RS-422, RS-485, or 20 mA current loops. Compared with other buses, MODBUS is relatively slower but has the decided advantage of very wide acceptance among control instrument manufacturers and users.

It is the industry's serial *de facto* standard since 1979. There is no formal way of certifying that a product is MODBUS compatible. It is the responsibility of the manufacturers to confirm that their products are compatible with other MODBUS devices. The protocol describes the manner by which a device accesses another, how information is received, and how queries are responded to. In case of error, the protocol provides a mechanism to send the corresponding command to the user. Communication may take place on a MODBUS network or on other networks (like Ethernet) by embedding the MODBUS protocol as data packets in the protocol of the other networks.

The MODBUS serial communication protocol is based on the master–slave principle, with the master initiating a transaction. The protocol provides for one master and up to 247 slaves.

Some characteristics of MODBUS are fixed and some others are selectable by the users. The fixed characteristics are frame format, frame sequence, handling of communication errors and exception conditions, and the functions performed. The selectable characteristics are transmission medium and transmission characteristics. The user characteristics, once set, cannot be changed when the system is in operation.

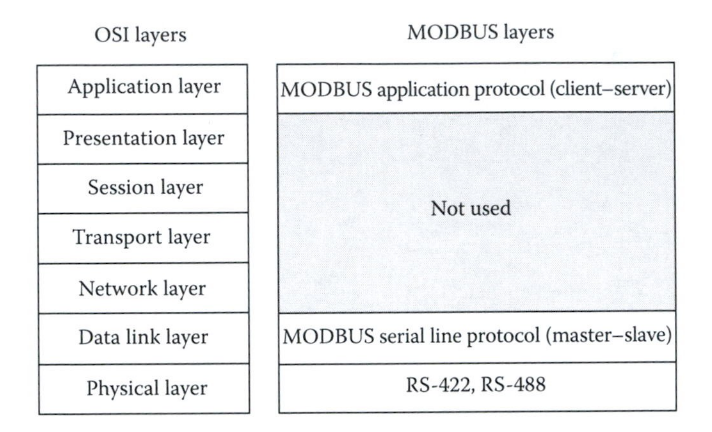

FIGURE 9.1 MODBUS protocol layers.

9.2 COMMUNICATION STACK

Presently, MODBUS is implemented using the following:

- TCP/IP over Ethernet
- Asynchronous serial transmission over different media, such as EIA/TIA-232E, EIA-422, EIA/TIA-485-A, fiber, and radio
- MODBUS Plus—a high-speed token-passing method

Figure 9.2 shows the implementation of a MODBUS communication stack using TCP/IP, master–slave, and MODBUS Plus physical layer.

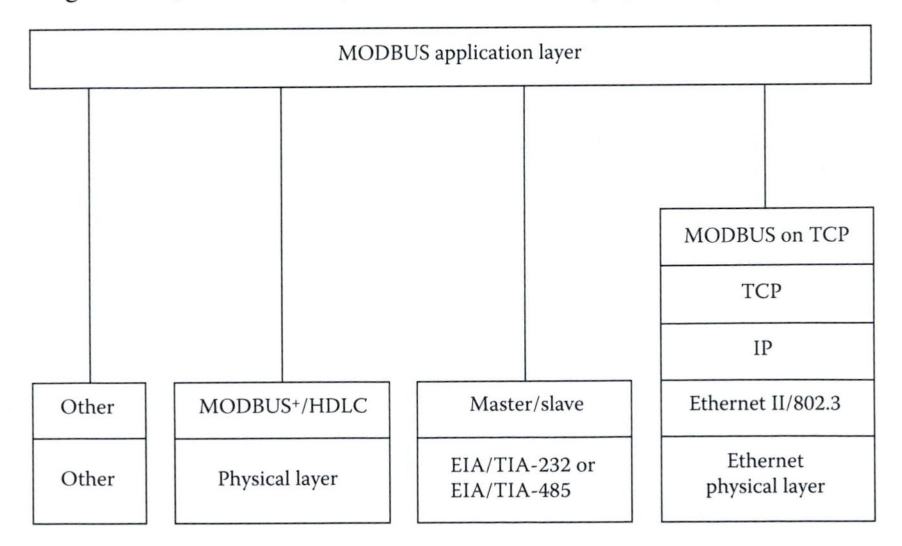

FIGURE 9.2 MODBUS communication stack. (From MODBUS, Application Protocol Specification, vol. 1.1b, December 28, 2006. Available at www.modbus.org/docs/Modbus_Application_Protocol_V1_1b.pdf.)

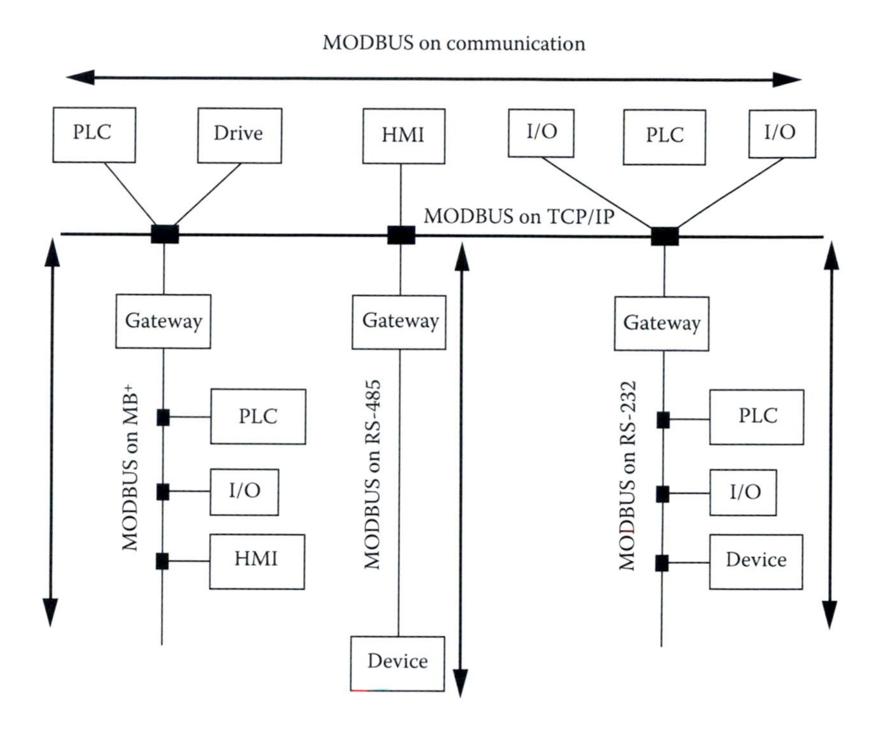

FIGURE 9.3 MODBUS network architecture scheme. (From MODBUS, Application Protocol Specification, vol. 1.1b, December 28, 2006. Available at www.modbus.org/docs/Modbus_Application_Protocol_V1_1b.pdf.)

9.3 NETWORK ARCHITECTURE

Figure 9.3 shows a scheme of MODBUS network architecture. The different devices such as PLC, HMI, and I/Os can be connected to the MODBUS TCP/IP via individual gateways. The different MODBUS protocols, viz., MODBUS on MB+, MODBUS on RS-232, and MODBUS on RS-485, initiate remote communication by using TCP/IP.

9.4 COMMUNICATION TRANSACTIONS

MODBUS serial communication uses the master–slave protocol. The master initiates the query and the slave responds by either providing the requisite data to the master or by taking the appropriate action as was requested for. The slaves respond by the following:

- By taking appropriate action
- By providing requisite data/information to the master
- By informing the master that the requisite action could not be carried out

An error message, termed *exception response*, is sent to the master when the slave is unable to carry out the required actions as requested by the master. The exception response to the master contains the following:

- The address of the responding slave
- The action that the slave was requested to carry out
- An indication of why the action could not be carried out

A slave ignores a message if it contains some error. In such cases, the master resends the query to the slave since it failed to receive a response from the slave.

9.4.1 Master–Slave and Broadcast Communication

A master can individually address the slaves one by one (called the *unicast* mode) or address all the slaves at the same time (called the *broadcast* mode). The slaves respond to unicast message, but do not respond to multicast messages. The master–slave communication and the broadcast communication modes are shown in Figures 9.4 and 9.5, respectively.

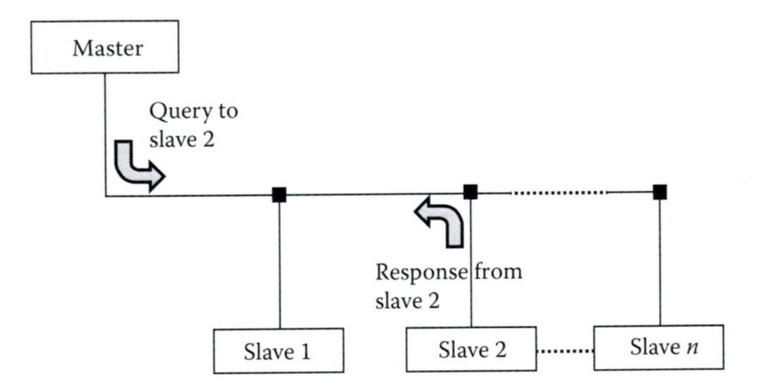

FIGURE 9.4 Master–slave communication model.

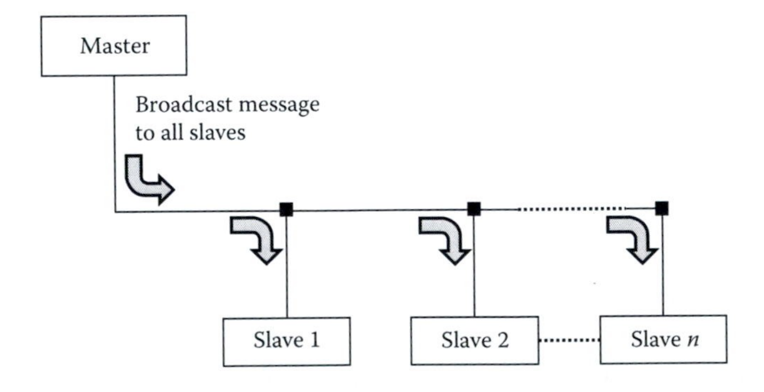

FIGURE 9.5 Broadcast communication mode.

9.4.2 Query–Response Cycle

Query is initiated by the master and the slave responds to it. The query–response cycle is the basis of all communication transactions on a MODBUS network. The query from the master contains four parts: device address, function code, data/message, and error check code. The response structure from the slave is identical to that of the master. Figure 9.6 shows query–response cycle.

9.4.2.1 Address Field

The address field sent by the master in its query contains the address of the slave to which the message is intended. Its value is in the range of 1–247, although practical limitations place a much lower value. When the slave sends its response, it places its own address in its address field so that the master can know that the correct slave is responding. Address "0" is earmarked for broadcasting. All slaves read them, but do not provide any response to such query from the master.

9.4.2.2 Function Field

Function codes are in the range of 1–255, although not all the function codes are supported by all the devices. When a function code reaches a slave from a master in its query, the slave then comes to know the actions that it will have to take. Examples of actions to be taken by the slave may include the following: read the input status, read register content, change a status within the slave, and operate a relay coil.

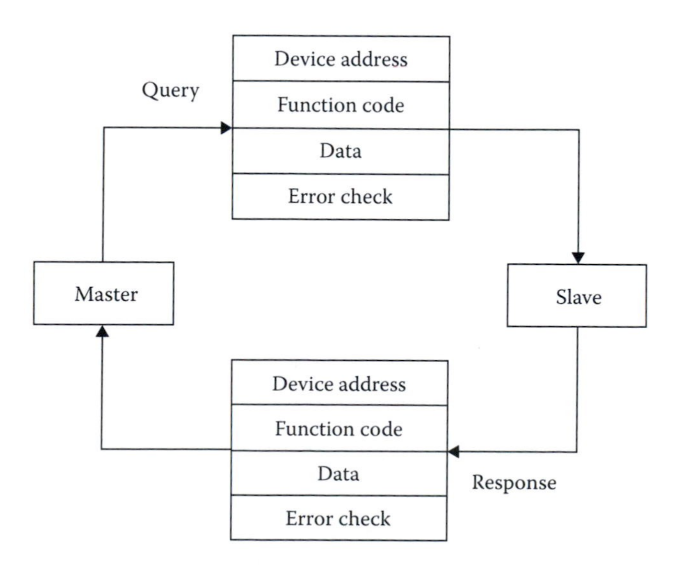

FIGURE 9.6 Query–response cycle.

When the slave sends its response to the master, it repeats the function received. It indicates that the slave has understood the query from the master and acted accordingly. If the instruction could not be carried out by the slave, it generates an "exception response" and the slave uses the function code and data field to inform the master the reasons for such exception.

In case the exception response is generated, the slave returns the original function code to the master, but with the MSB set to 1. The data field of the response message from the slave, in such cases, indicates to the master the nature of error that has occurred. Thus, the master can take appropriate actions based on this. The actions taken by the master may be either to repeat the original message or to try and diagnose the problem or to set an alarm, etc.

9.4.2.3 Data Field

The data field received by a slave in the query from the master may typically include a register value, a register address, or a register range. Some functions do not require the data field and, thus, the same is not included in the query from the master.

If no error has occurred, the data field of the response is used by the slave to pass data back to the master. When an error occurs, the data field from the slave passes on more information, informing the master about the nature of the error detected.

MODBUS does not encode the data and consequently a number of encoding schemes can be employed. This choice is left with the user of MODBUS.

9.4.2.4 Error Check Field

The error check field allows the master to confirm the integrity of the message received from the slave. The error check method employed depends on the transmission mode selected. It may be cyclic redundancy check (CRC) for the Remote Terminal Unit (RTU) mode or longitudinal redundancy check (LRC) for the ASCII mode of transmission.

On receiving the full message, the receiving device calculates the error check value and compares it with the error check value in the received message. If the two agree, no error has occurred and actions are taken accordingly. The received message is rejected if the two values differ.

9.5 PROTOCOL DESCRIPTION: PDU AND ADU

MODBUS can be implemented on different types of buses and networks, but the portion Protocol Data Unit (PDU) is an integral part in each of them. It consists of two fields: function code and data.

The two most common types of MODBUS implementations are Ethernet (TCP/IP) and serial, which may be RS-232, RS-422, or RS-485. The most

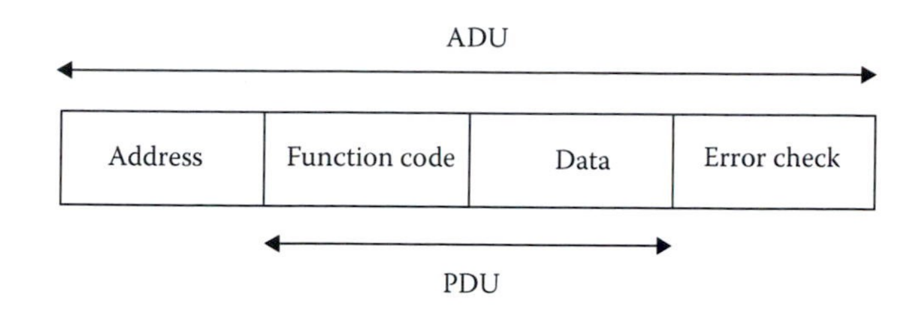

FIGURE 9.7 General MODBUS frame.

common MODBUS serial protocol is MODBUS-RTU. Irrespective of the type of protocol used for a particular application, application-specific addressing and error checking are attached to the PDU to give rise to Application Data Unit (ADU). It is shown in Figure 9.7. The ADU represents the MODBUS frame. Since error checking is handled by TCP protocol, it is omitted in the ADU of a MODBUS TCP.

9.6 TRANSMISSION MODES

Two types of transmitting modes are possible for transmitting serial data over a MODBUS network: RTU and ASCII. The two modes differ in a number of ways: the manner in which information is packed in the message field, the way the bit contents of the message are interpreted, the way the message is decoded, and the speed of operation at a given baud rate.

The two modes cannot be used together, and the user has the option of selecting the particular mode for a certain application. The RTU mode is faster and more robust than the ASCII mode. Thus, it finds more applications than the ASCII form for message transmission.

The RTU transmission mode is sometimes referred to as MODBUS-B, while the ASCII transmission mode as MODBUS-A. The typical message length in ASCII mode is roughly twice the length of the equivalent RTU message. While in the ASCII transmission mode each byte in the message is transmitted as two ASCII characters, the same is sent as one 8-bit binary number containing two hexadecimal digits in the RTU transmission mode.

MODBUS packets can also be transmitted over local area networks and wide area networks by encapsulating the MODBUS data in a TCP/IP packet.

9.6.1 ASCII Mode

In this, each 8-bit byte is sent as two ASCII characters. The format for each byte in ASCII mode is given below:

- Coding: hexadecimal, ASCII characters 0 to 9, A to F. One hex character comprises each ASCII character of the message.

- Bits per byte: 1 start bit, 7 data bits, LSB sent first, 1 bit for odd or even parity, no bit used for no parity, 1 stop bit if parity is used, 2 bits used for no parity.
- Error check field: LRC.

9.6.2 RTU Mode

In this, each 8-bit byte is sent as two 4-bit hex characters. The format for each byte in ASCII mode is given below:

- Coding: 8-bit binary, hex 0 to 9, A to F. Two hex characters comprise each 8-bit field of the message.
- Bits per byte: 1 start bit, 8 data bits, LSB sent first, 1 bit for odd or even parity, no bit used for no parity, 1 stop bit if parity is used, 2 bits used for no parity.
- Error check field: CRC.

9.7 MESSAGE FRAMING

For transmission of messages, a frame is constructed before its eventual transmission. A frame consists of start and end character(s), address of the device in the unicast mode or device addresses in the broadcast mode, function code, and data and error check code.

9.7.1 ASCII Framing

The MODBUS ASCII frame consists of six fields: Start, Address, Function Code, Data, LRC, and End. The frame begins with a "start," which is a colon (:) character, and its ASCII character 3Ah, h signifying hex. The frame ends with an "end," which is a "carriage return-line feed" and representing the two ASCII character pair 0Dh and 0Ah. The other remaining four fields in between have characters 0 to 9 and A to F in hex. The ASCII frame format is shown in Figure 9.8.

The ASCII mode allows an interval of up to 1 s between two successive transmissions without generating any error. All devices connected to the MODBUS network continue to monitor the colon character that would

Start 1 char	Address 2 char	Function code 2 char	Data N char	LRC 2 char	End 2 char

FIGURE 9.8 ASCII frame format.

signal the start of an ASCII character. If a particular device finds the address matching its own, then it would start decoding the function code and the rest fields and take actions accordingly. Otherwise, the device will ignore the same and continues its monitoring of the next colon character followed by the address and would take subsequent actions.

9.7.2 RTU Framing

In this mode, the message frame starts with a silent time gap of at least 3.5 times character length. The message ends with the same time gap of 3.5 times character length. After the start field, the address field is monitored by the receiving devices to know whether the message is meant for that device. The RTU frame format is shown in Figure 9.9.

The entire message frame must be transmitted in one continuous stream or else an error will be generated. If there is a gap in transmission of over 1.5 character times, then the receiving device assumes an error has occurred and removes any received information. Similarly, if a message starts within 3.5 character times after the end of a message, the receiving device assumes that the received information is part of the previous message. In either of the two cases cited above, a CRC error occurs.

9.8 MODBUS TCP/IP

The open MODBUS TCP/IP specification was first introduced in 1999. There are several advantages of using the MODBUS TCP/IP protocol, such as simplicity, use of standard Ethernet, and openness. Transfer rates in excess of 1 kB/s can very easily be achieved on a single station.

MODBUS TCP/IP is an Internet protocol. It is simply a MODBUS protocol with a TCP wrapper. Thus, MODBUS devices can communicate over MODBUS TCP/IP. A gateway device is simply needed to convert from the physical layer (RS-232, RS-485, or others) to Ethernet and to convert MODBUS protocol to MODBUS TCP/IP. Figure 9.10 shows the MODBUS TCP/IP protocol layers and also the MODBUS frame wrapped into the MODBUS TCP/IP frame.

The normal MODBUS frame contains the device address as either two ASCII codes (in ASCII mode) or an 8-bit hex byte (in RTU mode).

Start 3.5 char times	Address 8 bits	Function code 8 bits	Data N × 8 bits	CRC 16 bits	End 3.5 char times

FIGURE 9.9 RTU frame format.

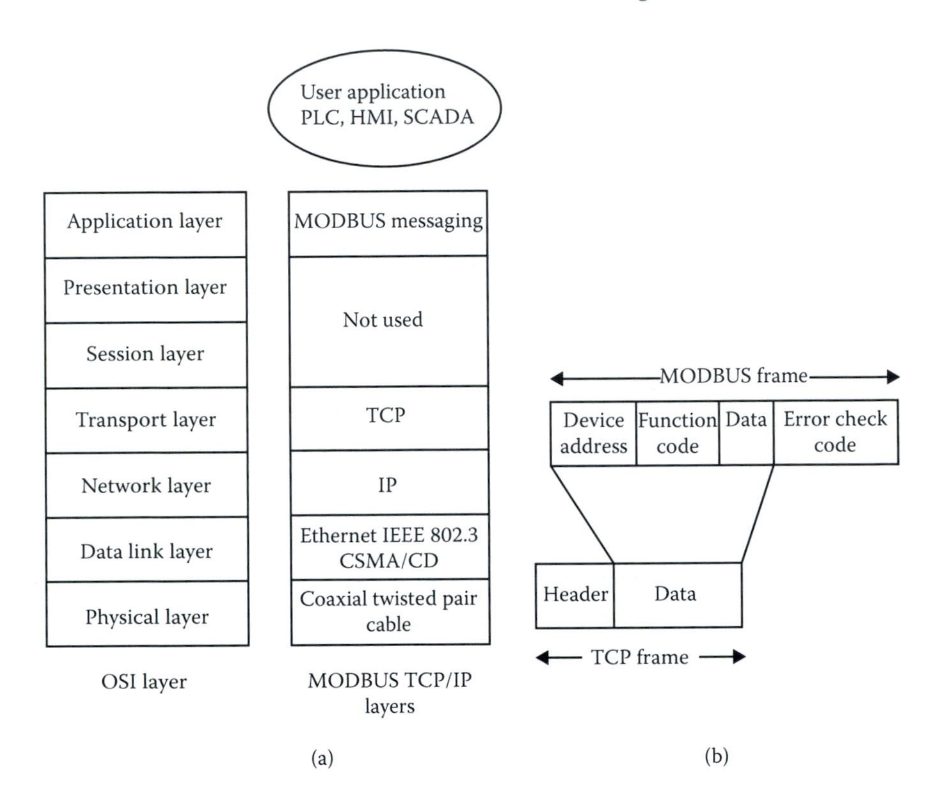

FIGURE 9.10 (a) MODBUS TCP/IP layers; (b) MODBUS TCP/IP frame.

The same is replaced by a combination of IP address and a unit identifier that identifies the device on the network. Also, the 16-bit checksum in a MODBUS frame is replaced by TCP's 32-bit CRC in the MODBUS TCP/IP frame.

The master–slave architecture of the MODBUS protocol is modified to client–server architecture in MODBUS TCP/IP. Since TCP is a connection-oriented protocol, for every query in MODBUS TCP/IP, there would be a response.

9.9 INTRODUCTION TO MODBUS PLUS

Apart from the standard MODBUS protocol, there are two others: MODBUS Plus and MODBUS II. MODBUS Plus is used because of additional cabling implementations; it is not an open standard protocol like the standard MODBUS.

The "single master" limitation of the MODBUS protocol has led to the development of MODBUS PLUS, which can share information and control strategy across various MODBUS networks. Thus, MODBUS Plus networks within its fold have the individual MODBUS communication networks. MODBUS Plus was one of the early token-passing protocols. It is a LAN in which the devices situated at geographically different positions can share information for measurement, control, and monitoring purposes.

MODBUS PLUS caters to individual MODBUS networks by bridging them together. The protocol allows a maximum of 64 devices on an individual protocol segment with each device assigned a network address in the range of 01 to 64. Messages can be routed from a device belonging to one MODBUS segment to another by both the network address of the device and internetwork address. The routing of a message may be five layers deep.

9.10 MESSAGE FRAME

The MODBUS Plus message frame is shown in Figure 9.11.

The frame begins with a preamble, followed by opening flag, broadcast address, data—whose length is variable in nature, an error check field, and lastly the closing flag. The variable data field itself has five fields: destination address, source address, MAC function, byte count, and LLC field.

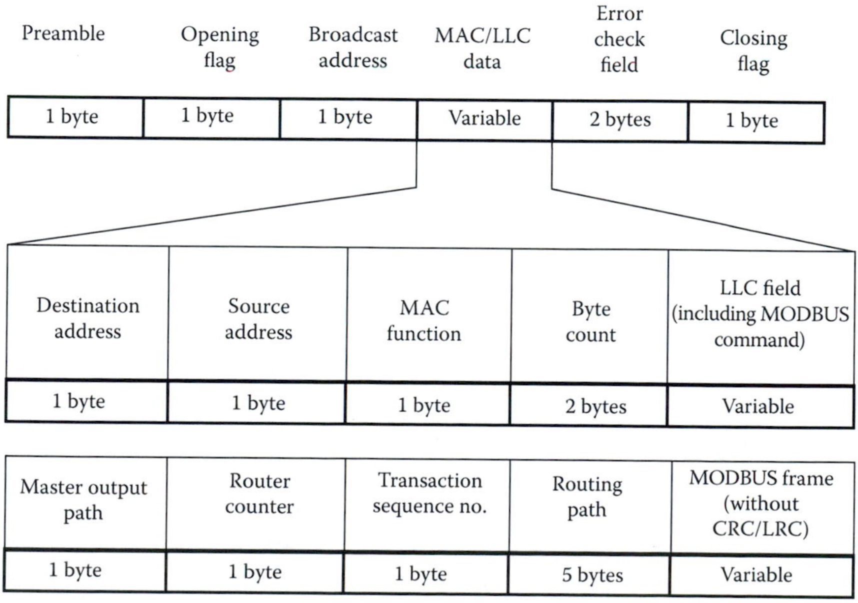

Preamble	Opening flag	Broadcast address	MAC/LLC data	Error check field	Closing flag
1 byte	1 byte	1 byte	Variable	2 bytes	1 byte

Destination address	Source address	MAC function	Byte count	LLC field (including MODBUS command)
1 byte	1 byte	1 byte	2 bytes	Variable

Master output path	Router counter	Transaction sequence no.	Routing path	MODBUS frame (without CRC/LRC)
1 byte	1 byte	1 byte	5 bytes	Variable

FIGURE 9.11 MODBUS PLUS message frame. (From S. Mackay et al., *Practical Industrial Data Networks, Design, Installation and Troubleshooting*. Newnes An Imprint of Elsevier, UK, p. 260, 2004.)

9.11 NETWORKING MODBUS PLUS

A single MODBUS PLUS network can support a maximum of 64 nodes. On a single section, up to 32 nodes (devices) can be connected with a maximum cable length of 450 m. The minimum cable length between any two successive pair of nodes is 3 m. Repeaters can be used to increase the node count to 64 in which case the maximum cable length becomes 1800 m. For longer cable distance requirements, fiber cables can be used. Figure 9.12 shows a single network consisting of two sections joined by a repeater. Thus, while a *repeater* is used to extend the length of a *single* network, a bridge is used to join *multiple* networks. A *bridge* cannot be used for deterministic timing of I/O processes.

A device on a network is assigned an address by the user of the network. No two device addresses can be the same. The devices in a network act as peer members of a logical ring, and a particular device identifies itself with the network on receiving the token frame. Each network maintains its own token rotation sequence. The token of a network is never passed on to another network. When a device or node holds a token, it can initiate message transmission with other nodes in the network.

When passing the token, a node writes into the global database, which is then broadcast to all the other nodes in the same network. The global data represents a field in the token frame. Nodes in the network monitor the global data and extract the same for its own use, like updating of alarms, set points, etc. Each network in a multiple network system has its own global database, as the token is not passed from one network to any other.

A typical global plus network with dual cabling is shown in Figure 9.13. The two cables are termed as cable A and cable B. Each cable will have a maximum length of 450 m. The maximum difference in length between cable A and cable B cannot be more than 150 m, measured between any pairs of nodes on the cable section.

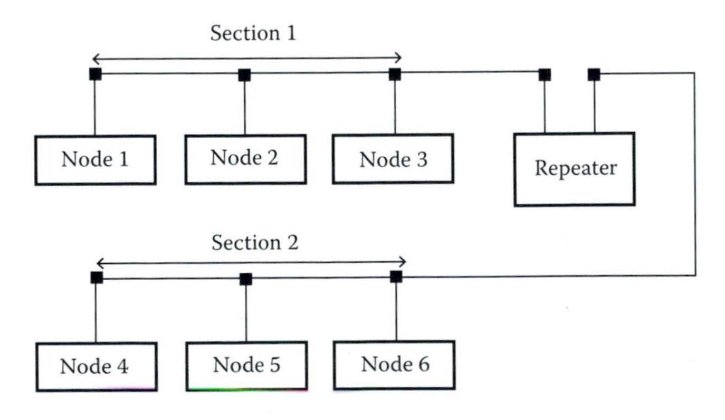

FIGURE 9.12 Two sections of a network connected by a repeater.

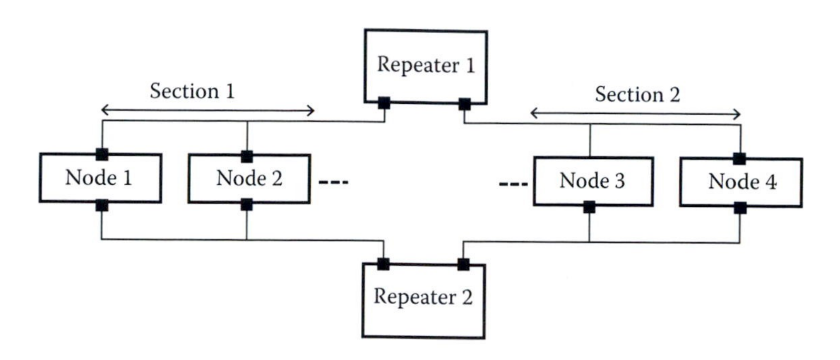

FIGURE 9.13 MODBUS PLUS network with dual cabling.

The token-passing mechanism works like this: the token rotation begins with the network's lowest addressed active node, and increases progressively until the highest active node address is reached. The token then begins at the lowest addressed active node again.

Addition and deletion of nodes to a network are transparent to the user. When a node leaves a network, a new token-passing sequence will be established in 100 ms. On the other hand, if a new node joins a network, it will be added to the newly formed address sequence in 5 s.

10 CAN Bus

10.1 INTRODUCTION

CAN, which stands for Controller Area Network, was developed by Bosch of Germany in 1986 to take care of growing demand of electronic control systems in automobile industries. CAN is a serial communication bus protocol standardized by the International Standards Organization (ISO). It does not use a master–slave or a token-passing method to access the bus. Instead it uses a unique bus access control method, called "nondestructive bitwise arbitration." It is a very simple, highly reliable, and prioritized communication protocol among sensors, actuators, and intelligent devices. A producer–consumer technique is applied to access the physical medium based on carrier sense multiple access with collision detection (CSMA-CD). It is a deterministic method to resolve collision/conflict on the bus by taking recourse to bus contention. The system thus uses the full bandwidth of the medium.

CAN refers to a network of independent controllers. It supports distributed real-time control with a very high level of security. The different controls in an automotive vehicle are of different data types requiring multibus lines to be sent to the controller. This resulted in many wires leading to various problems. CAN was developed to effectively address the above and became a standard for vehicle networking. It has since been applied to numerous fields for control purposes.

10.2 FEATURES

The units connected to the bus can send messages when the bus is free, i.e., the system is a multimaster type. When more than one unit starts sending messages at the same time, their priority is resolved by a message identifier residing in the data frame. Thus, a particular unit wins the bus contention and the message is sent by that unit. The other units that lost out can send their message when the bus goes into idle state. The units connected to the bus do not have any address, and any unit can be added/deleted at any time without any change in software or hardware. The speeds of the connecting units in the network are unique as they are already set depending on the network size.

Data from outside the network can be accommodated by sending a "remote frame" to the units. The CAN protocol has error detection, error notification, error recovery, and error confinement capabilities. The number of units that can be connected to the system has no logical limit, although it is dependent on latency and load on the bus. When more units are added to the system, speed decreases and vice versa.

10.3 TYPES

CAN has several ISO standards, such as ISO 11898 and ISO 11519-2. These two standards do not differ in data link layer but have differences in physical layer. ISO 11898 refers to high-speed CAN communication. Later on, it was divided into ISO 11898-1 and ISO 11898-2. The former stands for the standard for the data link layer, while the latter for the physical layer. The standard ISO 11519-2 refers to low-speed communication with speeds up to 125 kbps. The main differences in the physical layer between ISO 11898 and ISO 11519-2 are shown in Table 10.1.

10.3.1 Speed vs. Bus Length

Communication speed decreases with bus length. As the bus length increases, the communication speed decreases (as shown in Figure 10.1).

10.4 CAN FRAMES

There are five types of frames for communication in CAN: data frame, remote frame, error frame, overload frame, and interframe space. The first two frames are set by the user, while the rest are set in the hardware portion of CAN.

The data and remote frames come in two formats: standard and extended. The standard has an 11-bit ID, while the extended version has a 29-bit ID. Table 10.2 shows the roles the different frames play.

If a device wants to know the data associated with an identifier that it does not know, then the device sends a frame, called a remote frame, which has the desired identifier in its arbitration field with the RTR (remote transmission request) in set condition and the data field empty.

10.5 CAN DATA FRAME

The most important frame among the five frames in CAN is the data frame and is shown in Figure 10.2. The data transmit unit uses the data frame to send a message to the receive unit. The data frame consists of seven fields: start of frame, arbitration field, control field, data field, cyclic redundancy check (CRC) field, acknowledge (ACK) field, and end of frame. The number of bits in each field is shown in Figure 10.2. The data field is of variable length—it

TABLE 10.1

Differences in the Physical Layer between ISO 11898 and ISO 11519-2

Physical layer	ISO 11898 (high speed)						ISO 11519-2 (low speed)					
Communication speed	Up to 1 Mbps						Up to 125 kbps					
Maximum bus length	40 m/1 Mbps						1 km/40 kbps					
No. connected units	Maximum 30						Maximum 20					
	Recessive			**Dominant**			**Recessive**			**Dominant**		
Bus Topology	**Min**	**Nom**	**Max**	**Min**	**Nom**	**Max**	**Min**	**Nom**	**Max**	**Min**	**Nom**	**Max**
CAN_High (V)	2.00	2.50	3.00	2.75	3.50	4.50	1.60	1.75	1.90	3.85	4.00	5.00
CAN_Low (V)	2.00	2.50	3.00	0.50	1.50	2.25	3.10	3.25	3.40	0.00	1.00	1.15
Potential diff. (H–L) (V)	−0.5	0	0.05	1.5	2.0	3.0	−0.3	−1.5	–	0.3	3.0	–
	Twisted pair wire (shielded/unshielded)						Twisted pair wire (shielded/unshielded)					
	Loop bus						Open bus					
	Impedance (Z): 120 Ω (min. 85 Ω, max. 130 Ω)						Impedance (Z): 120 Ω (min. 85 Ω, max. 130 Ω)					
	Bus resistivity (r): 70 MΩ/m						Bus resistivity (r): 90 MΩ/m					
	Bus delay time: 5 ns/m						Bus delay time: 5 ns/m					
	Terminating resistance: 120 Ω (min. 85 Ω, max. 130 Ω)						Terminating resistance: 2.20 Ω (min. 2.09 Ω, max. 2.31 Ω)					
							CAN_L and GND capacitance: 30 pF/m					
							CAN_H and GND capacitance: 30 pF/m					

Source: Renesas Electronics America Inc. *Introduction to CAN*, REJ05B0804-0100/Rev. 1.00. Renesas Electronics America Inc., USA, 2006.

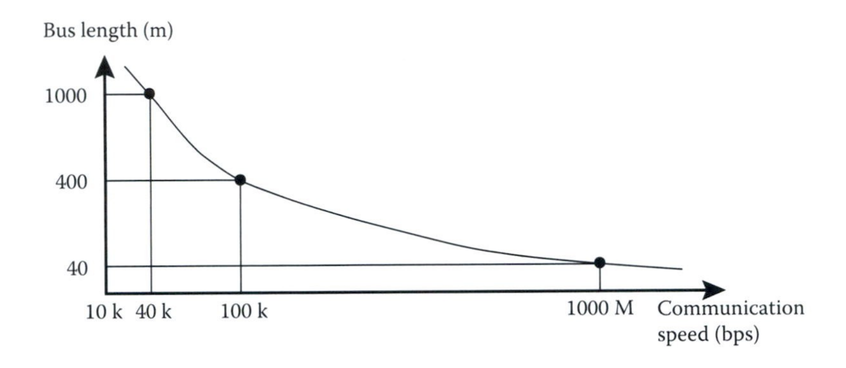

FIGURE 10.1 Communication speed vs. bus length. (From Renesas Electronics America Inc. *Introduction to CAN*, REJ05B0804-0100/Rev. 1.00. Renesas Electronics America Inc., USA, 2006.)

TABLE 10.2
Different Frames and Their Roles

Frames	Roles of Frames	User Settings
Data frame	This frame is used by the transmit unit to send a message to the receive unit	Necessary
Remote frame	This frame is used by the receive unit to request transmission of a message that has the same ID from the transmit unit	Necessary
Error frame	When an error is detected, this frame is used to notify other units of the detected error	Unnecessary
Overload frame	This frame is used by the receive unit to notify that it has not been prepared to receive frames yet	Unnecessary
Interframe space	This frame is used to separate a data or remote frame from a preceding frame	Unnecessary

Source: Renesas Electronics America Inc. *Introduction to CAN*, REJ05B0804-0100/Rev. 1.00. Renesas Electronics America Inc., USA, 2006.

can be a maximum of 8 bytes. This is sufficient to convey the information requirements from most of the devices. The CRC field in conjunction with the ACK field lends integrity to data as it is sent over the CAN bus.

10.6 CAN ARBITRATION

The bus arbitration principle about the device that gains control of the bus is shown in Figure 10.3. The device that first outputs a message on the bus during a bus idle state gains control of the bus and sends information

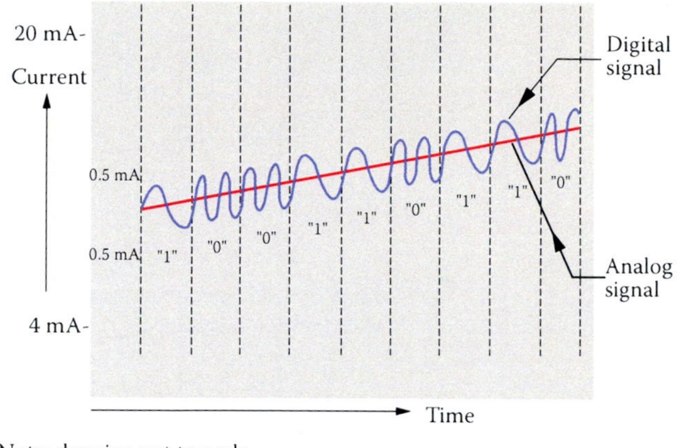

FIGURE 6.1 HART digital signal superimposed on 4–20 mA analog signal.

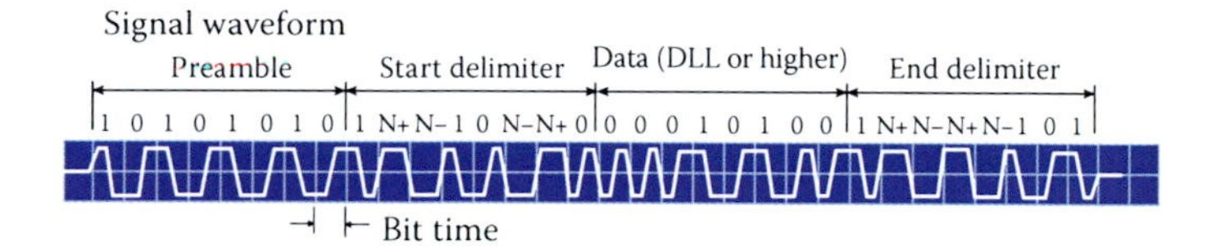

FIGURE 7.5 Physical layer signal waveform. (From Yokogawa Electric Corporation. *Fieldbus Book—A Tutorial*, Technical Information TI 38K02A01-01E, pp. 1–33, 2000.)

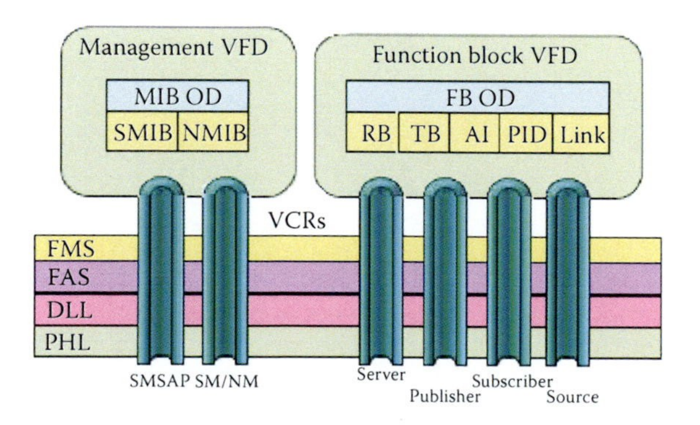

FIGURE 7.12 Block schematic showing management and function block VFDs. (From Yokogawa Electric Corporation. *Fieldbus Book—A Tutorial*, Technical Information TI 38K02A01-01E, pp. 1–33, 2000.)

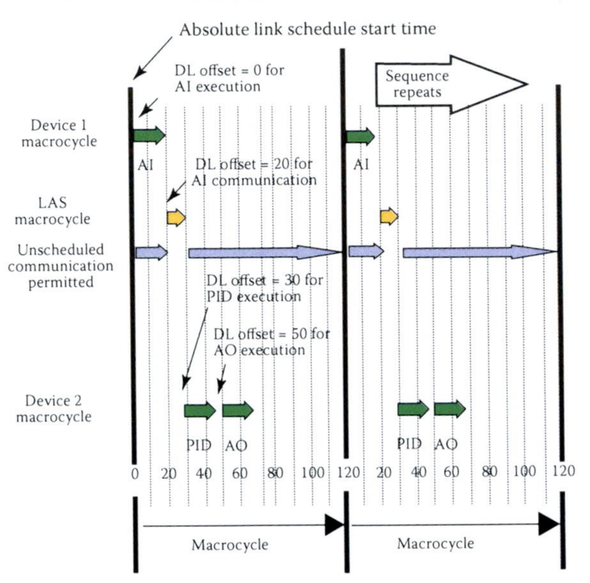

FIGURE 7.15 Scheduling of function blocks. (From SMAR. *Fieldbus Tutorial—A Foundation Fieldbus Technology Overview*. USA, pp. 1–29. Available at http://www.smar.com/PDFs/catalogues/FBTUTCE.pdf.)

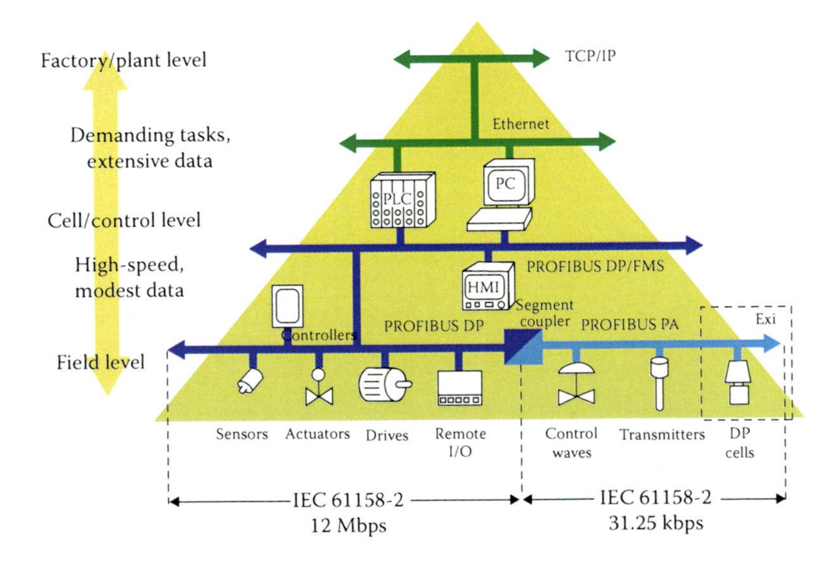

FIGURE 8.1 Typical PROFIBUS network includes both process and factory automation. (From A. A. Verwer. *Introduction to PROFIBUS*. Manchester Metropolitan University, Department of Engineering, Automation Systems Centre, PROFIBUS Competence Centre, 2005.)

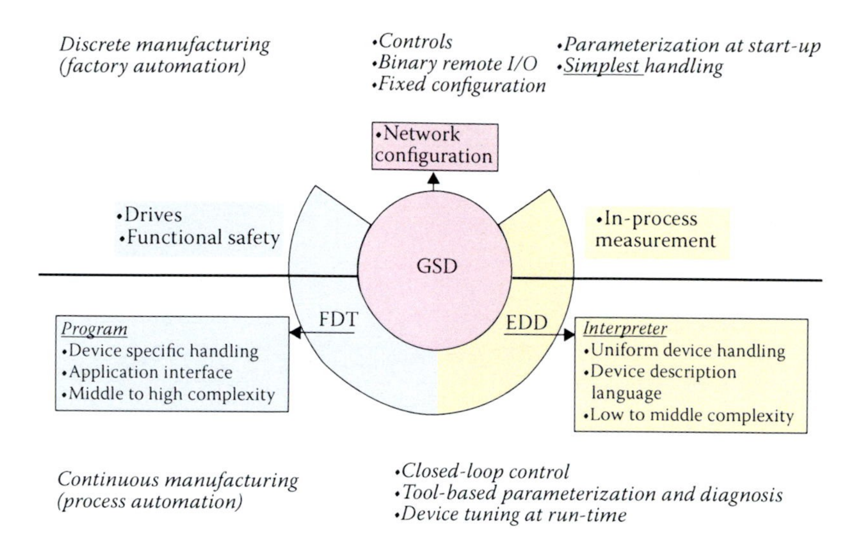

FIGURE 8.20 Technology integration by GSD, FDT, and EDD. (From PROFIBUS, Technology and Application: System Description, PROFIBUS International Support Center, Karlsruhe, Germany, Copyright by PNO 10/02, October 2002. Available at www.pacontrol.com/download/profibus-overview.pdf.)

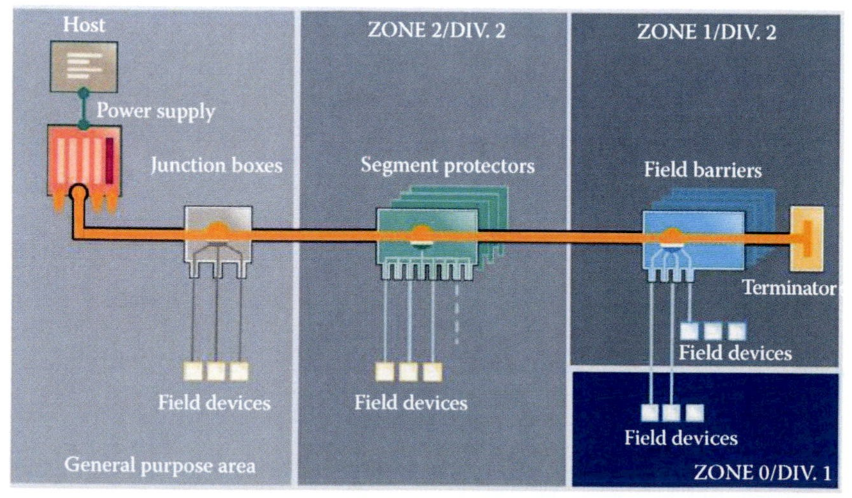

High-power trunk for any hazardous area. Segment protectors provide short-circuit protection and non-incendive energy limitation (Ex nL). Field barriers provide intrinsic safety (Ex i).

FIGURE 16.8 Block diagram of HPTC for hazardous area. (From A. Beck and A. Hennecke. *Intrinsically Safe Fieldbus in Hazardous Areas*, Technical White Paper, EDM TDOCT-1548_ENG. Pepperl+Fuchs GmBH, p. 5, 2008.)

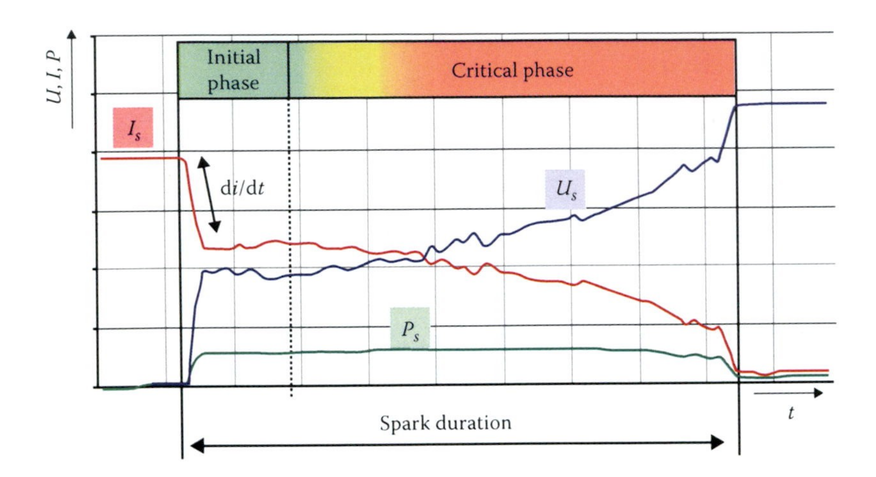

FIGURE 16.9 Electrical behavior of spark. (From A. Beck and A. Hennecke. *Intrinsically Safe Fieldbus in Hazardous Areas*, Technical White Paper, EDM TDOCT-1548_ENG. Pepperl+Fuchs GmBH, p. 6, 2008.)

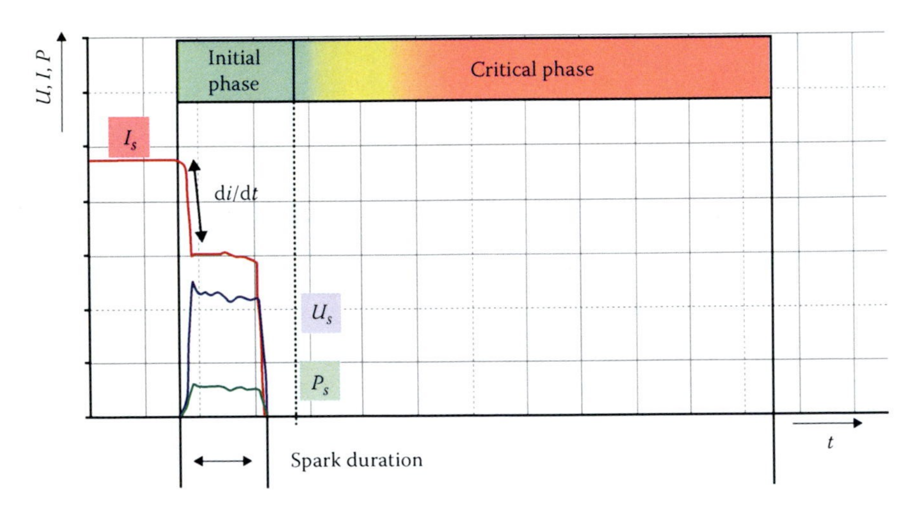

FIGURE 16.10 Highly characteristic di/dt associated with a spark. (From A. Beck and A. Hennecke. *Intrinsically Safe Fieldbus in Hazardous Areas*, Technical White Paper, EDM TDOCT-1548_ENG. Pepperl+Fuchs GmBH, p. 6, 2008.)

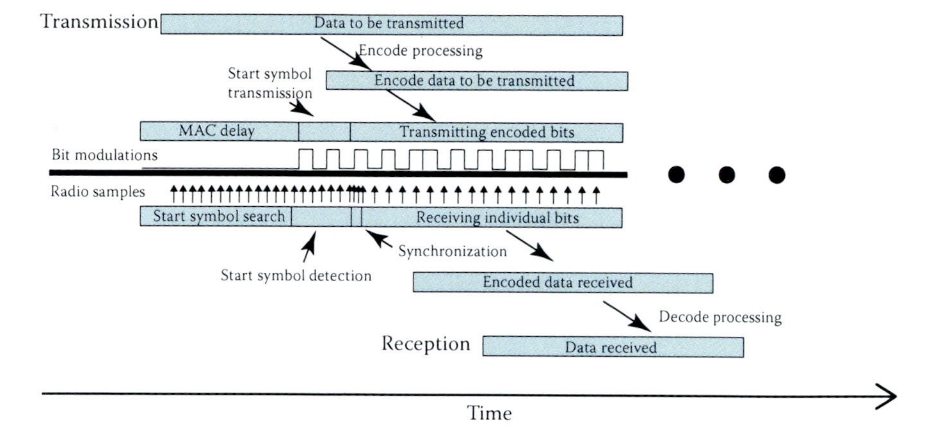

FIGURE 18.1 Wireless communication methodology. (Hill J. L. System architecture for wireless sensor networks. PhD Dissertation, Department of Computer Science, University of California, Berkeley, CA, Spring, 2003.)

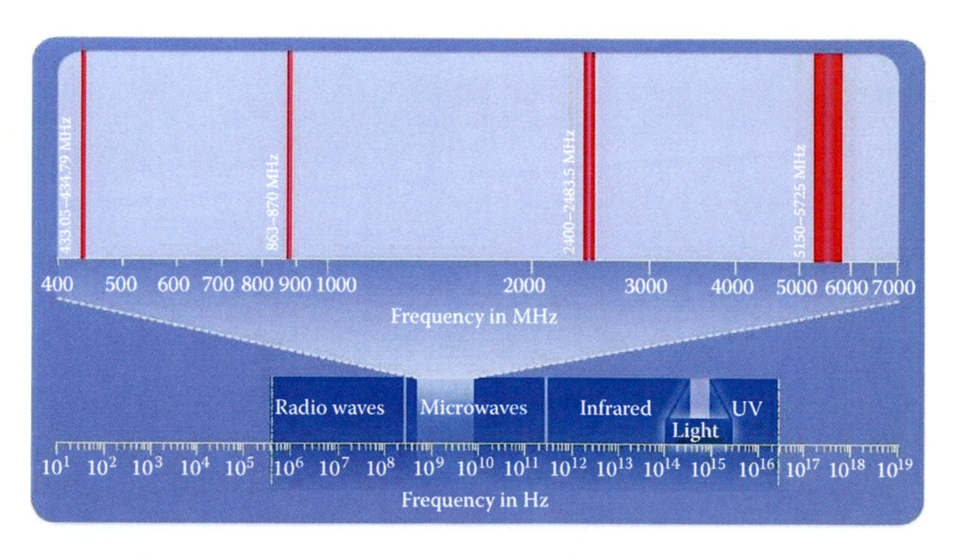

FIGURE 18.2 License-free frequency bands (in red). (Bentje H. et al. "Wireless in Automation" Working Group. *Coexistence of Wireless Systems in Automation Technology: Explanations on Reliable Parallel Operation of Wireless Radio Solutions*, 1st Edition. ZVEI-German Electrical Manufacturers' Association, Automation Division, Frankfurt, Germany, April 2009.)

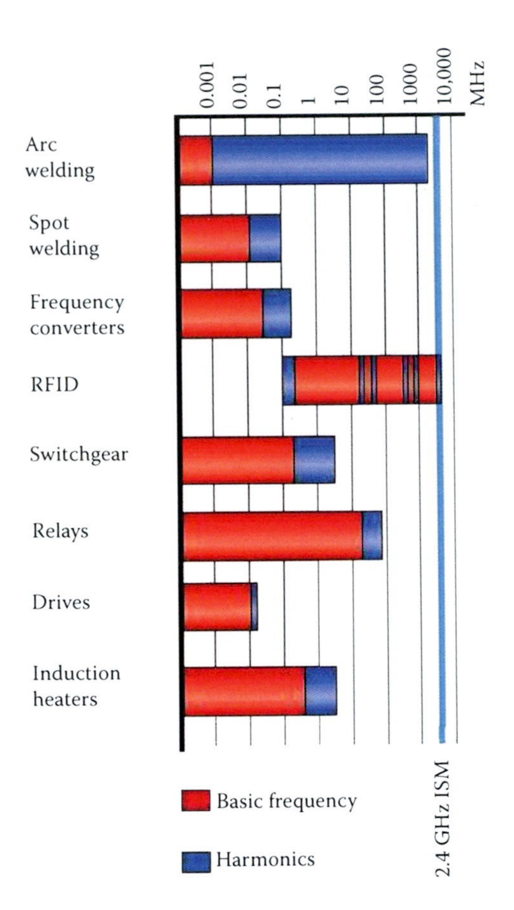

FIGURE 18.3 Interference spectra of some devices used in industries. (Bentje H. et al. "Wireless in Automation" Working Group. *Coexistence of Wireless Systems in Automation Technology: Explanations on Reliable Parallel Operation of Wireless Radio Solutions*, 1st Edition. ZVEI-German Electrical Manufacturers' Association, Automation Division, Frankfurt, Germany, April 2009.)

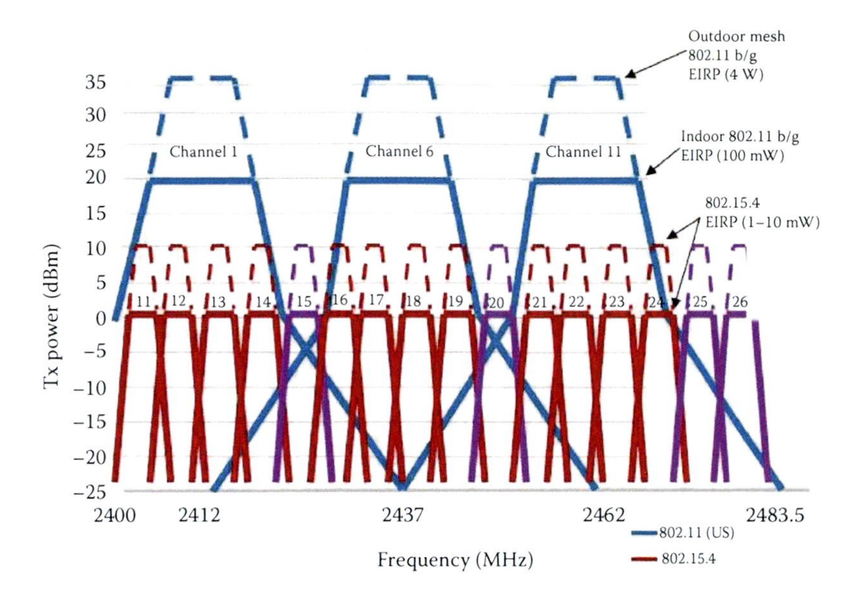

FIGURE 18.7 Response of 802.15.4 and 802.11 b/g in 2.4-GHz ISM band.

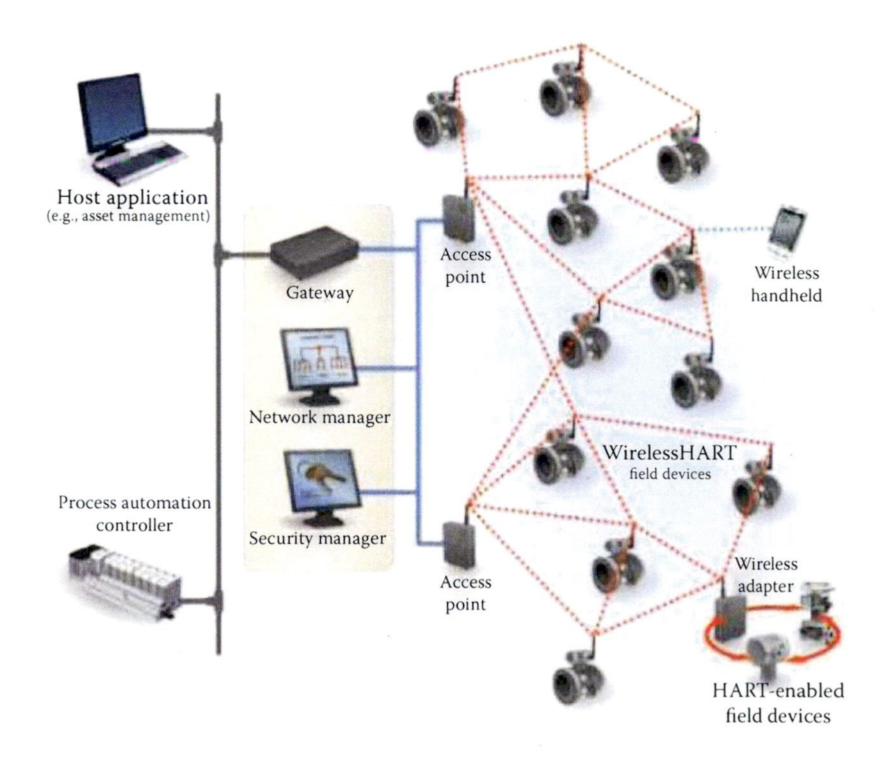

FIGURE 19.1 WHART network architecture. (Available at www2.emersonprocess.com/siteadmincenter/PM%20Central%20Web%20Documents/EMR_WirelessHART_SysEngGuide.pdf.)

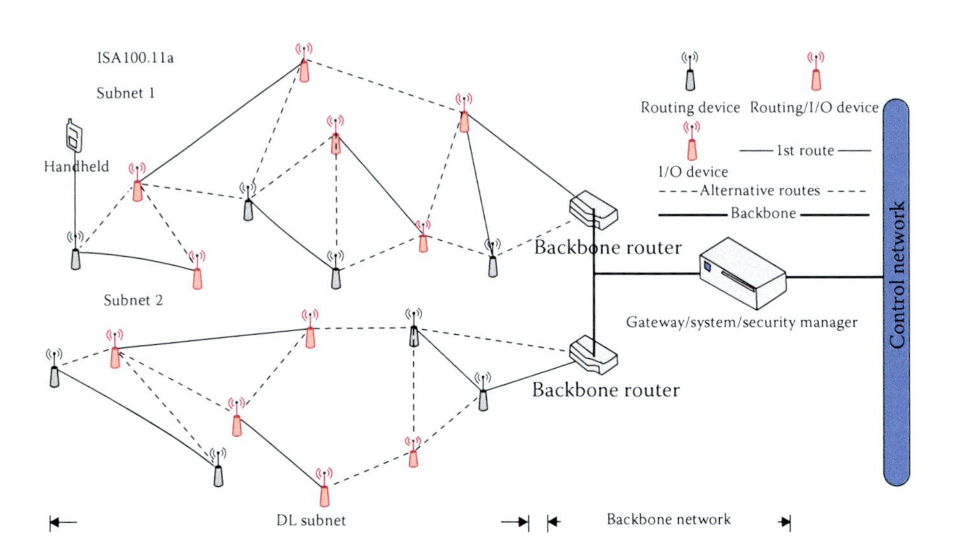

FIGURE 20.2 Detailed architecture of ISA100.11a. (From G. Wang. *Comparison and Evaluation of Industrial Wireless Sensor Network Standards: ISA100.11a and WirelessHART.* Master of Science Thesis, Communication Engineering, 2011.)

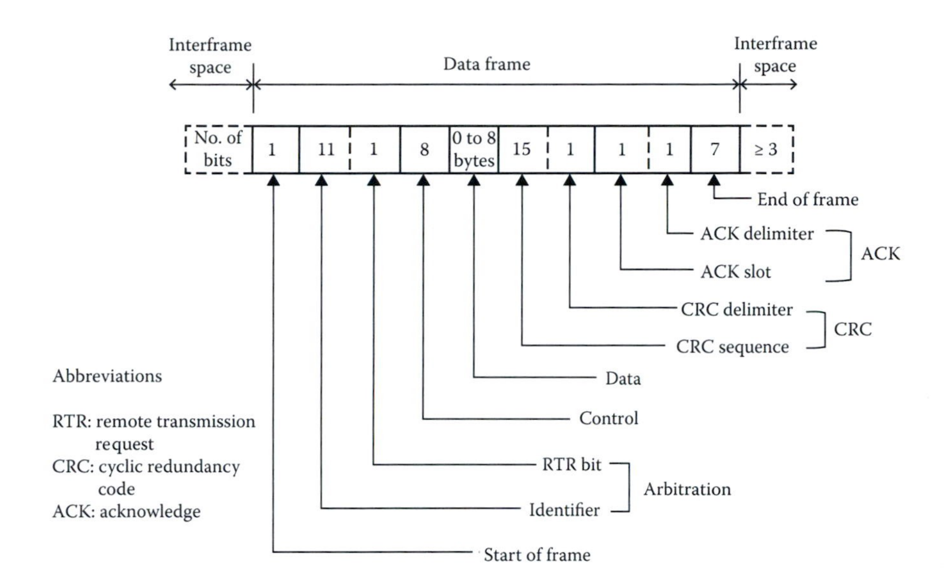

FIGURE 10.2 CAN bus data frame. (Reprinted from *Practical Data Communications for Instrumentation and Control*, J. Park et al., p. 266, Newnes, Copyright 2003, with permission from Elsevier.)

through it. In case more than one device wants to send data over the bus, a station identifier bit pattern for each device try to gain access of the bus. The priority regarding access of the bus by a device is determined by the addressing assignments during configuration of the network and allows the highest priority device to access the same. The devices lower in priority lose the contention and get another chance once the bus becomes idle.

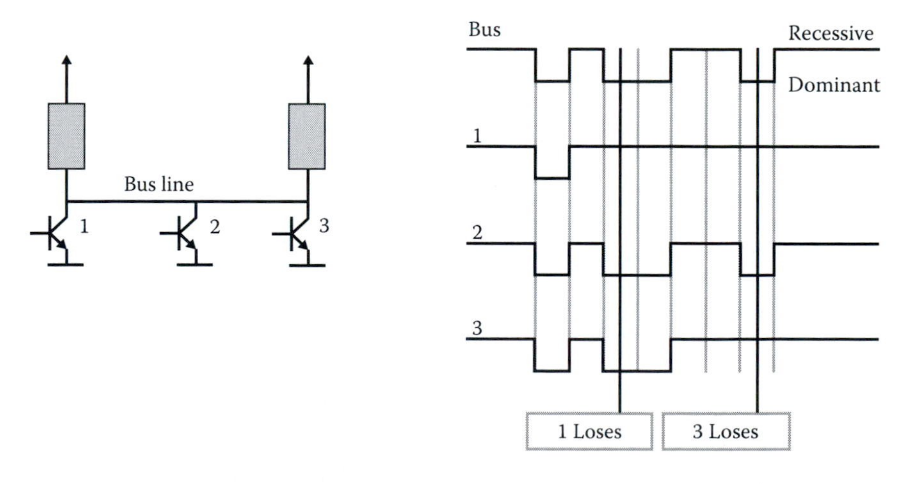

FIGURE 10.3 Bus arbitration principle. (Reprinted from *Practical Data Communications for Instrumentation and Control*, J. Park et al., p. 265, Newnes, Copyright 2003, with permission from Elsevier.)

The data frame consists of a station identifier field of 11 bits and a single RTR. The identifier field determines the priority of a device over others. In Figure 10.3, three devices try to gain control of the bus at the same time. State "0" is "dominant," while state "1" is "recessive." The "0" state dominates over the "1" state. The devices are connected to the network by an open collector stage. The bits of devices 1, 2, and 3, as configured during initialization, are put on the bus and shown in Figure 10.3. Device 1 first loses the bus arbitration due to both 2 and 3. Thereafter, device 3 loses out to device 2. The pattern on the bus thus corresponds to the identifier bit pattern of the device that wins the contention, which in this case is device 2. Devices 2 and 3 again get the next chance to transmit a message once the bus becomes idle. In conclusion, it can be said that if a device while transmitting a recessive bit detects a dominant bit from another device, the former stops transmitting. In this way, the less-prioritized devices lose the contention and the highest-prioritized device wins the same, thereby gaining the right of way over the bus.

10.6.1 CAN Communication

When not communicating, the CAN bus is idle and both the wires are tied at 2.5 V. Two dedicated wires are used for communication. The wires are called CAN high and CAN low. When a message is being transmitted, the CAN-high voltage assumes 3.75 V and the CAN-low voltage becomes 1.25 V, causing a voltage gap of 2.5 V between the two lines. This is shown in Figure 10.4. The CAN bus is robust in networked communication because it is insensitive to inductive spikes, electromagnetic noise, etc.

10.7 TYPES OF ERRORS

There are five types of errors that may occur in CAN communication, and more than one error may occur simultaneously. Table 10.3 shows the types of errors, content of error in each type, etc.

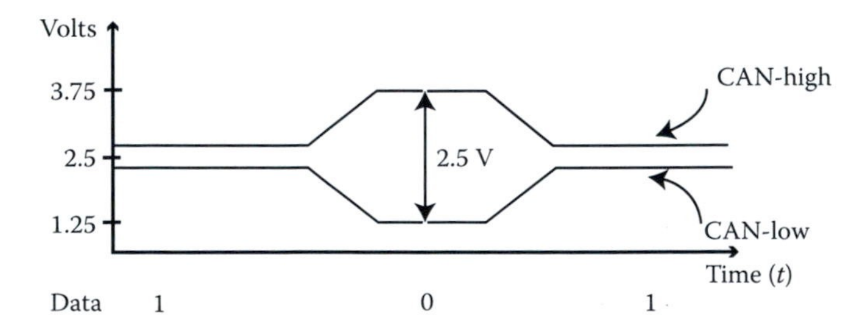

FIGURE 10.4 Voltage levels in CAN bus communication. (From Axiomatic Global Electronic Solutions. *Q&A—What is CAN?* Application Note, 2006. Available at www.axiomatic.com/whatisCAN.pdf.)

TABLE 10.3
Types of Errors

Type of Error	Error Content	Target Frame (Field) in which Errors Are Detected	Error Detection by Unit(s)
Bit error	This error is detected when the output level and the data level on the bus do not match when they are compared (level comparison: dominant output stuffing bits are compared, whereas arbitration fields and ACK bits during transmission are not compared)	Data frame (SOF to EOF) Remote frame (SOF to EOF) Error frame Overload frame	Transmit, receive
Stuffing error	This error is detected when the same level of data is detected for six consecutive bits in any field that should have been bit stuffed	Data frame (SOF to CRC sequence) Remote frame (SOF to CRC sequence)	Transmit, receive
CRC error	This error is detected if the CRC calculated from the received message and the value of the received CRC sequence do not match	Data frame (CRC sequence) Remote frame (CRC sequence)	Receive
Form error	This error is detected when an illegal format is detected in any fixed format bit field	Data frame (CRC delimiter, ACK delimiter, EOF) Remote frame (CRC delimiter, ACK delimiter, EOF) Error delimiter Overload delimiter	Transmit, receive
ACK error	This error is detected if the ACK slot of the transmit unit is found recessive (i.e., the error that is detected when ACK is not returned from the receive unit)	Data frame (ACK slot) Remote frame (ACK slot)	Transmit

Source: Renesas Electronics America Inc. *Introduction to CAN*, REJ05B0804-0100/Rev. 1.00. Renesas Electronics America Inc., USA, 2006.

10.8 ERROR STATES

A device remains in one of three error states: error-active state, error-passive state, and bus-off state. In the error-active state, a device can participate in communication on the bus in a normal manner. In case a device detects an error in the error-active state, it sends an active error flag. A device tends to cause an error in the error-passive state. In this state, the device does not communicate to other devices about the error in the receive mode. A passive error flag is transmitted when the device detects an error in the error-passive state. In the bus-off state, a device cannot participate in communication. It is disabled from all transmit/receive operations.

11 DeviceNet

11.1 INTRODUCTION

Developed by Allen–Bradley in 1994, DeviceNet is a device level or low-level industrial open network that communicates between the device controller and such sensors or actuators as limit switches, valve manifolds, motor starters, variable frequency drives, and remote I/Os. It is included in EN 50325 and IEC 62026 standards. DeviceNet is a CAN-based layer 7 application protocol and is maintained by Open DeviceNet Vendor Association Inc. (ODVA). The association issues specifications and ensure compliance with the stated specifications.

Devices from different manufacturers that comply with the DeviceNet standards can be connected together in the network. The DeviceNet specification is defined in two volumes: volume 1 and volume 2. Volume 1 pertains to the application layer, which uses Control and Instrumentation Protocol (CIP), and the data link and physical layers, which use the CAN protocol. Volume 2 is concerned with the device profiles to obtain interoperability and interchangeability among the different products.

11.2 FEATURES

DeviceNet supports up to 64 nodes with a maximum device count of 2048. The network topology used is trunk or bus line with drop cables that connect to the devices. On either end of the trunk line, terminating resistances of 121 ohm are each placed. It supports the use of repeaters, bridges, routers, and gateways.

DeviceNet is used when devices are mostly discrete in nature with an analog mix, and where motor control and variable frequency drives are present. It has moderate transmission speed and it depends on cable run lengths.

DeviceNet supports master–slave, peer-to-peer, and multimaster modes to transfer information in the network, utilizing the total bandwidth in the process. Slaves can be owned by one master only.

DeviceNet is robust in nature, offering some diagnostics at the same time. It has the capability to detect duplicate node addresses. It supports power and signal on the same cable. DeviceNet supports 8 A on the bus. Because of huge power handling, the system is not intrinsically safe. Devices can be added/withdrawn from the network under power.

The DeviceNet network is classified as a device bus network. Its characteristics are byte-level communication, high speed, and a lot of diagnostic power by the devices in the network.

Every DeviceNet device has a configuration file in its electronic device data sheet (EDS file). It maintains important data about the device that must be registered on the network configuration software. A configuration tool is used to input the EDS files and for configuring the devices. Devices that are modularly designed use one EDS file per component, i.e., module. This definitely is advantageous in configuring such devices.

11.3 OBJECT MODEL

A DeviceNet station or node is considered as a collection of objects. An object provides an abstract representation of a particular component in a product. Class is a group of objects that represent the same type of component, while the attributes provide the characteristics of objects.

Every device has mandatory and optional objects. Optional objects give a device the category (also called profiles) to which it belongs, such as pneumatic valve and AC/DC drive.

11.4 PROTOCOL LAYERS

Figure 11.1 shows the DeviceNet protocol along with the OSI protocol. It is seen that layers 3 to 6 of the OSI protocol are absent in this case.

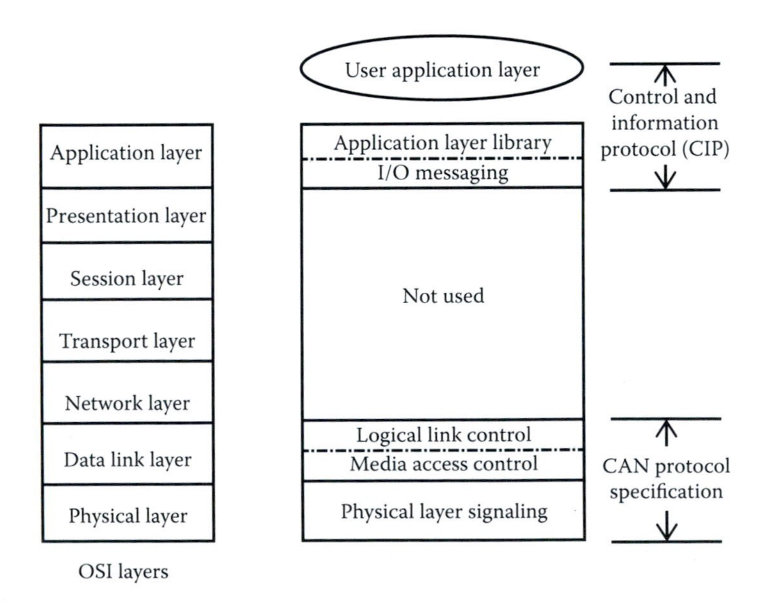

FIGURE 11.1 DeviceNet protocol layers.

DeviceNet uses the CAN protocol to implement the data link and physical layers, while the application layer conforms to CIP.

11.5 PHYSICAL LAYER

DeviceNet uses the CAN protocol for its physical layer implementation, and it offers a great advantage because CAN chips are very easily available. Each node in the network is assigned a unique address in the range 0–63. The same is set by using switches on the devices and a configuration tool existing in the system.

The devices can be powered from the bus line or else the devices can be powered separately. Each device power should at least be 11 V or higher. For greater protection, optical isolation can be used.

11.5.1 Data Rate

The DeviceNet protocol offers variable data rate, trunk distance, and drop length. This relationship is shown in Table 11.1. The total length of the trunk line is dependent on data rate, cable type, and the number of devices. Devices can be connected by means of screw terminals, sealed type screw, tight connectors, etc.

11.6 DATA LINK LAYER

The data link layer comprises of logical link control (LLC) and media access control (MAC), and follows the data link layer of CAN specification. As already mentioned, carrier sense multiple access with collision detection (CSMA-CD) is used to access the physical medium. Contention on the bus is resolved by a priority-based deterministic procedure. This avoids collision on the bus and at the same time utilizes the total bandwidth of the transmitted medium.

TABLE 11.1
Relationship between Data Rate, Trunk Distance, and Drop Length

		Drop Length (m)	
Data Rate (K baud)	**Trunk Distance (m)**	**Maximum**	**Cumulative**
125	500	3	156
250	200		78
500	100		39

It should be mentioned here that although DeviceNet uses the CAN data frame, encoding of identifier and the data packets for DeviceNet is left to the application layer developer. A producer–consumer model is used for communication on the bus, where a station places data on the bus (producer) and this is read off by another station on the network (consumer).

11.7 APPLICATION LAYER

DeviceNet categorizes CAN frames into four groups, from group 1 to group 4, with decreasing priority levels. This is achieved with the help of an 11-bit CAN identifier field. It includes a massage ID and a MAC ID. The MAC ID can be either a source or a destination.

A predefined master–slave connection set defines the group via which data transfer is to take place. Such an action is done to considerably reduce the software load to establish a connection. To establish a connection, the master sends a service request to the designated slave(s). A subsequent request by the master establishes the connection, which then takes the help of the predefined master–slave connection set.

The CAN protocol, used by data link and physical layers, does not interpret the fields in the CAN message frame. In the case of DeviceNet, software developed and residing in the application layer distinguishes between two types of messages: cyclic I/O and explicit type. Cyclic I/O is of four types: bit strobe command/response message, poll command/response message, change of state message, and cyclic message. The type is dependent on the manner in which data is exchanged. These are now discussed.

Bit/strobe command is used for exchange of small amounts of data. Here, the message gives out 8 bytes, i.e., 64 bits of data, one for each slave. If a slave is absent or not included in the scan list, the corresponding bit does not carry any meaning. The response from the slaves can contain a maximum of 8 bytes of data.

In the *poll command* type, the master sends a message to a single slave. The slave may also return data to the master.

In the *change of state method* of communication, data exchange may be initiated by either master or slave. This takes place if the values monitored or controlled change within a set time limit. This time limit is set in the network configuration program.

In the *cyclic method*, data exchange takes place at regular time intervals regardless of being altered or not. These intervals are adjustable on the network configuration program.

Explicit message type is a general purpose one. This is used in asynchronous jobs, e.g., to transfer the values of some attributes, parameterization, and configuration of equipment.

11.8 POWER SUPPLY AND CABLES

A single four-conductor cable in bus topology configuration can be used in DeviceNet. Two wires are used for supplying power to the devices and two for supporting communication. Both the pairs have a foil shield. The cable has an overall braiding. ODVA recommends the use of different types of cables as follows: ODVA type I thin cable for drop wires, ODVA type II thick cable for trunk wires, while ODVA type III is used when flexible drop cable is the requirement. A 24-V DC power is provided on the power lines and supports 8 A on the thick cable and 3 A on the thin cable. The thick and thin cables are connected together by "T" junctions.

11.9 ERROR STATES

DeviceNet devices assume any one of the following error states.

Nonexistent	The device has shut down owing to an internal error or some remote command.
Unallocated	The device has successfully joined the network, but is not currently owned by a DeviceNet Master device. The LED network status indicator for an unallocated device flashes green.
Timed out	Messages have failed to arrive at one or more connections with the master device. This is typically a recoverable error. The network status LED indicator on a device will flash green in this state.
Faulted	The device has detected an internal error or received a duplicate MAC ID response message. This is not a recoverable error. The network status LED indicator on a device will typically be solid red in this case.
Bus off	In the bus-off state, the device has detected significant network errors and has removed itself from network operation. This is typically a hardware failure in the device circuitry. The network status LED will typically be solid red in this case.

12 AS-i

12.1 INTRODUCTION

Actuator sensor interface (AS-i) is a bit-oriented master–slave type open system fieldbus designed to be used at the device level of process control systems. It is included in both EN 50295 and IEC 62026 standards. AS-i is designed to connect binary sensors and actuators (i.e., slaves) that require very small number of bits to convey device status. Thus, this protocol is very efficient for this type of slaves. AS-i is not good enough to be connected to intelligent controllers that require information beyond the limited capacity of such a protocol.

The main design of AS-i is based on modular components. They act as bridges between the network and the binary sensors. Nodes can be added or taken out in "live" condition, i.e., while the system is running, without any kind of interruption. AS-i data capacity ranges from 1 to 16 bits per device.

12.2 FEATURES

It is a cyclic polling, single-master multislave type of system where the slaves have specific addresses and sensors/actuators from different manufacturers can be connected together. The topology used can be ring, linear, star, or tree with a cable length of 100 m, which can be extended to 300 m by using repeaters. The "alternating pulse modulation" technique used during transmission of information reduces the bandwidth and also the "end of line reflection" as is very common in networks that use square wave pulse techniques. Synchronizing information is passed on to the receiver, making the AS-i protocol very reliable one as far as data integrity is concerned. It is an open standard in which error detection and retransmission of incorrect data are possible. It eliminates the need to employ programmable logic controller input and output (I/O) modules; hence, cost is low per slave basis. Data and power for the devices are sent through two unshielded, nontwisted cables. Each slave typically consumes 200 mA with a maximum of 8 A per bus. Cycle time is a maximum of 5 ms for the 2.0 version and 10 ms for the 2.1 version.

12.3 DIFFERENT VERSIONS

The AS-i protocol is available in three versions: the original version 2.04 (1994); version 2.11 (1998), which is an enhanced version of the earlier one; and version 3.0 (2005/2007), featuring some additional capabilities. These versions are also known as AS-i 2.0 or AS-i 1 specification, AS-i 2.1 or AS-i 2 specification, and AS-i 3.0 or AS-i 3 specification, respectively.

In version 2.04, a maximum of 31 slaves could be employed with each slave linking four digital inputs and four digital outputs, resulting in 124 inputs and 124 outputs on a single network. A feature of this version is the automatic substitution of a network module. Update time is approximately 150 μs. This is calculated by multiplying the number of I/O nodes with the deterministic update time for each node.

In version 2.11, the number of slaves increased to 62 from 31, the number of I/Os to 434, and the update time to 10 ms. The number of slave profiles increased to 225 from 15 and the peripheral error could be taken care of by adding a bit in the status record.

In version 3.0, some of the extra capabilities added are full duplex bit serial data channel and configurable fast analog channel of 8, 12, or 16 bits.

12.4 TOPOLOGY

Communication in AS-i is controlled by a single master, with which the slaves are connected in various configurations, such as star, line, branch, or tree. These are shown in Figure 12.1.

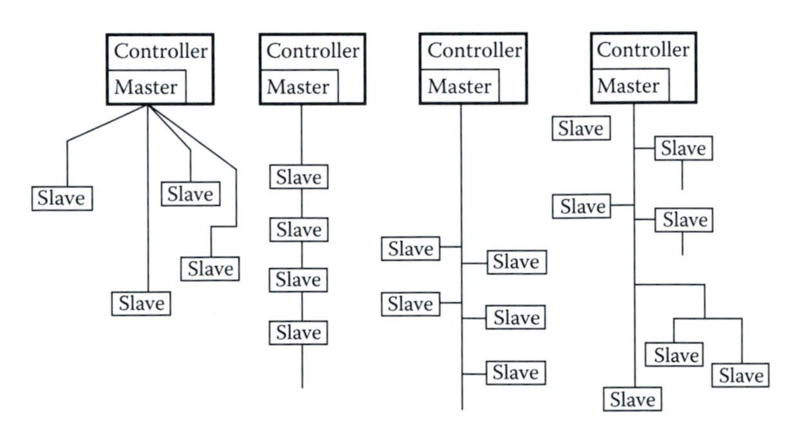

FIGURE 12.1 Physical network topologies.

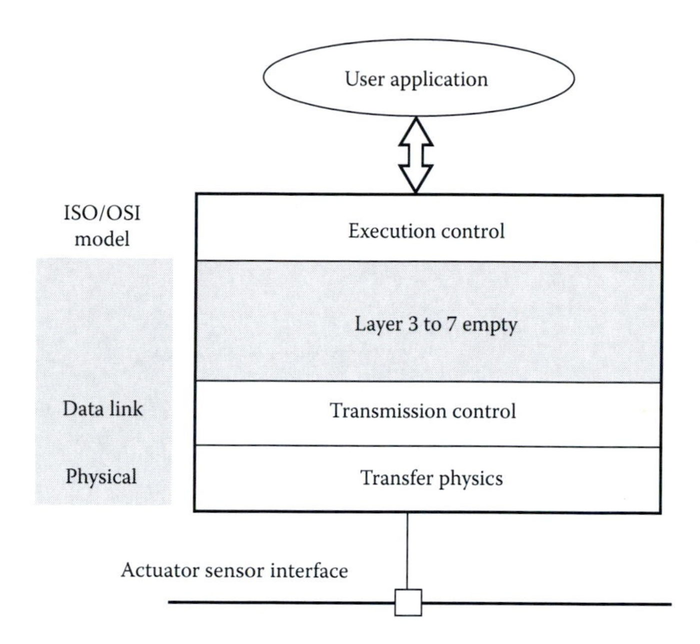

FIGURE 12.2 AS-i protocol. (From B. G. Liptak. *Instrument Engineers' Handbook*, 3rd Edition. CRC Press, Boca, Raton, FL, p. 621, 2002.)

12.5 PROTOCOL LAYERS

Figure 12.2 shows the AS-i protocol along with the OSI protocol. Layers 3 to 7 of OSI are nonexistent in this case. Layer 1 is termed as "transfer physics," while layer 2 is called "transmission control." There is a layer called "execution control" and placed above layer 7 of OSI, which is responsible for overall operation of AS-i.

12.6 PHYSICAL LAYER

As already mentioned, the physical layer of AS-i is called "transfer physics" and is responsible for physically establishing the connection between master and slaves. It uses a two-wire untwisted and unshielded cable, which is available in two versions and is used for both communication and power supply to the devices. The general cable version is used to connect devices by normal screw terminals, while the special cable can connect stations directly by contacts that penetrate the cable isolation. A specially shielded cable is used when the environment is encountered.

12.7 DATA LINK LAYER

It consists of a master call-up and a slave response. It is a bit-oriented protocol with the master call-up, shown as the master request protocol

Master request PDU

Start bit	Control bit	Address	Information	Parity bit	End bit
1 bit	1 bit	5 bits	5 bits	1 bit	1 bit

Slave response PDU

Start bit	Information	Parity bit	End bit
1 bit	4 bits	1 bit	1 bit

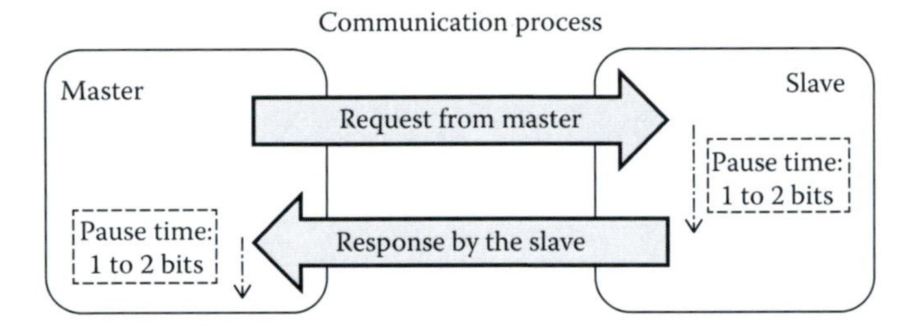

FIGURE 12.3 AS-i master, slave PDUs, and communication sequences. (From B. G. Liptak. *Instrument Engineers' Handbook*, 3rd Edition. CRC Press, Boca, Raton, FL, p. 622, 2002.)

data unit (PDU) in Figure 12.3. It is 14 bits in length and has six fields. The slave response, shown as the slave response PDU in the same figure, consists of 7 bits. When the master request PDU (to slave) is over, there is a pause of 3–10 bits, while it is 1–2 bits pause for the slave request PDU (to master). This communication sequence between request–response is shown in the lower part of the same figure. The pause between each transmission is used to ensure synchronization, error detection, and error correction.

Various combinations are possible in the 5-bit information field of the master PDU, the content of which tells the operation that is to be executed on the slaves. Some of them are Address_Assignment, Data_Exchange, and Write_Parameter. When the code corresponding to Address_Assignment is set, the master sets the address of the slave, which is possible only if the slave has a default address that does not allow for data exchange.

12.8 EXECUTION CONTROL

The execution control layer exists at the top of layer 7 of OSI. It is necessary for the management and overall proper operation of AS-i. The user application layer requests the execution control layer for different functionalities. The main jobs performed by the execution control unit are initialization, start-up, and normal operation.

Initialization involves setting some parameters offline, such as setting some parameter values for the master or testing the power supply to ensure it can adequately supply the load to the slaves.

In the start-up phase, the master detects and activates the slaves. There are two means to achieve the above: the protected operation mode and the configuration mode. In the former, the user prepares a list beforehand, called the "list of projected slaves," and the master activates the slaves from that list only. In the latter, all the slaves detected by the master are activated.

The jobs carried out by the normal operation phase include cyclic data transfer between master and slave, acyclic management tasks, and inclusion function. In cyclic data transfer, the master updates each slave with output data cyclically and acquires data from each slave in an identical fashion. This cyclic data transfer automatically takes place without user intervention. User requests are taken care of in management tasks. Inclusion function allows inclusion of some slaves to the cyclic data transfer. For this, the master first detects a new slave and then only the slave can be activated for cyclic data transfer to take place with this new slave.

12.9 MODULATION TECHNIQUE

The "alternating pulse modulation" (APM) technique is used for data transmission in case of AS-i, shown in Figure 12.4. It produces a baseband signal that is superimposed on a DC power supply. As the information field of AS-i is of very limited in size, APM is used to ensure data integrity. The modulated signal is "sin squared," which is similar to Manchester II coding. This coding scheme reduces the bandwidth required in transmission and end-of-line reflections. As in Figure 12.4, each data bit has

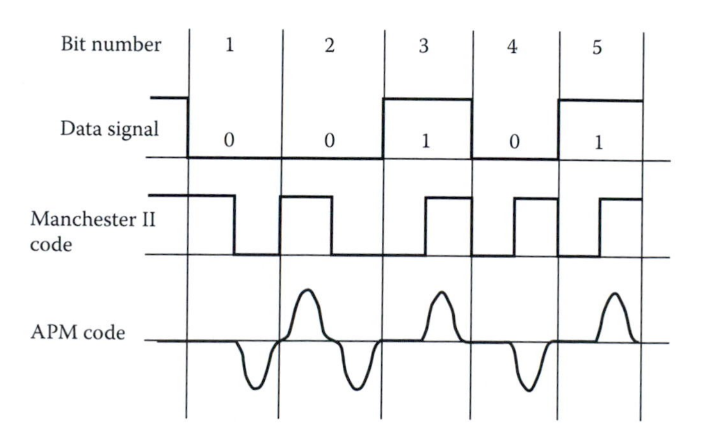

FIGURE 12.4 APM technique used in AS-i.

an associated pulse in the second half of the bit period, which helps in bit-level error checking. Also, this scheme helps in synchronization at the receiving end. The start or first bit of the APM is a negative pulse, while the stop bit is a positive pulse. Even parity and a recommended frame length are included in the APM technique. These inclusions offer more credence to data integrity in the received signal.

13 Seriplex

13.1 INTRODUCTION

Seriplex Control Bus, developed in 1987 by Automated Process Control Inc., is a control network specifically intended for simple industrial control applications. Seriplex is most favored for simple devices such as push buttons, valves, contactors, and limit switches to the network. A single cable connects all the devices to save considerably on cabling costs. The total cable run is about 1500 m, while for other networks that is about 300 m.

Seriplex control schemes employ intelligent modules that provide a link between inputs and outputs, much like logic gates. This obviates the need for a supervisory processor. For supervisory control systems, the network is to be connected to a host processor via interface adapters.

Various topologies are available for connecting the intelligent modules to the Seriplex network via cables that provide for power for data and clock signals. Without address multiplexing, Seriplex can support 255 input bits and 255 output bits; sixteen 16-bit analog inputs and sixteen 16-bit analog outputs. With address multiplexing, the system can support 3840 input bits and 3840 output bits; two hundred forty 16-bit analog inputs and two hundred forty 16-bit analog outputs.

Seriplex bus can be configured to operate in both master–slave and peer-to-peer versions.

13.2 FEATURES

A lot of features have gone into the development of Seriplex Control Bus. It is highly cost-effective for networking simple control devices. The Seriplex protocol is very efficient for data length ranging from 1 to 64 bits, while other protocols are efficient for data length in excess of 16 bits. Thus, for very simple control systems, the Seriplex protocol is mostly used. The bus is inherently deterministic, implying that the time at which response is available is known. Seriplex employs a very highly efficient protocol with less overhead. Hence, its response is very fast, although its baud rate is less than that of other protocols.

Without multiplexing, up to 510 bits can be transmitted in frames continually with very short synchronization periods. The data bits located within the frame corresponds to the addresses assigned to the I/O devices.

The data and clock signals operate at 12 V with a 4.5 V hysteresis. This gives very good noise immunity. The network power cable is separate from the I/O power cable. Additionally, two other wires are earmarked for data and clock signals. The four conductors are enclosed in a shield.

The efficiency of Seriplex systems can be as high as 98%. It means that for 98% of the time, the system transports actual data bits. The setup tool available in Seriplex systems is meant for configuring the master–slave modes, and it takes around 15 s to configure them.

Seriplex Control Bus updates all I/O data on a regular basis; thus, there is no need to have a collision resolution or message prioritization.

13.3 PHYSICAL LAYER

The physical layer, belonging to layer 1 of the OSI, connects the Seriplex network with each individual device. It is a four-conductor cable, with two thicker wires of AWG #16 meant for power and common. The two relatively thinner wires of AWG #18 carry data and clock signals. The maximum clock rate is 200 kHz.

First-generation I/O devices are powered by 12-V DC, while the second-generation ones are powered by either 12-V or 24-V DC; the value employed depends on the user requirements. Connections are made through Seriplex modules placed near the field devices.

Cable capacitance plays a vital role in determining the maximum distance that data can be delivered. The lower the cable capacitance, the greater the distance. A cable capacitance of approximately 50 pF/m allows data rates up to 100 kHz for a maximum distance of 150 m, while a cable capacitance of 65 pF/m would enable data transmissions at 100 kHz for a maximum distance of 100 m. The number of digital and analog I/Os depend on address multiplexing.

13.4 DATA LINK LAYER

Seriplex Control Bus can be operated in two modes: peer-to-peer mode and master–slave mode. In the former, a host controller is not needed and data can be interchanged between devices. For the latter, a host controller along with an interface card controls the activities on the bus following some preloaded software.

A peer-to-peer timing diagram is shown in Figure 13.1. This is called mode 1. Each input and output device are assigned the same address. Data with an input address 7 appears as output with the same address 7. A separate clock is required in this case, as there is no host to provide the same.

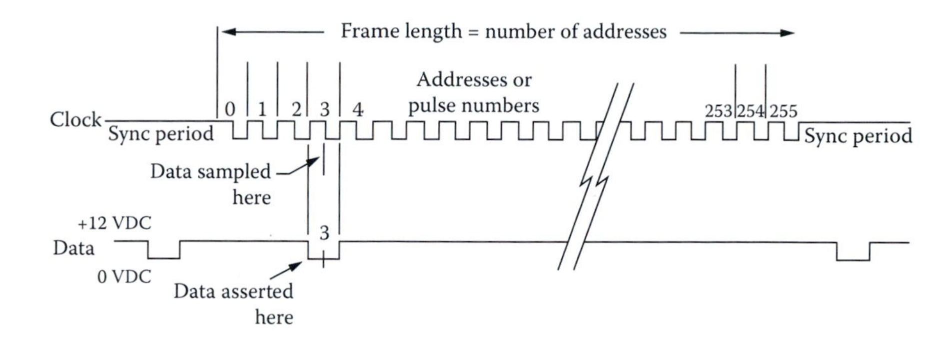

FIGURE 13.1 Seriplex peer-to-peer timing diagram. (Square D. SERIPLEX: Design, Installation and Troubleshooting Manual, Bulletin no. 30298-035-01A, Raleigh, NC, January, 1999. Available at www.guilleviniag.com/downloads/Products/Schneider/Seriplex.pdf.)

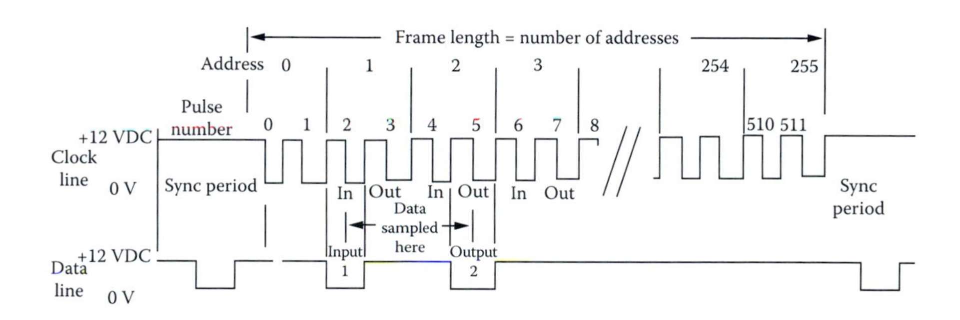

FIGURE 13.2 Seriplex master–slave timing diagram.

In the master–slave mode, also known as mode 2, there are two clock pulses per address; the timing diagram is shown in Figure 13.2. During the first clock pulse for an address, input data is transferred to the interface card and from there it is retransferred to the host. During the second clock pulse for the same address, the host gives out output data to remote devices. This method thus facilitates the separation of input and output data for the same address. In this mode, the host takes all the control decisions regarding all data transfers on the bus.

13.5 DATA INTEGRITY

Seriplex systems offer a fair degree of data integrity in the form of "digital debounce" and "data echo" features. These two methods efficiently verify data for its integrity than either the parity or the CRC method of data verification for its authenticity. The second-generation Seriplex ASIC2B provides the above two features.

Devices such as valves and contactors, which receive data, check the data stability for multiple readings before acting to filter out spurious data.

Data echo mechanism can determine whether some device is actually connected or not and whether the ASIC is operating properly or not. For output data from output devices, data can be echoed and monitored. If the data is properly echoed, it can be concluded that the device is operating faithfully and the ASIC is responding properly to data transmissions.

Multibit data, such as obtained in case of analog signals, can be verified by employing the Complementary Data Retransmission (CDR) feature, which is supported by Seriplex version 2.

14 Interbus-S

14.1 INTRODUCTION

Interbus-S is an open architecture, serial bus, mater–slave type fieldbus communication system that operates in an active ring topology. It consists of one master and up to 511 slave devices. Each slave has an input and an output connector. It was developed by Phoenix Contact in 1984 to interchange data between control systems such as personal computers (PCs), programmable logic controllers (PLCs), and distributed input/output (I/O) modules that communicate with sensors/actuators. A maximum of 4096 digital I/O points for a maximum distance of 400 m can be connected via this network. These points can be updated in 14 ms by employing a variable frame transfer protocol. This time can further be reduced by employing fewer I/Os. The data transfer rate on this bus is 500 kbps. The basic scheme of process control via Interbus is shown in Figure 14.1.

A bus cycle begins with the master passing a bit stream to the first slave. The first slave then transfers the same to the second slave in the ring, and at the same time data from the first slave is transferred to the master.

14.2 FEATURES

The protocol is efficient, deterministic, cyclic, and full duplex in nature. The network system is manufacturer independent and can itself adapt to future modifications and expansions. Because the devices are connected in a ring fashion to transfer data from each device to the next in the line, Interbus works as a network-based shift register. Interbus can handle both single-bit data (process data) and data records for intelligent field devices (parameter data). Process data is transmitted in fixed and cyclic time cycles in real-time mode, while parameter data is taken care of in acyclic transmission involving large volumes of data as and when required. Data with a width of between 1 bit to 64 bytes per data direction is permitted in a single Interbus device.

14.3 OPERATION

Interbus is a master–slave type communication system. The master acts as a controller with all the devices connected in a ring fashion. Each slave,

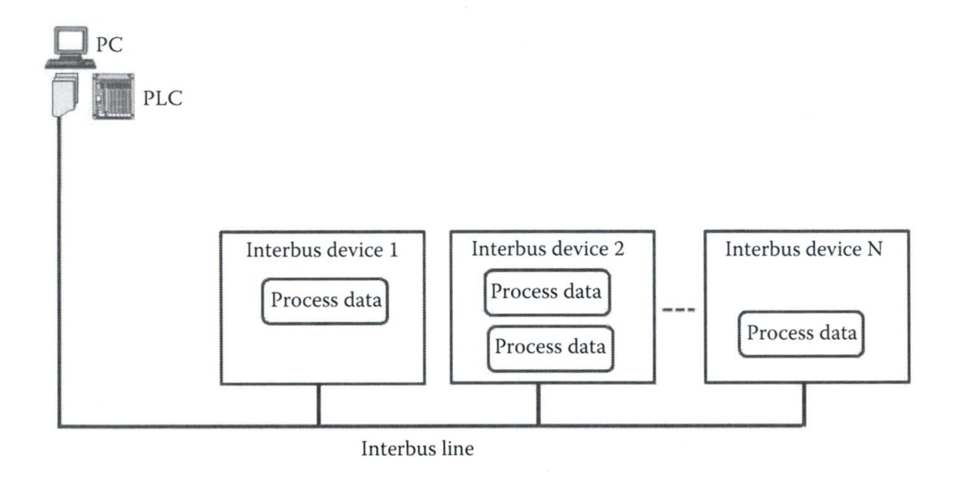

FIGURE 14.1 Basic scheme of process control via Interbus. (From Prof. Dr.-Ing. Reinhard Langmann. *INTERBUS Basics*. Process Control Laboratory, University of Applied Sciences, FH Dusseldorf.)

i.e., device, has two distinct lines for data transmission: one for forward data transmission and the other for return data transmission.

Figure 14.2 shows the basic structure of an Interbus system. It consists of one main ring and subrings—also known as bus segments. Each subring is connected by a bus coupler (or bus terminal module).

The local bus is connected to local bus devices, while the remote bus is connected to remote devices and bus couplers. It is controlled from the controller board. Each bus coupler is connected to a subring. The remote bus branch is connected to remote devices spread over a wide geographical area. It is suitable for complex networking for processes distributed over a wide area, separated by large distances.

The process of sending data from the master to the slaves (OUT data) and from the slaves to the master (IN data) is executed in full duplex transmission mode and is shown in Figure 14.3, in which (a) and (b) correspond to data positions before and after a data cycle. The master provides a data packet, contained in a summation frame, to the send shift register. The data registers in each device (slave) contains the data meant for the master. In a data cycle, the OUT data is transferred from the master to the devices and the IN data is transferred from the devices to the master. The loopback word, shown in Figure 14.3, pulls the OUT data along behind it while pushing the IN data along in front of it.

A bus cycle begins with the master passing a bit stream to the first slave. The first slave then transfers the same to the second one in the ring and at the same time data from the first slave is transferred to the master.

The master knows the physical position of a slave in the ring and so the slaves need not be addressed separately. The master also knows the

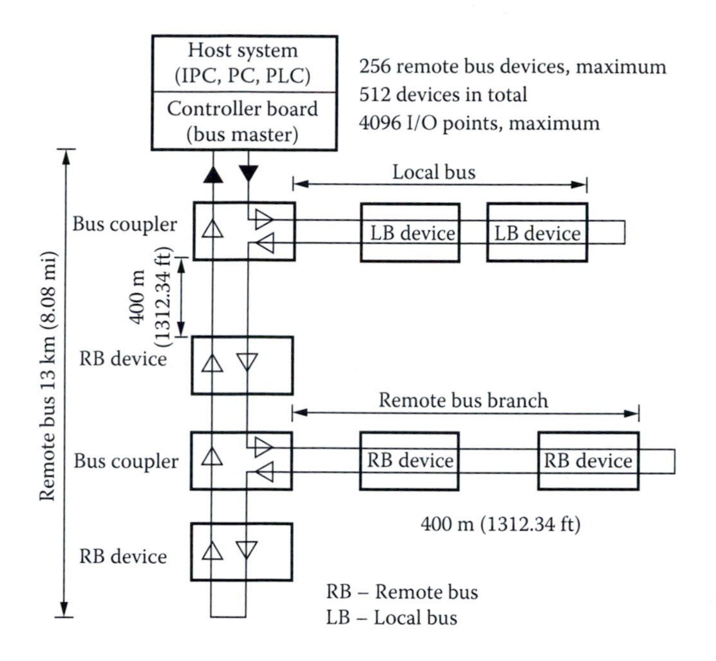

FIGURE 14.2 Structure of an Interbus system. (From Prof. Dr.-Ing. Reinhard Langmann. *INTERBUS Basics*. Process Control Laboratory, University of Applied Sciences, FH Dusseldorf.)

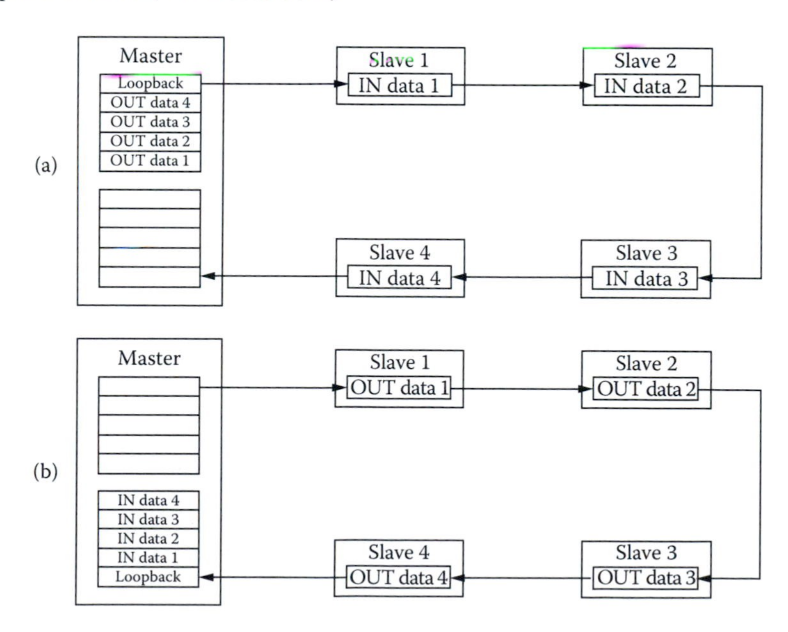

FIGURE 14.3 Principle of data transmission on Interbus: (a) distribution of data before a data cycle; (b) distribution of data after a data cycle. (From Prof. Dr.-Ing. Reinhard Langmann. *INTERBUS Basics*. Process Control Laboratory, University of Applied Sciences, FH Dusseldorf.)

amount of information to be delivered to each device, which is ultimately included in the summation frame.

14.4 TOPOLOGY

The system comprises the communication bus master, bus coupler, and bus devices (slaves), and connecting them are the local bus as well as the remote bus, shown in Figure 14.4.

The bus master, normally available as a controller board, performs the following tasks: transferring data between the host (normally an industrial PC or a PLC) and the bus device, management of the bus, and communication between the devices. The management functions include error detection and configuration.

The bus coupler, also known as the bus terminal device, divides the ring connection into bus segments. It functions as a slave and connects the incoming (i) and outgoing (o) interfaces. Figure 14.5 shows two types of bus couplers normally required in an Interbus system.

The bus coupler, along with the bus devices, configures the ring system. The bus devices are connected to the process signals and transfer communication signals as per the requirements, i.e., either in analog or digital from. The bus coupler, under the command of bus master, can activate or deactivate either the incoming or outgoing interfaces to support system configuration as well as error diagnostics. The same is shown in Figure 14.6.

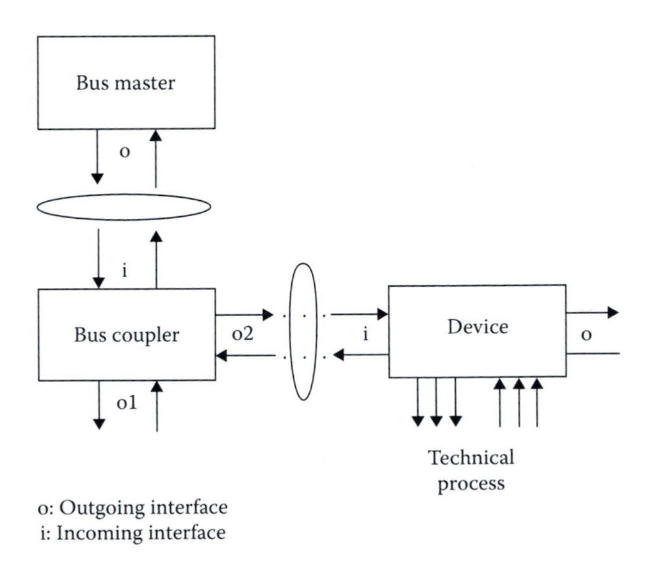

FIGURE 14.4 Components in an Interbus system. (From Prof. Dr.-Ing. Reinhard Langmann. *INTERBUS Basics*. Process Control Laboratory, University of Applied Sciences, FH Dusseldorf.)

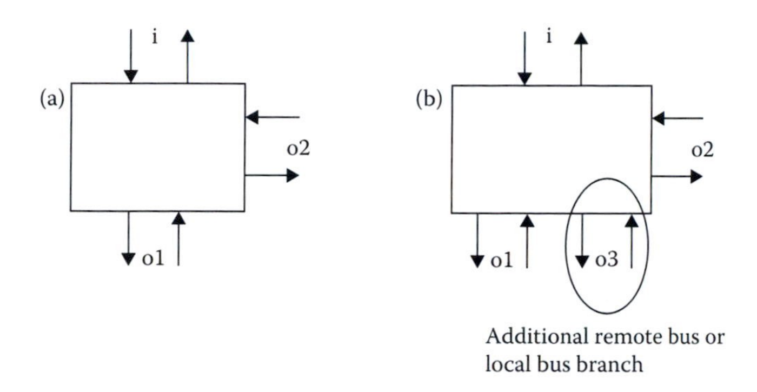

FIGURE 14.5 Bus couplers: (a) standard; (b) with additional interface. (From Prof. Dr.-Ing. Reinhard Langmann. *INTERBUS Basics*. Process Control Laboratory, University of Applied Sciences, FH Dusseldorf.)

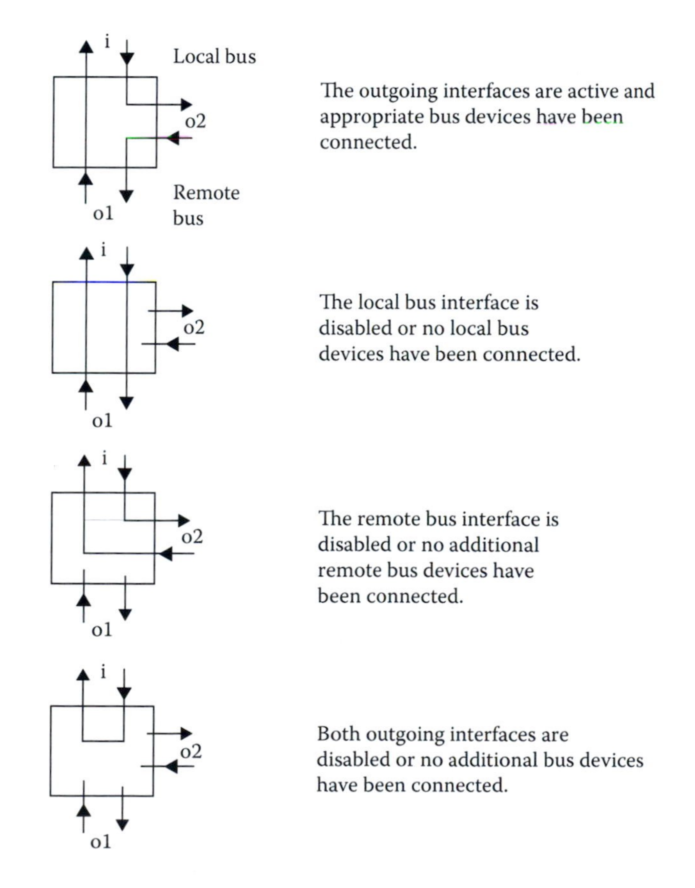

FIGURE 14.6 Techniques or activating/deactivating interfaces. (From Prof. Dr.-Ing. Reinhard Langmann. *INTERBUS Basics*. Process Control Laboratory, University of Applied Sciences, FH Dusseldorf.)

14.5 PROTOCOL STRUCTURE

The protocol structure follows the OSI reference model and comprises layers 1, 2, and 7 and shown in Figure 14.7. Some functions of layers 3 to 6 are included in the application layer.

The application layer helps in accessing data from the slaves (Interbus devices), the physical layer determines the baud rate and performs data encoding, and the data link layer ensures data integrity. Layers 1 and 2 adhere to DIN 19 258.

14.5.1 Physical Layer

Data transmission on the physical layer takes place at a rate of 500 kbps, while the data line is scanned 16 times faster by the slaves. Data is encoded in NRZ (nonreturn-to-zero) form before transmission. There is a clock generator in each device, and the individual clocks are synchronized internally by a common synchronization marker.

The format for the line encoding in the physical layer is shown in Figure 14.8; it comprises the status telegram and the data telegram. The status telegram transmits the status of the SL (select) signal. It is 5 bits in length and generates activity on the bus during pauses in transmission. The data telegram is 13 bits in length, with 5 bits of header and 8 bits of user data. It also contains the status of the CR (control) signal.

14.5.2 Data Link Layer

The summation frame telegram is shown in Figure 14.9, and it corresponds to the cyclic Interbus protocol in the data link layer. The methodology used

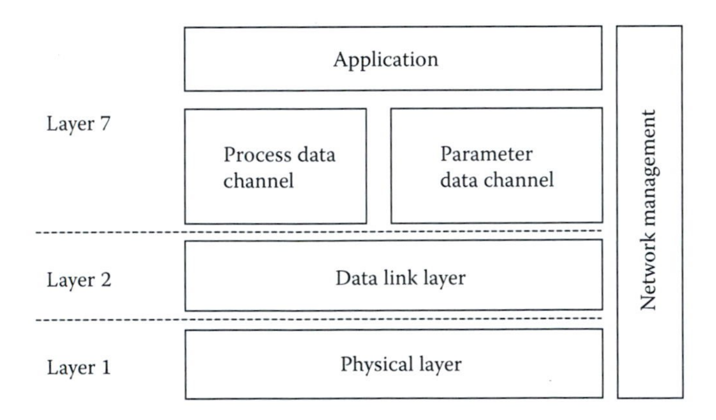

FIGURE 14.7 Interbus protocol structure. (From Prof. Dr.-Ing. Reinhard Langmann. *INTERBUS Basics*. Process Control Laboratory, University of Applied Sciences, FH Dusseldorf.)

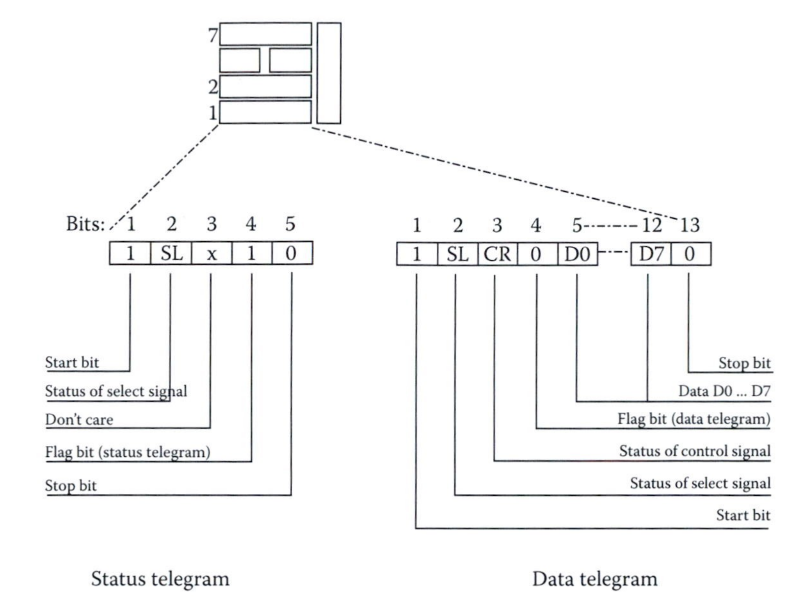

FIGURE 14.8 Line encoding format in the physical layer. (From Prof. Dr.-Ing. Reinhard Langmann. *INTERBUS Basics*. Process Control Laboratory, University of Applied Sciences, FH Dusseldorf.)

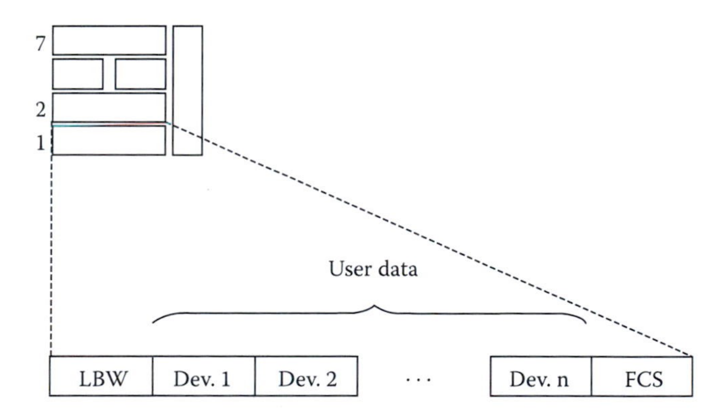

FIGURE 14.9 Summation frame telegram format in the data link layer. (From Prof. Dr.-Ing. Reinhard Langmann. *INTERBUS Basics*. Process Control Laboratory, University of Applied Sciences, FH Dusseldorf.)

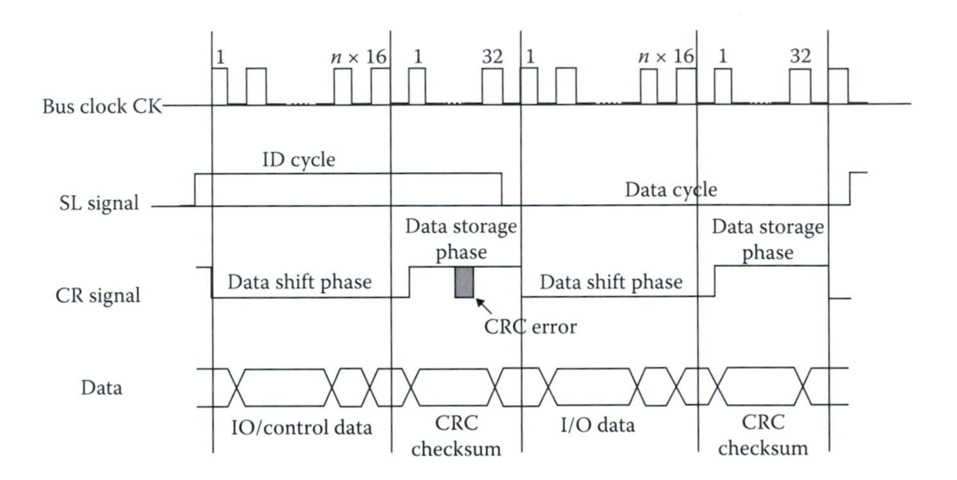

FIGURE 14.10 Different operating phases in the Interbus protocol. (From Prof. Dr.-Ing. Reinhard Langmann. *INTERBUS Basics*. Process Control Laboratory, University of Applied Sciences, FH Dusseldorf.)

in this frame is Time Division Multiple Access (TDMA) with collision-free transmission. Each device is allocated a time in tune with its function, and thus the total transition time can easily be calculated. The LWB, or loopback word, is used by the master at the beginning of each cycle to ascertain the amount of data and consequently the number of shift registers in the ring. It is a 16-bit word and is used in the identification cycle (ID cycle) to detect the end of data shift phase.

The Interbus protocol in the data link layer has different operating phases, and the different activities are governed by the SL (select) and CR (control) signals. This is shown in Figure 14.10. The SL signal is characterized by the ID cycle and data cycle, while the CR signal by the data shift phase and data storage phase.

In the ID cycle within the SL signal, the devices switch the ID registers in the Interbus ring and the master can thus identify the devices individually. The status of the SL signal in the data cycle is made zero by the master, and thus the Interbus ring is closed by the data shift registers. When the CR signal is asserted, the data already transmitted during the SL signal is now stored in the data storage phase or FCS. CRC checksum is employed in this phase of data storage and saving.

14.5.3 Application Layer

Interbus devices are accessed for data via two channels: the process data channel and the parameter channel. The former is employed for process data and the latter for parameter data. Cyclic data exchange between the sensors and the higher-level control systems takes place via the process data channel, while the parameter channel helps accessing the connection-oriented message exchange. Message transmission takes place in the parameter channel in the client–server model, and a large volume of data is exchanged between the communicating devices. Each Interbus device has a process data channel, while the parameter channel requirement is optional. The performance of the application layer is dependent on how seamlessly it can handle the two types of data efficiently.

The process data for a device is in the range of a few bits and comes from the set point, limit switches, control signals, etc. The process data for a device is almost always uniquely identified by its address and/or the sensor/actuator it represents. The collective total of I/O data in the summation frame telegram during a data cycle represents the process data.

Interbus devices with a parameter channel are servo amplifiers, operating and display units, etc., and may be in the range of 10–100 bytes. Parameter data is acyclic in nature and transmitted only when it is required between two Interbus devices.

15 ControlNet

15.1 INTRODUCTION

ControlNet is an open industrial network protocol belonging to the Common Industrial Protocol (CIP) family. It was initially supported by ControlNet International. Since 2008, the activities and management of ControlNet is being handled by ODVA. The CIP is also used in DeviceNet and Ethernet/IP.

ControlNet is included in the EN 50170 and IEC 61158 standards. In ControlNet fieldbus systems, only the session layer is empty when it is seen in the context of the OSI model. It is designed to be used both at the device and cell levels of industrial control systems. Applications of ControlNet include batch control systems, automotive industries, and process control systems. Figure 15.1 shows the ControlNet communication profile, which shows that layer 5 of OSI, i.e., the session layer, is absent in ControlNet.

15.2 FEATURES

ControlNet communication is a high-speed deterministic network used for the transmission of time-critical applications and can be scheduled as per requirement. Common network sniffers cannot sniff into ControlNet packets. In peer-to-peer communication mode, it provides real-time control and messaging services. It supports the producer–consumer network model, allowing data from a node to be multicast, resulting in increased efficiency and higher system performance.

It offers high-throughput—about 5 Mbits/s for improved I/O performance. Some of the other features of ControlNet include deterministic and highly repeatable data delivery, and multiple controllers controlling I/Os independent of each other belonging to the same link.

ControlNet supports both coaxial and fiber optic cables in tree, bus, or star topology. Repeaters can extend the reach of the network to distances more than 30 km. It has media layer redundancy for increased network reliability.

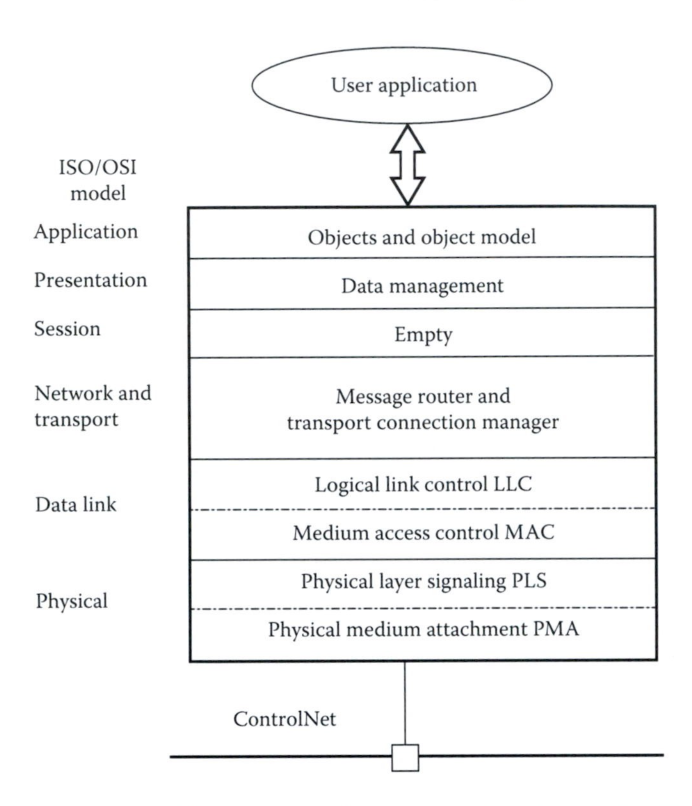

FIGURE 15.1 ControlNet communication profile. (From B. G. Liptak. *Instrument Engineers' Handbook–Process Software and Digital Networks*, 3rd Edition. CRC Press, Boca Raton, FL, p. 614, 2002.)

15.3 PRODUCER–CONSUMER MODEL

The producer–consumer model is followed in ControlNet in which all the nodes access data from a single source at the same time, i.e., synchronously. Synchronism is achieved because data arrives at each node at the same time. This results in higher performance and increased efficiency. In traditional legacy source–destination model, data is received by different destinations at different times and requires multiple packets to deliver the same data to multiple nodes. It results in reduced efficiency because of extra network traffic. Also, a legacy system requires different networks for sending messages and time-critical I/Os. Compared with this, the producer–consumer model is fully synchronized—the same network can send messages and time-critical I/Os, and bandwidth is optimized for higher performance. This system supports master–slave, multimaster, peer-to-peer communication, and also hybrid systems—i.e., a mix of the above three types. Figure 15.2 shows a legacy source–destination model and a producer–consumer model.

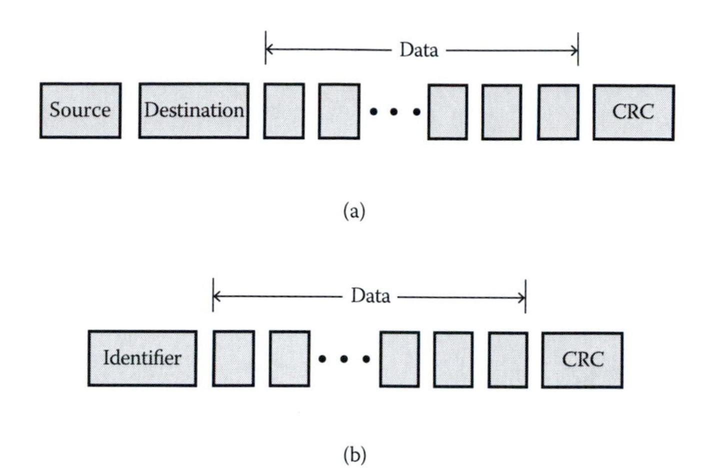

FIGURE 15.2 (a) Legacy source–destination model. (b) Producer–consumer network model.

15.4 CONTROLNET MEDIA

ControlNet media consists of the physical components that make up a ControlNet network. The components are connectors, taps, repeaters, bridges, cables, terminators, etc. (as shown in Figure 15.3).

The ControlNet media shown in Figure 15.3 has two links, with one link having two segments and the other one segment only. A link is a collection of nodes having unique addresses in the range of 1 to 99. Figure 15.3 also shows a repeater that replicates the signal from one segment to another. The nodes connect the physical devices to the cable system via drop

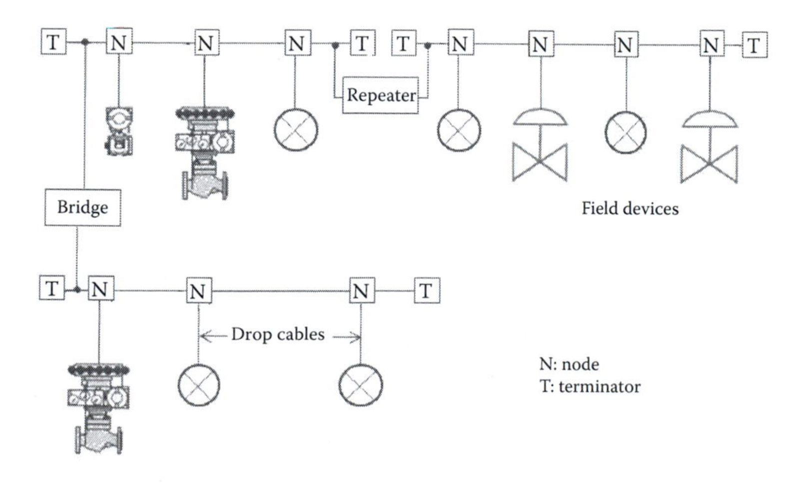

FIGURE 15.3 ControlNet media.

cables. Two terminators (75 Ω each) are inserted on either side of a segment. Terminators prevent reflections from the ends of a cable. The bridge connects one segment to another. Taps connect devices to the trunk cable. A tap is required for each and every node and on either side of a repeater.

15.5 PHYSICAL LAYER

The physical layer of ControlNet is split into two sublayers: Physical Medium Attachment (PMA) and Physical Layer Signaling (PLS), as shown in Figure 15.1. PMA comprises the circuits that aid in delivering/accepting data to/from the bus to which the devices are attached. PLS is responsible for appropriate timing and bit representation of data generated at the devices. It also acts as an interface to the next higher layer—the data link layer.

The physical media can be of three types: coaxial cable, fiber optic, and network access port or NAP. The coaxial cable is RG-6 quad shield. The maximum cable length is 1000 m without repeaters, and the maximum number of nodes is 99. The media can support a maximum of five repeaters (10, if redundancy is used). Physical layer signaling uses Manchester code with a bit rate of 5 Mbits/s.

When used with fiber optic cables, the system can have two lengths: up to 300 m for short-range systems and up to 7000 m for medium-range systems. The fiber optic type can employ both active star and active hub topologies.

NAP has eight conductors with an overall shield. It is used to establish a connection between a programming unit and a station that is already attached to ControlNet. For a point-to-point temporary connection between two nodes, NAP is used.

15.6 DATA LINK LAYER

The data link layer of ControlNet consists of two sublayers: medium access control (MAC) and logical link control (LLC).

When a node sends data over the network, it is packed into a MAC frame. The form of the MAC frame is shown in Figure 15.4. A single MAC frame may contain several LPackets (link packets). A maximum of 510 bytes of data can be sent in a single frame. The format of the LPacket is also shown in Figure 15.4, which contains a field called connection ID (CID), apart from the other fields. The CID can be of two types: a 2-byte fixed CID and a 3-byte general CID. The two types of CIDs are shown in Figure 15.5. The 2-byte CID contains a service code and destination network address byte.

The code byte is used to indicate the type of service required, while the destination byte indicates the network address to which it is to be delivered.

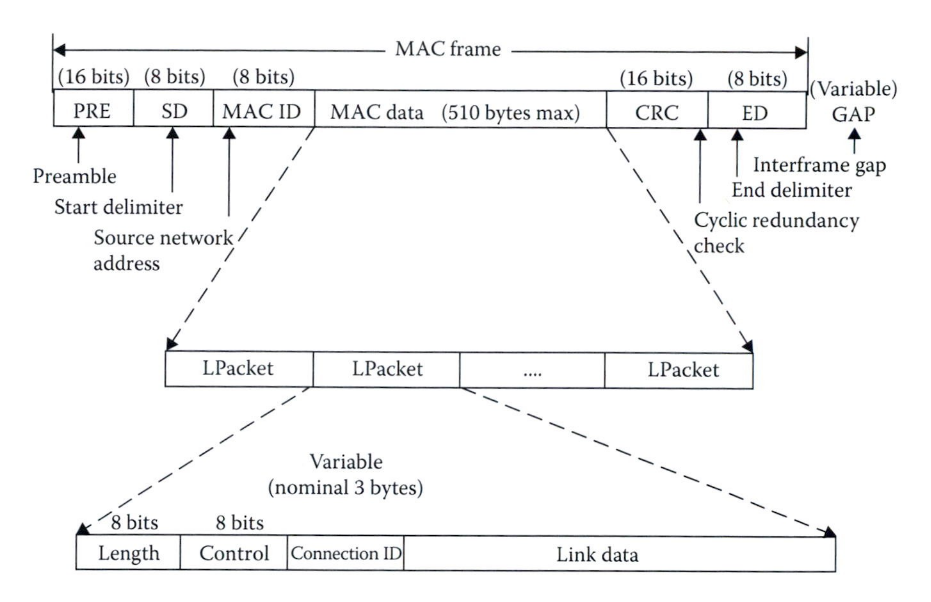

FIGURE 15.4 MAC frame and LPacket formats. (From Rockwell Automation—Allen–Bradley. *An Introduction to the ControlNet Network*, System Overview, Release 1.5, 1997.)

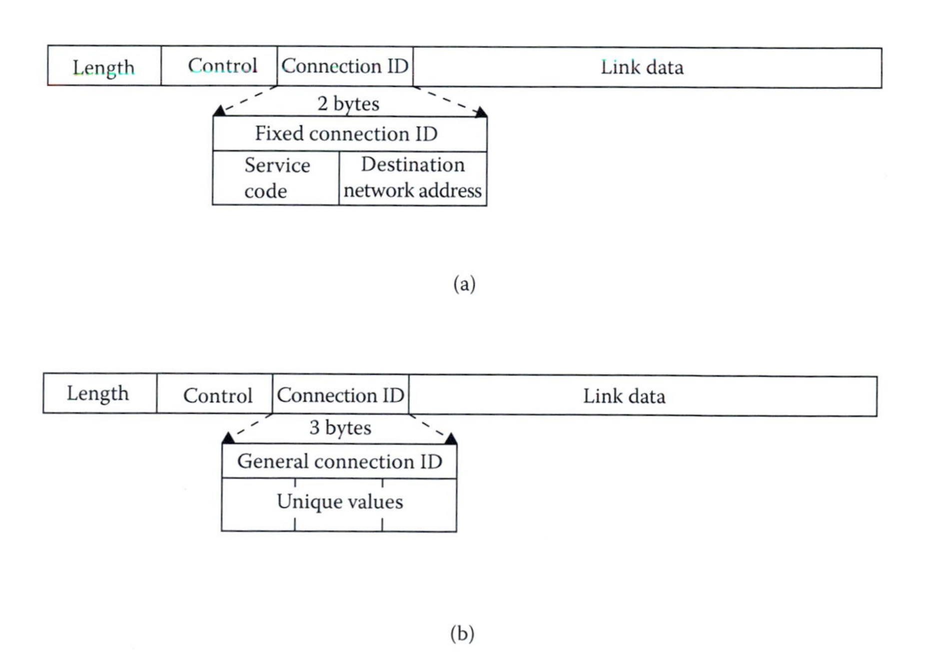

FIGURE 15.5 Two CID formats: (a) 2-byte fixed CID and (b) 3-byte general CID. (From Rockwell Automation—Allen–Bradley. *An Introduction to the ControlNet Network*, System Overview, Release 1.5, 1997.)

The 3-byte general CID contains a number that identifies a specific data packet. On any given ControlNet network, this number is unique.

Access to the network is based on a time slice algorithm, based on MAC. This algorithm is called Concurrent Time Domain Multiple Access (CTDMA). This regulates the data traffic on the network. Figure 15.6 shows ControlNet's media access mechanism. It is apparent from Figure 15.6 that the network update time (NUT) is fixed to where all the nodes are synchronized. For the network, NUT is indefinitely repeated. A NUT is divided into three sections: scheduled, unscheduled, and guard band, also known as network maintenance.

Real-time data, for example, both the cyclic data exchange and asynchronous data traffic, are sent in the scheduled interval. In this, the message delivery is fully deterministic and repeatable.

The bandwidth in the scheduled portion is reserved and configured beforehand to support real-time data transfers. The types of messages supported in this interval are analog data, digital data, and peer-to-peer interlocking data.

Each station or node is granted only a single frame transmission in a NUT. Each station is assigned an address called its MAC ID. When the NUT begins, the station with the lowest MAC ID is granted permission to transmit a frame. Every station has an implicit register, which is its own MAC ID. When a frame transmission is over, the MAC ID is automatically incremented by one. Each station compares this new ID with the contents of the implicit register. The station for which these two match each other gets the right to transmit its own frame. The scheduled transmission technique is shown in Figure 15.7.

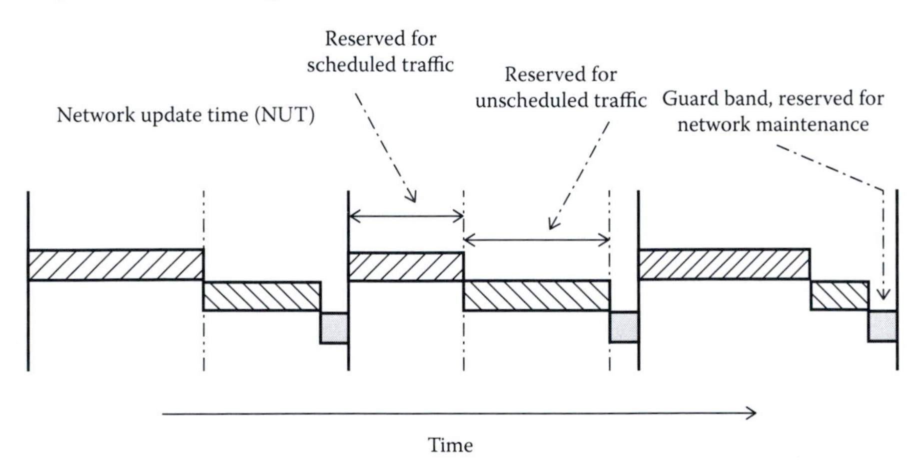

FIGURE 15.6 Media access mechanism for ControlNet. (From Rockwell Automation—Allen–Bradley. *An Introduction to the ControlNet Network*, System Overview, Release 1.5, 1997.)

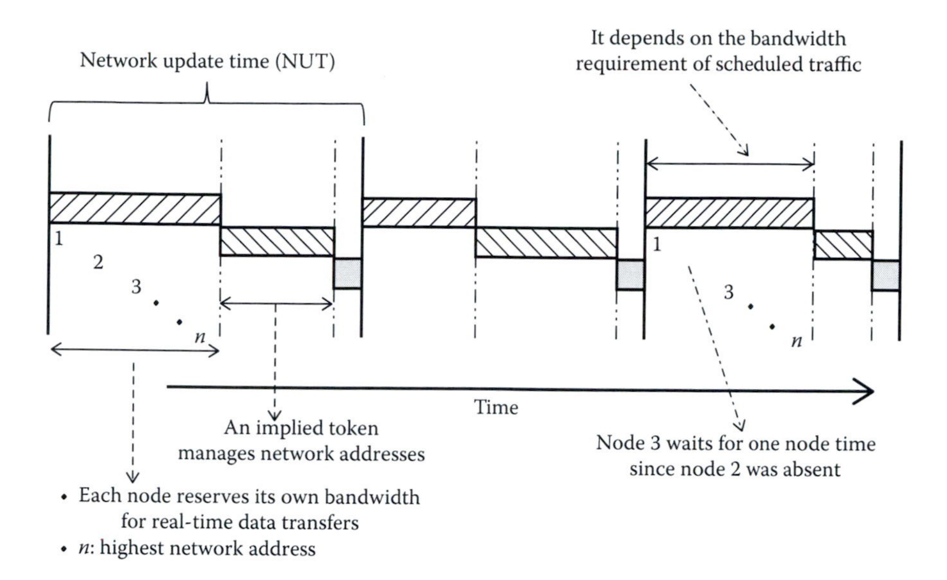

FIGURE 15.7 Media access mechanism in scheduled time in ControlNet. (From Rockwell Automation—Allen–Bradley. *An Introduction to the ControlNet Network*, System Overview, Release 1.5, 1997.)

Any station having a network address between 1 and SMAX (scheduled maximum) will be granted exactly one opportunity to transmit in a single NUT. Up to SMAX bandwidth is reserved in the scheduled portion of NUT. Network address 0 is not allowed for a station, which is reserved for future use. Slot time is the time duration that a node or station would wait for a missing network address before initiating transmission on the network bus.

Once the data traffic in the scheduled interval is over, the unscheduled portion of the NUT takes over, which does not have any time-critical constraints. Unscheduled transmission in NUT is neither deterministic nor repeatable. In this, the implicit token is still circulated among the stations in a round-robin scheme. This continues for network addresses between 0 and UMAX (unscheduled maximum) until the beginning of guard band. The scheduled transmission technique is shown in Figure 15.8.

A node may have the opportunity to transmit more than once in the unscheduled portion of NUT, but transmission is not always guaranteed. Unscheduled data types include connection establishment, peer-to-peer messaging data, and programming data—both upload and download. Some important characteristics of unscheduled communication are as follows: (a) Unscheduled service is from 0 to UMAX; UMAX can never be less than SMAX; (b) Stations with addresses greater than SMAX and less than or equal to UMAX may only send unscheduled messages; (c) Stations with addresses less than or equal to SMAX may send both scheduled and

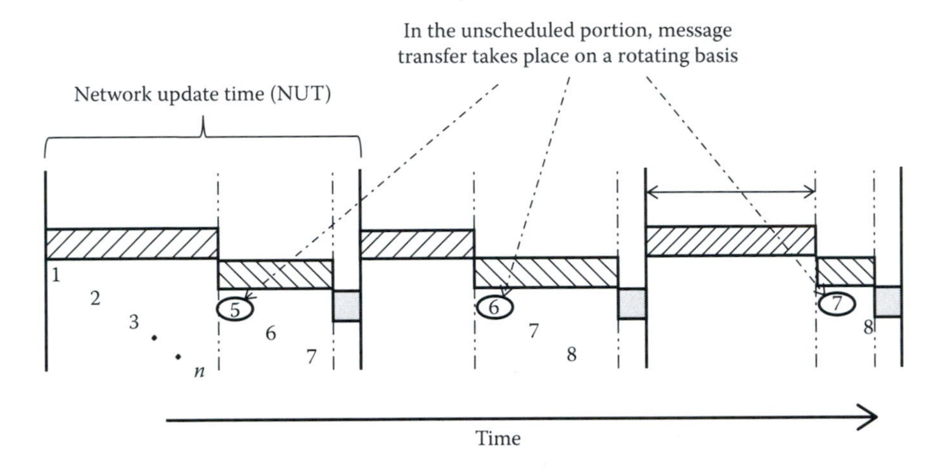

FIGURE 15.8 Media access mechanism in unscheduled time in ControlNet. (From Rockwell Automation—Allen–Bradley. *An Introduction to the ControlNet Network*, System Overview, Release 1.5, 1997.)

unscheduled messages; and (d) Stations having addresses greater than UMAX cannot communicate through the ControlNet network.

The scope of transmitting first in the unscheduled part of NUT passes on a rotating basis. In the first NUT, node 7 transmits first in the unscheduled portion. In the second NUT, it is node 8 that transmits first irrespective of which node finished the last interval. In conclusion, it can be said that the right to transmit first in the unscheduled portion rotates by one node per NUT.

In each NUT, once the scheduled and unscheduled portions are over, the guard band interval takes over. Transmission on the network is stopped, and the station with the lowest network address (MAC ID) is granted access. This station or node is called moderator. In the guard band, this node transmits only the moderator frame, which keeps all the nodes in synchronism and also a set of parameters necessary for correct operation of the network.

15.7 NETWORK AND TRANSPORT LAYERS

The three major basic modules existing on these two layers are Unconnected Message Manager (UCMM), Message Router (MR), and Connection Manager (CM).

UCMM can send a message without an established connection. It is used to transmit nonrepetitive, non-time-critical data on a single link, which is always sent in the unscheduled portion of NUT. It facilitates execution of unconnected messages. In this case, each and every data transfer is independent of others and the messages carry the descriptions of destination and source applications, which are provided by the CM.

As shown in Figure 15.9, the message coming from the ControlNet network is passed on to MR via the UCMM. The MR analyzes the same message and sends it to the specified function or object. UCMM functions include receiving incoming UCMM messages, and sending and receiving unconnected messages to and from UCMM objects on other nodes and its local MR.

The MR guides the correct dispatch of services to the addressed applications within the node. Figure 15.10 shows the functioning of MR.

The MR analyzes the message to determine the service to be executed by the identified object. The message is forwarded to the destination object, which in turn sends its response to the requesting object via the MR.

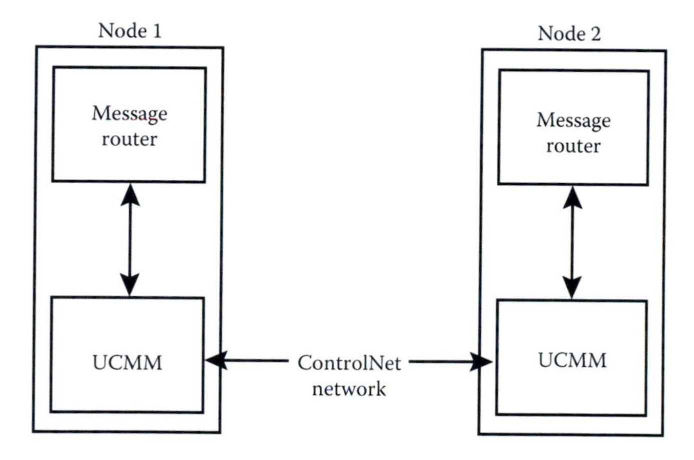

FIGURE 15.9 UCMM and message router interconnection. (From Rockwell Automation—Allen–Bradley. *An Introduction to the ControlNet Network*, System Overview, Release 1.5, 1997.)

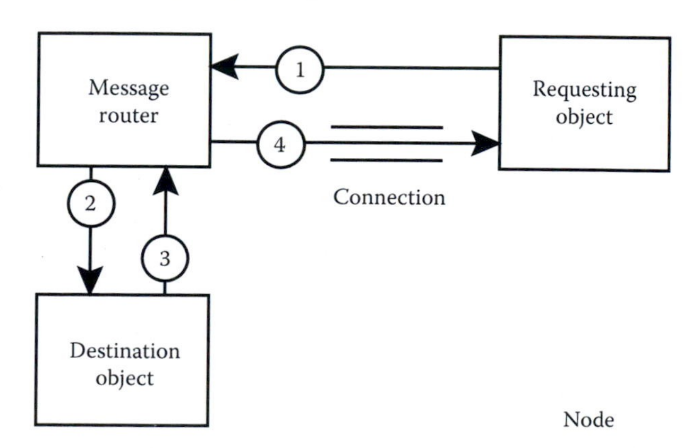

FIGURE 15.10 Functioning of a message router. (From Rockwell Automation—Allen–Bradley. *An Introduction to the ControlNet Network*, System Overview, Release 1.5, 1997.)

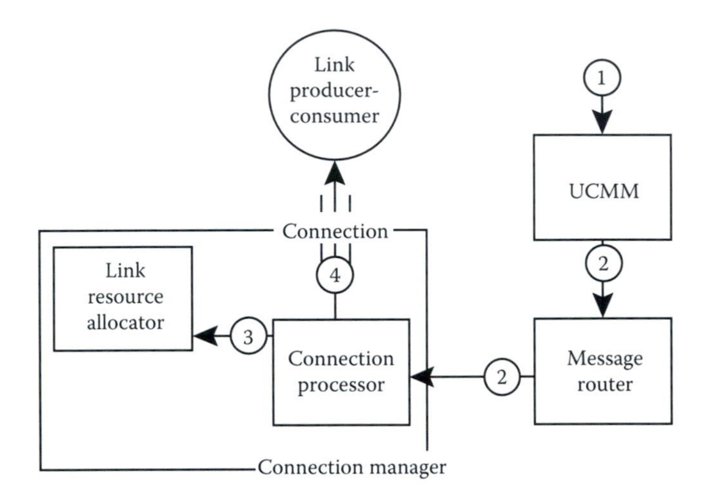

FIGURE 15.11 Functioning of a connection manager. (From Rockwell Automation—Allen–Bradley. *An Introduction to the ControlNet Network*, System Overview, Release 1.5, 1997.)

Internal resources required for a connection establishment is provided by the CM. The request for connection may come from another node via the UCMM or else an application existing on a node. Figure 15.11 shows the operation of a CM.

First, the UCMM of the source tries to establish a connection with the UCMM of the destination with a connection request. This is routed through the MR of the target to the CM, which then allocates the requisite resources. Finally, the needed connection is established.

Seven classes (classes 0 to 6) of connections are defined for the CM. They are distinguished by their features, complexities, and nature of services required of them.

15.8 PRESENTATION LAYER

The presentation layer is based on the IEC 1131-3 standard. The elementary and derived data types are defined by this standard. Data management is managed by this layer. It specifies the format of the data to be handled by the application layer.

15.9 APPLICATION LAYER

The ControlNet network application layer is based on object modeling. Object modeling refers to organizing related data and procedures into one entity: the *object*. An object can be thought of as a collection of related *services* and *attributes*. Application objects share their resources and

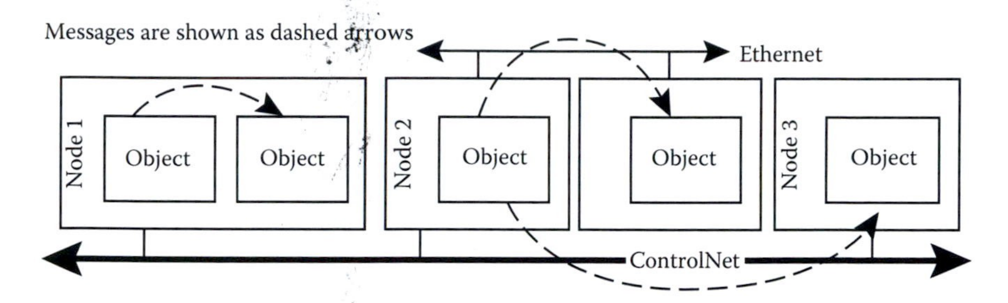

FIGURE 15.12 Message paths for application objects. (Department of Informatics and Automation. An Introduction to the ControlNet Network. Available at www.dia.uniroma3.it/autom/Reti_e_Sistemi_Automazione/PDF/ControlNetDetails.pdf.)

information by sending messages. Messages can be sent between two objects of a node or across the ControlNet network. The message paths of such objects are shown in Figure 15.12.

A standard is an object library that contains all the details of the objects used in a station. Sometimes, vendor-specific objects are included in the standard by incorporating their device profiles so that devices from different manufacturers can be used interchangeably.

16 Intrinsically Safe Fieldbus Systems

16.1 INTRODUCTION

Fieldbus standard IEC 61158-2 defines the maximum dimensions for a fieldbus segment and its proper operation for the case of a safe area. These operations and calculations are for a network based on Foundation Fieldbus H1 or PROFIBUS-PA (MBP). Some additional constraints are imposed for proper operation of the network segment in the hazardous area for explosion protection as defined in IEC 60079.

16.2 HAZARDOUS AREA

In chemical industries in general and petrochemical industries in particular, which deal with oil and gases, a high degree of caution is always maintained because of the presence of flammable gases and oils. Heat and electrical arcing are the main reasons for fire/explosion in these hazardous areas. Safety of plant and personnel is of prime importance in such industries. Intrinsic safety, explosion-proof enclosures, and purging are some of the measures undertaken to overcome fire hazards.

16.3 HAZARDOUS AREA CLASSIFICATION

Classification is based on the identification and quantity of flammable materials, their characteristics, and the chances of their coming into contact with an electrical arc, etc. The basis includes types of flammable gases/vapors, liquids, combustible dust, and fire-prone fibers.

Depending on the classification, equipments are employed in hazardous areas to match and conform to the safety standards. The standards are classified in National Fire Protection Agency (NFPA) Standard 70—the National Electrical Code. Normally, two classifications are followed: the conventional Division Classification System and the Zone Classification System.

16.3.1 Division Classification System

In this system, a hazardous area is identified with a class, division, and group. *Class* deals with whether the flammable material is a gas or dust or fiber, while *division* deals with the probability of the presence of flammable material by defining how it is used. Lastly, *group* deals with the physical properties of the flammable material.

There are three classes: I, II, and III. Class I deals with the presence of flammable gases, class II with combustible gases, and class III with ignitable gases.

Classes are further subdivided into groups. The groups assign categories of chemical and physical properties to the hazardous materials. These are found in National Fire Protection Association NFPA 497 and NFPA 325.

Class I is subdivided into four groups: A, B, C, and D. Group A consists of acetylene, while groups B, C, and D are concerned with hydrogen, ethylene, and propane gas, respectively.

Groups B, C, and D are flammable gases/vapors produced by flammable/combustible liquids. Maximum Experimental Safe Group and Minimum Igniting Current Ratio are the two factors that differentiate groups B, C, and D.

Class II is subdivided into three groups: E, F, and G. Information about the class II group is found in NFPA 499. Group E is concerned with metal dust and its properties (such as abrasiveness, particle size, and conductivity) in the hazardous area's atmosphere, while group F pertains to carbon-based dust (coke, coal, graphite, charcoal, etc.) in the industrial atmosphere. Group G includes flour, grain, and wood. Class III does not have any subgroup.

Division is concerned with the probability of flammable material present and its use in the hazardous area. It is subdivided into division 1 and division 2.

16.3.2 Zone Classification System

This system classifies a hazardous area into three zones: zone 0, zone 1, and zone 2. These were introduced in 1999 by the National Electrical Code. Zone 0 area refers to the presence of flammable mixtures of gaseous vapors continuously or for long periods of time. In zone 1, flammable mixtures are expected to occur under normal circumstances, while in zone 2, flammable mixtures are unlikely under normal circumstances and would quickly disperse if they do occur.

In passing, it can be mentioned here that "divisions" are normally used in North America, while "zones" are used in the rest of the world.

16.4 EXPLOSION PROTECTION TYPES

Figure 16.1 shows both division-wise and zone-wise hazardous area classifications.

Tables 16.1 and 16.2 show the division 2/zone 2 and division 1/zone 0,1 methods of protection for different categories of fire hazards, respectively.

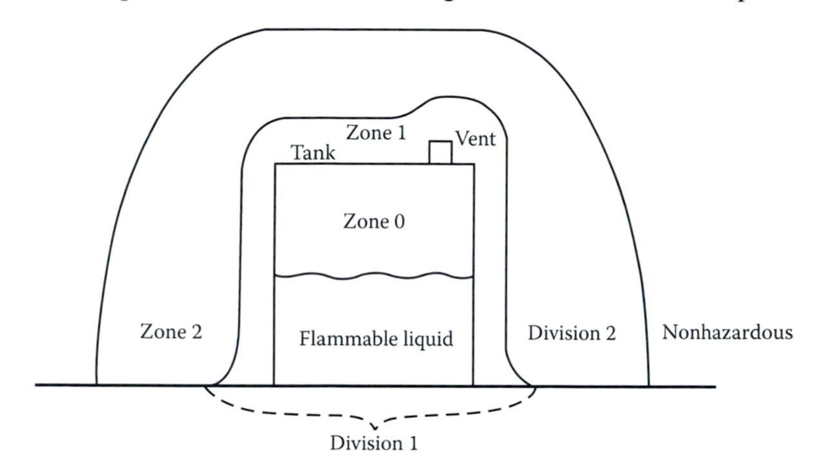

FIGURE 16.1 Hazardous area classifications. (From Fieldbus Wiring Guide, 4th edition, Austin, TX, Doc. No. 501-123, Rev.: E.0. Available at www.relcominc.com/pdf/501-123FieldbusWiringGuide.pdf.)

TABLE 16.1
Division 2/Zone 2 Methods of Protection

Method of Protection	Trunk	Spurs
Non-arcing	Non-arcing	Non-arcing
High-energy trunk	Non-arcing	Energy limited
FNICO (FISCO ic)	Energy limited	Energy limited

Source: From Fieldbus Wiring Guide, 4th edition, Austin, TX, Doc. No. 501-123, Rev.: E.0. Available at www.relcominc.com/pdf/501-123FieldbusWiringGuide.pdf.

TABLE 16.2
Division 1/Zone 0,1 Methods of Protection

Method of Protection	Trunk	Spurs
Increased safety	Non-arcing	Non-arcing
High-energy trunk	Non-arcing	Intrinsically safe
FISCO	Intrinsically safe	Intrinsically safe

Source: From Fieldbus Wiring Guide, 4th edition, Austin, TX, Doc. No. 501-123, Rev.: E.0. Available at www.relcominc.com/pdf/501-123FieldbusWiringGuide.pdf.

16.5 INTRINSIC SAFETY IN FIELDBUS SYSTEMS

A regular fieldbus network and an intrinsically safe fieldbus network do not differ much. Topologically, they are identical. Differences of the latter include fewer devices per wire and a safety barrier is placed instead of regular power supply impedance. Field instruments for intrinsically safe systems are designed such that they do not inject power into the bus.

A terminator is placed on either side of a fieldbus network. The network contains a capacitor, which must be certified to be intrinsically safe. Also, the terminators must be totally passive, which would draw no current. This is a mandatory requirement so as not to reduce either length or device count.

Again, for each intrinsically safe segment, only one power supply source is employed—i.e., no redundancy is allowed. The barriers are placed in safe areas so that they are cheaper. Sometimes, exigencies force the barriers to be mounted in hazardous areas, in which case they must be housed in flame-proof enclosures with flame-proof seals.

The barriers employed can be a zener barrier or a galvanic barrier. Fieldbus devices operate between 9 and 32 V. Thus, to limit power, lesser

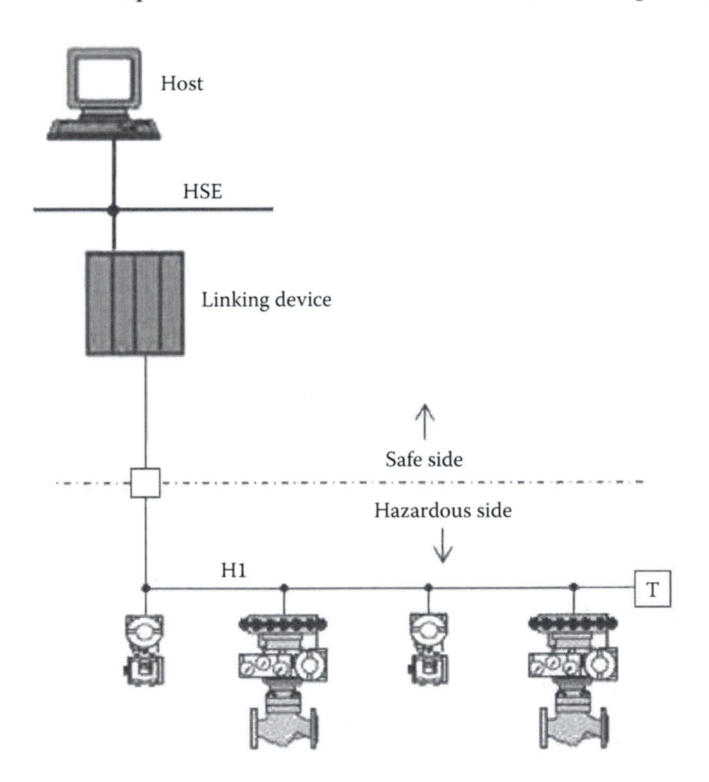

FIGURE 16.2 Safety barrier separates safe area from hazardous area. (From B. G. Liptak. *Instrument Engineers' Handbook—Process Software Digital Network*, 3rd Edition. CRC Press, Boca Raton, FL, p. 161, 2002.)

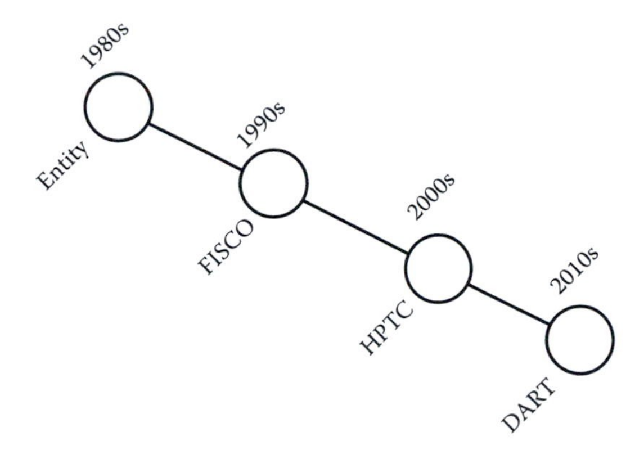

FIGURE 16.3 Development of different intrinsically safe fieldbus systems. (From A. Beck and A. Hennecke. *Intrinsically Safe Fieldbus in Hazardous Areas*, Technical White Paper, EDM TDOCT-1548_ENG. Pepperl+Fuchs GmBH, Mannheim, Germany, p. 8, 2008. Available at files.pepperl-fuchs.com/selector_files/navi/productInfo/doct/tdoct1548a_eng.pdf.)

voltage should be chosen for powering field devices in order to limit power consumption. It is a good investment if field devices are chosen that consume the least power. In such cases, fewer barriers are needed, which keeps the system less clumsy. Figure 16.2 shows a safety barrier that separates a safe area from a hazardous area.

Intrinsically safe fieldbus barriers have been around since the 1960s, and different barriers have been introduced over a period of time as shown in Figure 16.3. These are Entity concept, Fieldbus Intrinsically Safe Concept (FISCO), High-Power Trunk Concept (HPTC), and Dynamic Arc Recognition and Termination (DART).

16.6 ENTITY CONCEPT

Entity concept is defined in IEC 60079-11 and NEC 515, and is a method of validating an intrinsically safe installation by employing intrinsically safe parameters. The parameters to be considered are voltage, current power, capacitance, and inductance. The cable capacitance and inductance for the hazardous side of the segment are considered lumped and must be considered along with the capacitance and inductance of all the devices connected to the segment. The total capacitance and inductance must be within the limits of the barrier as envisaged in the recommendation. Several devices are normally multidropped off a single barrier, and hence the entity parameters of all the devices must match those of the barrier.

A linear output characteristic for the barrier is used in Entity concept, shown in Figure 16.4. The output power for Exia IIC is roughly 1.2 W at

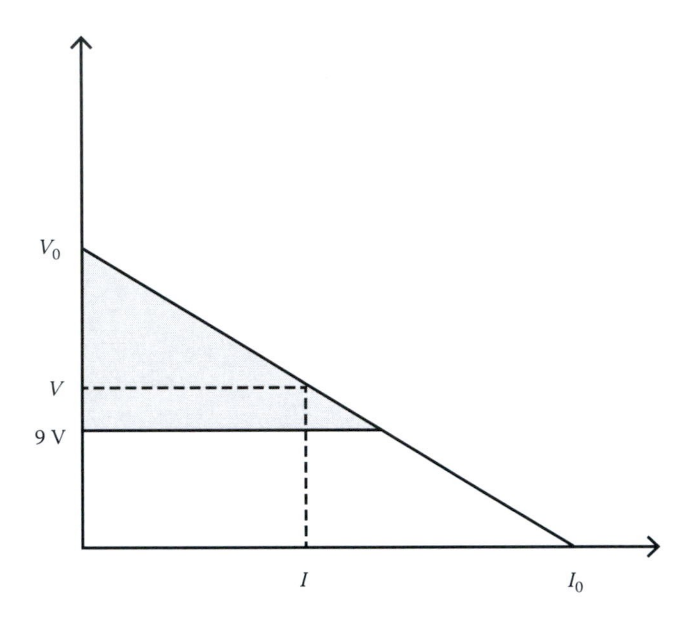

FIGURE 16.4 Linear output characteristic. (From J. Berge. *Fieldbuses for Process Control: Engineering, Operation and Maintenance*. ISA, USA, p. 103, 2004.)

11-V DC; the current available is a maximum of 60 mA. This thus limits the maximum number of devices to a few for each barrier that can be accommodated; otherwise, the available voltage at the devices would become less than the minimum recommended. Application of small voltage (11-V DC) limits the total cable run, since a small voltage drop along the cables would bring down the voltage available at the devices to a value less than the acceptable recommended. A barrier allows a maximum L/R ratio (inductance/resistance) that should be compared with that of the cable. Again, cables with shielding have higher value of capacitance, which lowers the total cable run compared with cables without shielding.

Fault disconnection electronics automatically disconnects the power supply to the barrier in case of malfunctioning/short circuit of any device.

The low power of 1.2 W allows a maximum of two to three devices per barrier for the IIC gas connection. This practically limits adopting the Entity concept in hazardous areas. Again, the complex and time-consuming calculation efforts to validate an installation do not find favor in employing the Entity concept.

16.7 FISCO MODEL

FISCO was developed by Physikaliseh-Technische-Bundesanstalt (PTB) and has trapezoidal characteristics, shown in Figure 16.5. It provides

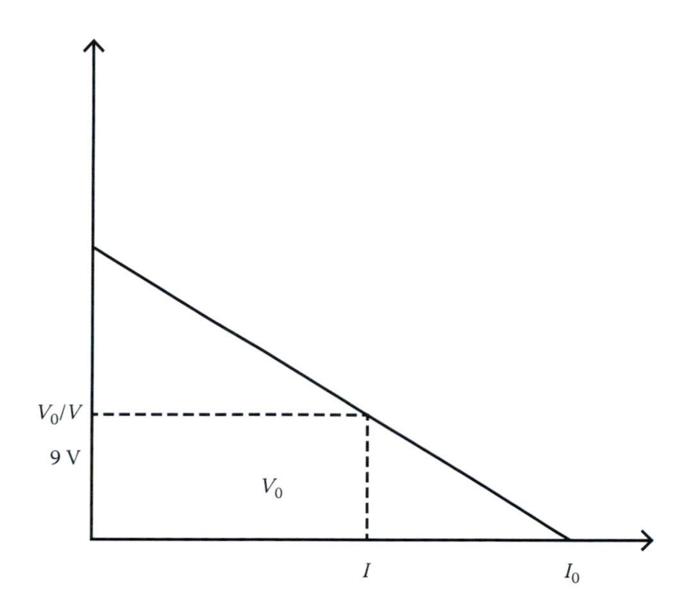

FIGURE 16.5 Trapezoidal output characteristic. (From J. Berge. *Fieldbuses for Process Control: Engineering, Operation and Maintenance.* ISA, USA, p. 106, 2004.)

an output power for Exia IIC. Thus, more devices can be connected per segment compared with the Entity model. FISCO offers the following advantages compared with the Entity model: (a) increased available power; (b) simplification in calculations; and (c) installation parameters are standardized.

FISCO recommends a single power supply per fieldbus segment, and other devices present in the segment are only power drains and no power feedback in the cable is allowed. Power supplies, devices, and cables require FISCO validation and certification, which in 2005 became a standard—IEC 60079-27.

Cable capacitance and inductance are not considered concentrated until they are within specified limits. Cable quality restrictions recommend the permitted FISCO cable parameters, shown in Table 16.3. In such cases, cable lengths up to 1 km with a maximum spur length of 30 m can be used. If shield is used, effective cable capacitance should be calculated.

Some FISCO models offer 1.2-W output power and are effective for devices with lower power ratings. Such FISCO equipment is often referred to as "small FISCO" or "fisco." It must be remembered that in the FISCO model, the barrier, devices, and cables must be FISCO compliant. FISCO restricts the use of a single power supply per network; thus, power supply redundancy is not possible.

TABLE 16.3
FISCO Cable Parameters (Permitted)

R(Loop)	15–150 ohm/km
L	400–1000 μH/km
C	80–200 nF/km

Source: J. Berge. *Fieldbuses for Process Control: Engineering, Operation and Maintenance.* ISA, USA, p. 106, 2004.

FISCO does not require any calculations and thus offers one of the easiest methods for validation of explosion protection. It shifts the onus of electric design from the planner and operator of process plants to the equipment manufacturers. For practical situations involving short cable lengths, four to eight devices can be connected per segment depending on the gas group. On the flip side, FISCO requires expensive power supplies and low MTBF (mean time between failure) due to complicated circuit designs.

The ratings for FISCO power supplies are around 250 mA (group C, D, IIB) and 110 mA (group A, B, IIC).

16.8 REDUNDANT FISCO MODEL

The difference between FISCO and redundant FISCO lies in power supplies for the latter. Two arbitration modules, shown in Figure 16.6, ensure

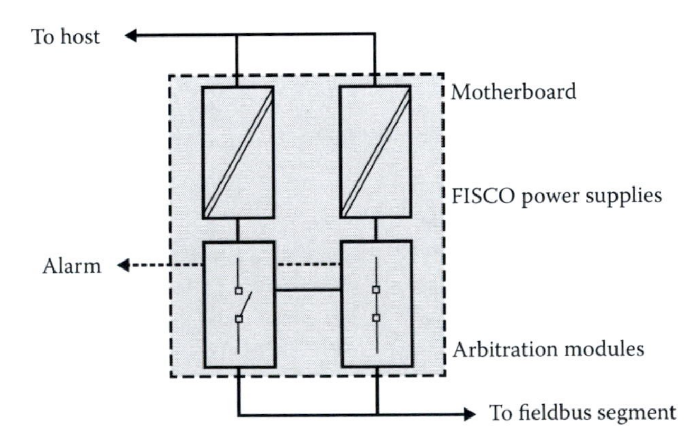

FIGURE 16.6 Redundant FISCO power supply arbitration principle. (From A. Beck and A. Hennecke. *Intrinsically Safe Fieldbus in Hazardous Areas*, Technical White Paper, EDM TDOCT-1548_ENG. Pepperl+Fuchs GmBH, Mannheim, Germany, p. 4, 2008. Available at files.pepperl-fuchs.com/selector_files/navi/productInfo/doct/tdoct1548a_eng.pdf.)

that only one power supply out of the two intrinsically safe power supplies is active at any given instant. The two modules always monitor the output voltage of both the power supplies.

When output voltage of one falls below a specified recommended value, the module ensures the switchover of the power supply to the other one. During the switchover, the bus loses power and the segment voltage drops. As a rule of thumb, a maximum of 100 μs is allowed to avoid field devices resetting due to power loss. However, due to redundancy, communication telegrams are lost due to power drop.

Redundant FISCO, although ensuring power supply to the segment, has drawbacks, such as higher capital cost, more cabinet space, probability of communication errors occurring during redundancy power transfer, and the two arbitration modules requiring their own power putting constraints on cable length and number of devices per segment.

16.9 MULTIDROP FISCO MODEL

A multidrop FISCO power supply configuration is shown in Figure 16.7. In this, several FISCO power supplies are multidropped to different segments—with each power supply connected to a particular segment. In situations where the number of field devices is limited due to intrinsic safety considerations, connecting several segments, as shown, results in utilizing the full capacity of the distributed control system's (DCS's) interface card. Thus, considerable cost savings is effected in the DCS hardware.

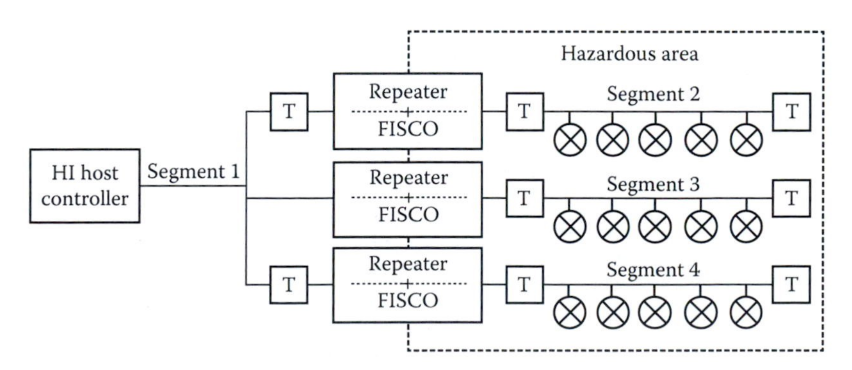

FIGURE 16.7 Multidrop FISCO power supply configuration. (From Fieldbus Wiring Guide, 4th edition, Austin, TX, Doc. No. 501-123, Rev.: E.0. Available at www.relcominc.com/pdf/501-123FieldbusWiringGuide.pdf.)

16.10 HPTC MODEL

HPTC was introduced and developed in 2002 by Pepperl+Fuchs. It removed the limitations with regard to segment length and number of devices. In HPTC, more power is available to the segment, which requires that the trunk cable be protected in the form of armoring or putting the same in a duct—i.e., mechanical methods of explosion protection is effected in the hazardous area. Field devices are connected to the trunk cable via intrinsic safety barriers to reduce power to intrinsically safe levels, as shown in Figure 16.8.

Compared with the other models, standard power supplies can be employed in HPTC. These power supplies are easily available and less costly. The power supplies are employed in redundant configuration. The two power supplies are in parallel and share the load evenly. In case of one power supply failing, the other immediately takes over. Also, since under normal circumstances both the power supplies share the load, their life expectancy obviously increases. The attributes of HPTC are (a) the highest possible overall cable length; (b) redundancy in power supply; (c) easy validation of intrinsic safety requiring no calculation; (d) Entity- and FISCO-compliant devices can be mixed in a segment; (e) field devices can be serviced without "hot work" permit; and (f) standard power supplies can be employed.

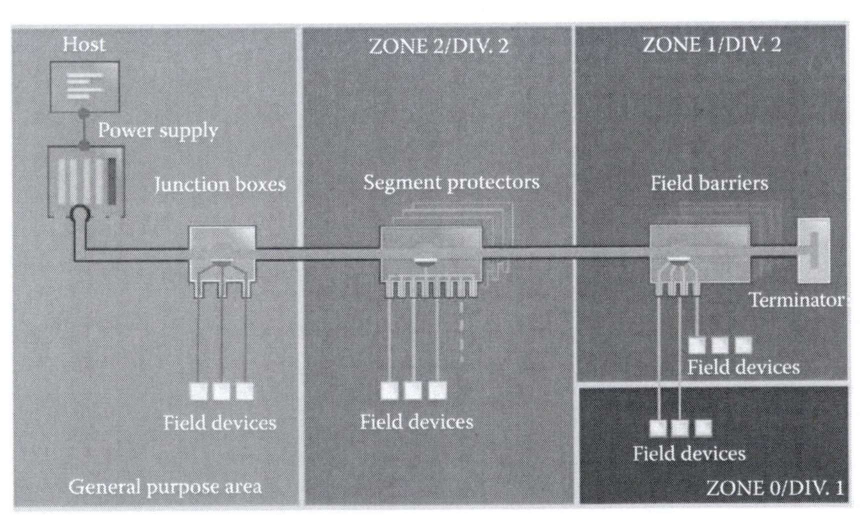

High-power trunk for any hazardous area. Segment protectors provide short-circuit protection and non-incendive energy limitation (Ex nL). Field barriers provide intrinsic safety (Ex i).

FIGURE 16.8 **(See color insert.)** Block diagram of HPTC for hazardous area. (From A. Beck and A. Hennecke. *Intrinsically Safe Fieldbus in Hazardous Areas*, Technical White Paper, EDM TDOCT-1548_ENG. Pepperl+Fuchs GmBH, Mannheim, Germany, p. 5, 2008. Available at files.pepperl-fuchs.com/selector_files/navi/productInfo/doct/tdoct1548a_eng.pdf.)

16.11 DART MODEL

DART attacks the power availability in intrinsic safety applications with a completely new approach compared with its predecessors. DART allows considerably higher available power during normal conditions, while conforming to intrinsically safe energy limitations via rapid disconnection. DART realizes long cable runs and many devices per segment without the need for protected installation, as in the case of HPTC.

Sparks in wiring of electrical systems can cause explosions of hazardous gases. Sparks may occur during "make" or "break" of electrical circuits. Enough energy in the spark may cause the existing gas to attain ignition temperature, leading to explosions. The typical behavior of spark is shown in Figure 16.9, which shows that it remains non-incendive during the initial phase, but reaches a critical phase within several microseconds and then becomes incendive.

All the attributes of HPTC are present in DART. Along with that, the trunk is intrinsically safe, which allows maintenance along the trunk without a hot work permit.

DART does not limit the power during normal operation, as is the concept prevalent in the other models. On the other hand, it detects a fault condition by the characteristic rate of change of current and disconnects the power before it becomes incendive, leading to catastrophic consequences.

A DART-compliant power supply feeds the system with the full 8–50 W, compared with approximately 2 W as in the Entity concept. When a fault occurs, DART detects the resulting rate of change of current, d*i*/d*t* (that

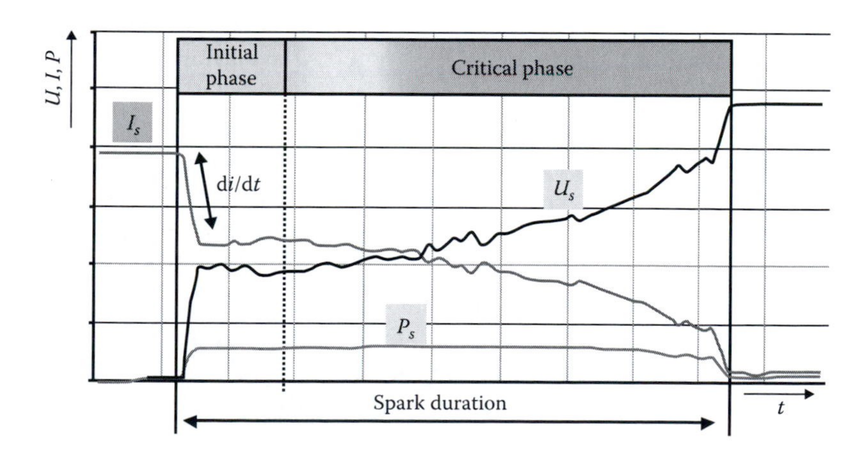

FIGURE 16.9 **(See color insert.)** Electrical behavior of spark. (From A. Beck and A. Hennecke. *Intrinsically Safe Fieldbus in Hazardous Areas*, Technical White Paper, EDM TDOCT-1548_ENG. Pepperl+Fuchs GmbH, Mannheim, Germany, p. 6, 2008. Available at files.pepperl-fuchs.com/selector_files/navi/productInfo/doct/tdoct1548a_eng.pdf.)

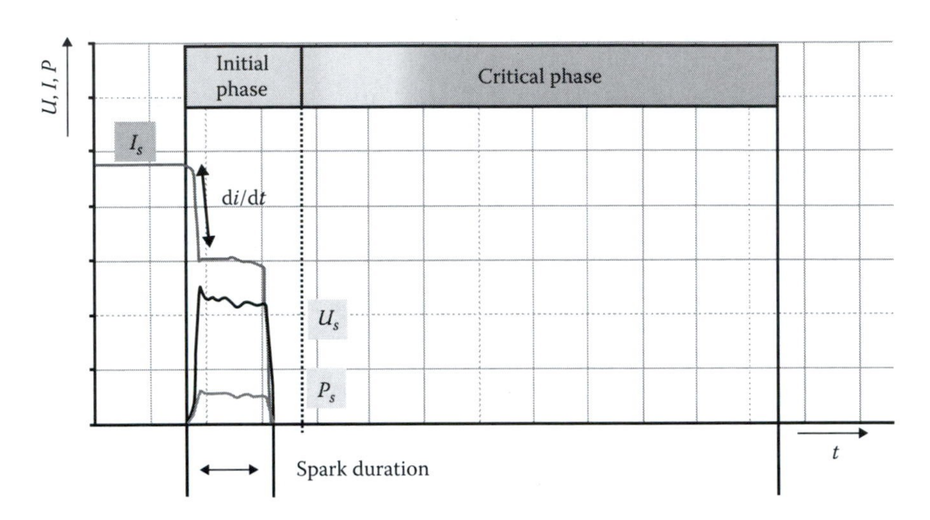

FIGURE 16.10 **(See color insert.)** Highly characteristic d*i*/d*t* associated with a spark. (From A. Beck and A. Hennecke. *Intrinsically Safe Fieldbus in Hazardous Areas*, Technical White Paper, EDM TDOCT-1548_ENG. Pepperl+Fuchs GmBH, Mannheim, Germany, p. 6, 2008. Available at files.pepperl-fuchs.com/selector_files/navi/productInfo/doct/tdoct1548a_eng.pdf.)

occurs in the initial phase of the spark), and switches off the power supply to the system in approximately 5 μs, before the critical phase or incendive condition arises. This is shown in Figure 16.10. Thus, power to the system is reduced to a safe level, spark is robbed of its energy, and the system does not become incendive.

The rate of change of current, d*i*/d*t*, is highly deterministic—thus, a fault condition is easily detected and the system is saved from catching fire.

DART fieldbus is very simple and straightforward, requiring no additional training for field personnel. During the planning phase, a potential user will only have to keep in mind the maximum trunk and spur lengths permissible. DART service personnel need not have to distinguish between "black" and "blue" cables—as is the common practice with field barriers. DART's value proposition lies in its ability to maintain normal power levels to field devices with a single fieldbus topology.

DART is very simple to install and maintain, entailing great savings both in terms of capital and operating costs. DART fieldbus is certified by PTB as per IEC 60079-11 with an IEC-Ex certificate.

DART supports 32 devices per segment, which is the maximum permissible number. This thus reduces excessive infrastructure and the capital cost. Thus, fewer junction boxes, power supplies, barriers, etc., are required to implement a DART fieldbus system, entailing a simpler network topology.

TABLE 16.4
General Performance Summary of Different IS Modules

	Entity	FISCO (Redundant)	High-Power Trunk	DART
Available power	–	0	+	+
Validation of explosion protection	–	+	+	+
Power supply redundancy	–	– (+[a])	+	+
Long-term physical layer diagnostics	–	–	+	+
Segment design mix	–	–	+	+
Cabinet space requirement	–	– (– –)	+	+
Power supply initial cost	–	–	0	0
Trunk live working	+	+	–	+

Source: A. Beck and A. Hennecke. *Intrinsically Safe Fieldbus in Hazardous Areas*, Technical White Paper, EDM TDOCT-1548_ENG. Pepperl+Fuchs GmBH, Mannheim, Germany, 2008. Available at files.pepperl-fuchs.com/selector_files/navi/productInfo/doct/tdoct1548a_eng.pdf.

[a] Currently not certified.

TABLE 16.5
Performance Parameters for Different IS Modules

		FISCO			
Performance Indicator	**Entity**	**IIC**	**IIB**	**High-Power Trunk**	**DART**
Maximum output voltage	10.9 V	14 V	14.8 V	30 V	24 V
Output voltage under load	10.6 V	12.4 V	13.1 V	28.5 V	22 V
Maximum output current	100 mA	120 mA	265 mA	500 mA	360 mA
Effectively available current	55 mA	66 mA	177 mA	360 mA	248 mA
Real-life trunk length	180 m	570 m	290 m	670 m	670 m
(theoretic trunk length)	(1900 m)	(1000 m)	(1900 m)	(1900 m)	(1000 m)
Real-life spur length	30 m	60 m	60 m	100 m	100 m
(theoretic spur length)	(120 m)	(60 m)	(60 m)	(120 m)	(120 m)
Max. no. of field devices	2	4	8	12	12

Source: A. Beck and A. Hennecke. *Intrinsically Safe Fieldbus in Hazardous Areas*, Technical White Paper, EDM TDOCT-1548_ENG. Pepperl+Fuchs GmBH, Mannheim, Germany, 2008. Available at files.pepperl-fuchs.com/selector_files/navi/productInfo/doct/tdoct1548a_eng.pdf.

16.12 PERFORMANCE SUMMARY

In DART, Table 16.4 shows the basic merits and demerits of the four mostly used intrinsically safe fieldbus systems, viz., Entity, Redundant FISCO, HPTC, and DART. It can be concluded from the table that both HPTC and DART have almost identical performance levels, but live working on the trunk is possible only in DART.

Again, Table 16.5 shows the effectiveness of the above four intrinsic barrier fieldbus systems based on several performance indicators.

It is seen from the table that the available current in case of HPTC is greater compared with DART, while on other counts, they are almost identical. However, with the gradual acceptance of DART and its ability for live working on the trunk, DART is increasingly being seen to be the frontrunner for solutions to intrinsic barrier fieldbus systems.

16.13 CONCLUSION

Several measures, such as intrinsic safety and explosion-proof enclosures, are employed for the safety of plant and personnel in hazardous industries. Uninterrupted operation of fieldbus-based instrumentation systems is a must to keep the downtime to a minimum and enhance profitability. Over the years, different types of intrinsically safe fieldbus systems have been introduced, the latest one being DART. HPTC and DART offer almost comparable benefits to the users of fieldbus systems. DART offers around 8 W per segment, which is about four times more than that provided by FISCO. Since live working on the trunk is possible only with DART, coupled with its ability to provide higher levels of available power, it is increasingly becoming the most sought-after technology where intrinsic safety is a must for proper fieldbus system operation.

17 Wiring, Installation, and Commissioning

17.1 INTRODUCTION

The field devices are to be installed, wired together, and commissioned properly before they are put into operation. Electrical installation wise, Foundation Fieldbus and PROFIBUS-PA are identical since they conform to IEC 61158-2. Again, Highway Addressable Remote Communication (HART), Foundation Fieldbus, and PROFIBUS-PA use existing wires for their proper operation. HART devices are discussed separately since their installations differ on several counts.

It is recommended to consult the manufacturer's data sheet before installation and to perform commissioning of any fieldbus system. Since fire hazards in fieldbus systems are a major area of concern in certain plant operations, it has been discussed separately in Chapter 16.

17.2 HART WIRING

HART is an open access process control network protocol and supported by HART Communication Foundation. In HART, 4–20 mA current signals carry the process measurement and control information, while other process parameters, calibration, device configuration, and diagnostic information are transmitted as the HART signal over the same pair of wires and at the same time. The HART protocol uses the Frequency Shift Keying (FSK) technique to superimpose the digital signal on the conventional 4–20 mA current loop, and hence sometimes termed as a hybrid communication. A 0.5 mA sine wave of two frequencies, viz., 1200 Hz and 2200 Hz, representing binary 1 and 0, respectively, are superposed on the 4–20 mA DC signals. Thus, it adds no DC component as the average value of the sine wave over a complete cycle is zero. Digital data is transferred at a rate of 1200 bps.

HART uses different topologies for connecting the field devices to the controllers. It can be point-to-point, multiplexed, or multidrop types, shown in (a), (b), or (c) of Figure 17.1. Irrespective of topology, an FSK modem is used at the output of a field device and also at the input to the computer, which may act as a host or a primary master.

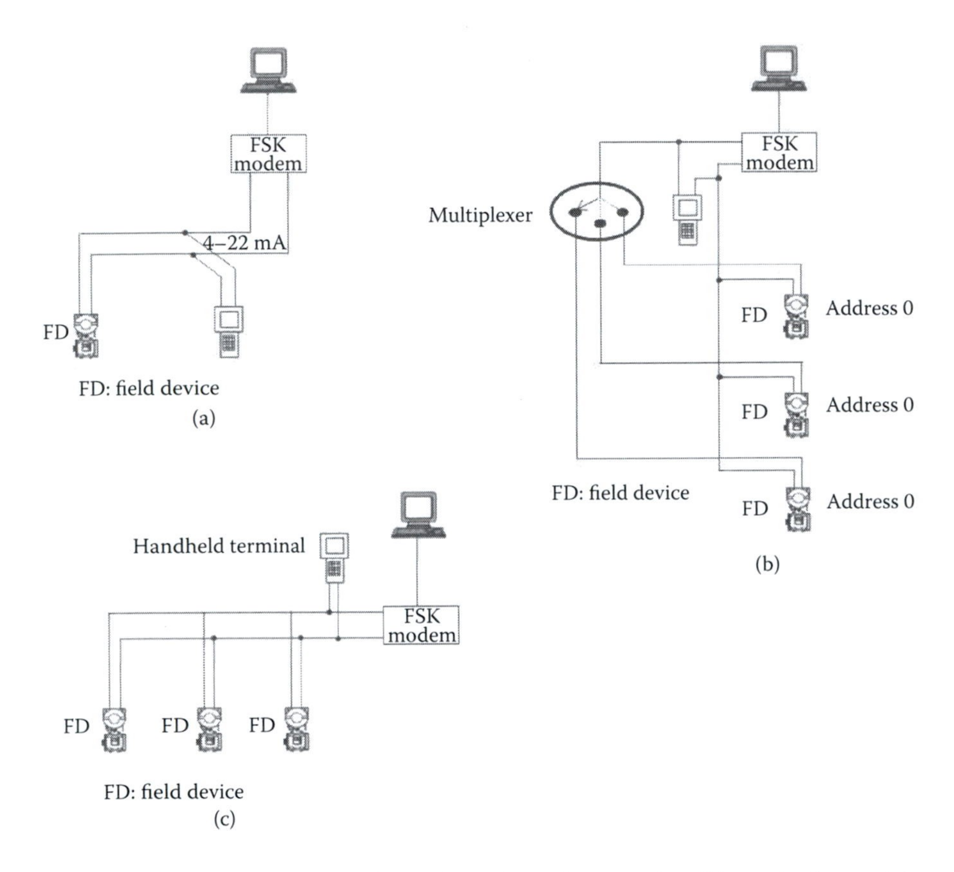

FIGURE 17.1 HART network topology: (a) point-to-point; (b) employing multiplexers; (c) multidrop mode.

The HART protocol uses master–slave communication. In this, a master initiates communication and the slave merely responds to it. The masters can be a PC-based network controller, handheld terminal, distributed control system (DCS), or programmable logic controller. There can be two masters in a HART network at the same time. A remote terminal unit or a multiplexer can act as an interface for the primary master. A handheld terminal is a secondary master and is connected only when working in the field. Sensors, actuators, and transmitters act as slave devices.

A multiplexer allows several fieldbus devices to be connected to the host one after the other, with all the field devices having the same address. The multiplexer continuously scans the process variables as per the written program and also receives their status information. They are then put into the host for processing. Several multiplexers can be cascaded so that many devices can be accessed by the host—all the connected devices having the same address.

An important functionality of a multiplexer is its *pass through* capability. It allows device data and field status information to be directly communicated between host the field device. Thus, data and status information from the HART network can be multiplexed on to a single network—such as PROFIBUS-DP and MODBUS—to be connected to the host controller. Thus, a multiplexer acts as a gateway.

17.2.1 Surge Protection

In industries where large motors are occasionally started and stopped, and places known for recurrent lightning strikes (such as the top of high towers or remote locations like pumping stations and well sites), induced voltage is of serious concern and must be addressed. Fieldbus systems installed in those areas must be protected from such lightning strikes.

Surge voltages are in the order of tens of KVs and last only a fraction of a second. A low-capacitance device is installed at the device that acts as a surge suppressor. This low-capacitive device acts as an open circuit to the segments and the spurs and, hence, does not affect communication. Surge protection methods and the corresponding equipment employed for their prevention are identical for HART and Fieldbus Foundation. A direct lightning strike to equipment is quite rare, but offers challenges when it strikes the ground. In such situations, a current of the order of kAs flows through the ground. This poses no problem in a confined area, but of serious concern when two locations are physically separated because of large ground potential differences between them.

Transmitters that are prone to higher surge strikes should be placed in separate segments with no final control elements. In the event of a surge strike and the transmitter getting affected, the consequent data loss will not affect the system much more than a process upset resulting from both the transmitter and the final control element going offline due to such a surge strike.

When surge protectors are to be used, they must be installed at each field device of a segment and at the host. Also, the number of field devices per segment should be kept at a low value so that in the event of a strike, the consequent process upset can be kept to a minimum. It should also be kept in mind that installation of surge suppressors must not attenuate the original fieldbus signal. Surge protectors are normally in the deactivated condition but get activated when surge current exceeds a certain threshold. The extra current flows into the earth—thus, a solid earth connection is a must. A surge protection method is shown in Figure 17.2, which shows employing the protectors at either end of the system.

Another important issue is the use of current-limiting couplers in conjunction with surge suppressors. In the event of a surge strike, the

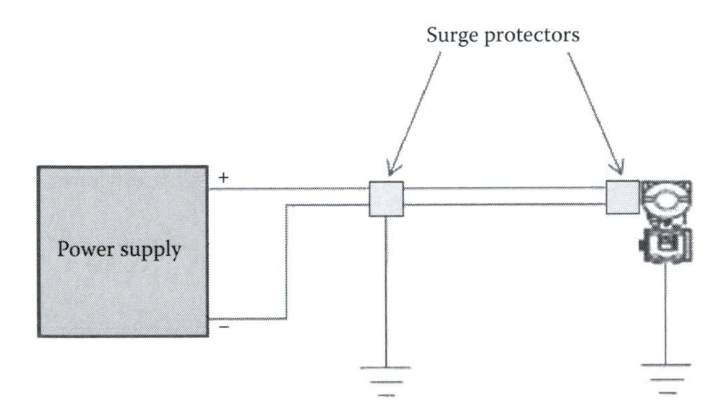

FIGURE 17.2 Surge protectors installed at either end of a network. (From J. Berge. *Fieldbuses for Process Control: Engineering, Operation and Maintenance.* ISA, Research Triangle Park, NC, p. 72, 2004.)

suppressor would cause failure of the coupler, resulting in data loss and control concerning that segment.

17.2.2 Device Commissioning

Two addresses identify a HART device: a unique hardware address and a polling address. The unique hardware address consists of the manufacturer's code, the device type code, and a unique identifier. The combination of these three codes ensures that no two hardware addresses are same. This eliminates any chances of conflict. The hardware address can never be changed.

The polling address is set to zero for point-to-point mode, while it is from 1 to 15 for multidrop mode. For the former, the transmitter is 4–20 mA, while for the latter it is fixed at 4 mA. For a multidropping case, each field device is to be configured separately with individual polling addresses before they are connected to the network. This address assigning is done for each device by connecting them point to point with the help of a handheld terminal. Communication is not possible in case duplication of address is inadvertently done even for two devices.

17.3 BUILDING A FIELDBUS NETWORK

Fieldbus uses a pair of wires to power the field devices and also carry the process signals to the local controller. Different types of topologies are employed to carry the bidirectional (from devices to host and vice versa) digital signals in a fieldbus system, which constitutes a local area network (LAN). Among the different fieldbuses available for different uses, Foundation Fieldbus and PROFIBUS-PA are the mostly used ones for process control purposes.

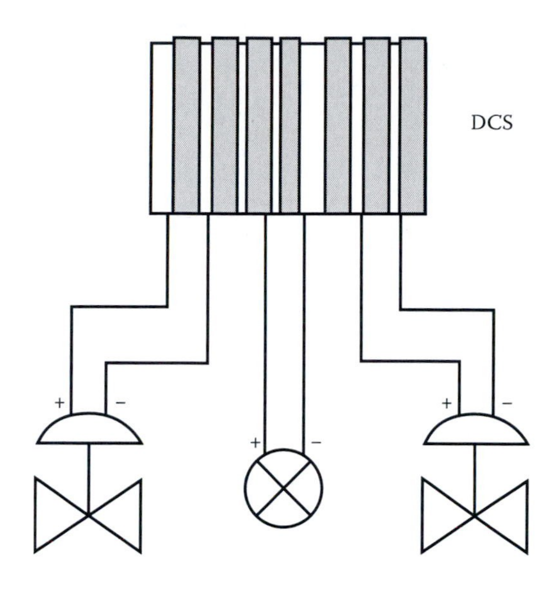

FIGURE 17.3 Conventional DCS system. (From Fieldbus Wiring Guide, 4th edition, Austin, TX, Doc. No. 501-123, Rev. : E.0. Available at www.relcominc.com/pdf/501-123FieldbusWiringGuide.pdf.)

They share identical networking and powering schemes, although differing significantly in communication protocol strategy.

For a conventional DCS, process parameters such as temperature, pressure from sensors, and control signals to actuators, valves, etc., are carried via a shielded twisted pair cable in the form of 4–20 mA current signals via I/O cards. Figure 17.3 shows several field devices connected to a DCS system via I/O cards.

The wiring diagram for a fieldbus system is shown in Figure 17.4. The upper part of the diagram shows several field devices, connected by two wires, while the lower part of the diagram shows the wiring in a simplified manner.

Figure 17.5 shows a single fieldbus device connected to a fieldbus network. Two terminators, one at each end of the wire pair, are connected.

17.3.1 Multifieldbus Devices

Figures 17.6 and 17.7 show a number of field devices added to the network with "star" connection for the former and "chained" network for the latter. In both Figures 17.6 and 17.7, only two terminators are attached, identical to Figure 17.5.

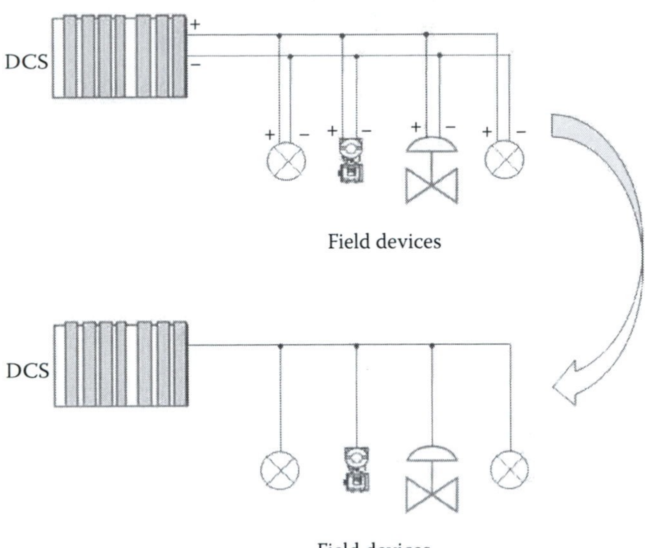

FIGURE 17.4 Fieldbus wiring diagram.

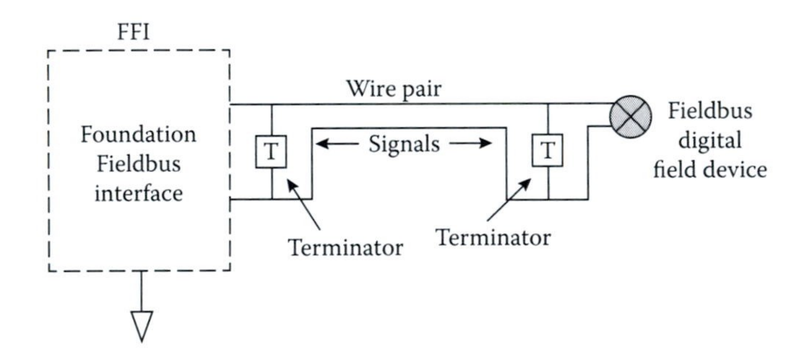

FIGURE 17.5 Fieldbus network with a single device.

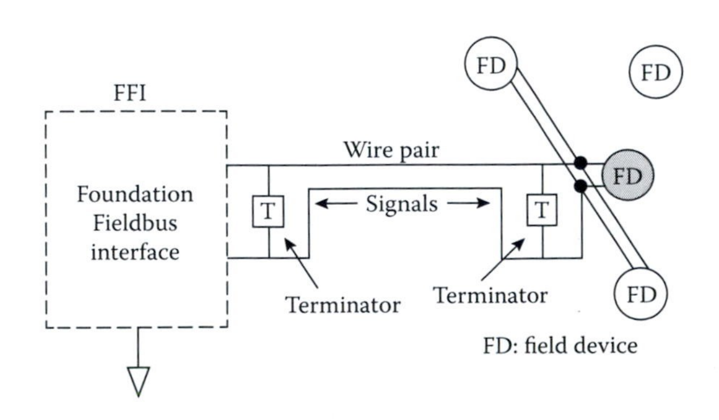

FIGURE 17.6 Star-connected fieldbus network.

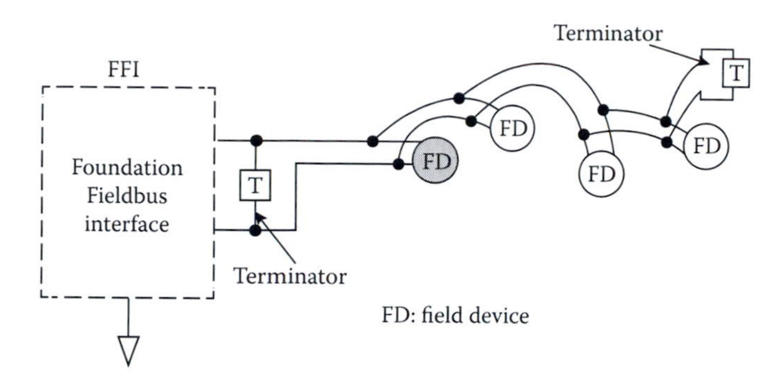

FIGURE 17.7 Chained fieldbus network.

In Figure 17.6, four field devices are joined at the junction with the terminator placed at the junction itself. It is tacitly assumed that all four devices are about the same distance from the junction. Had one device been placed at a larger distance from the junction, then the terminator would have been shifted to the end of that device. In Figure 17.7, as the devices are continued to be added to the chain, the terminator is shifted and connected across the last field device.

17.3.2 Expanding the Network

It is necessary to extend the reach of a network when the plant to be controlled is physically spread over a considerable area. In such cases, network traffic regulators (also called network devices) such as hubs, repeaters, switches, bridges, routers, and gateways are used to realize the above.

For the traffic to flow in an orderly manner in a network, the system follows some protocols comprising hardware and software. Network interface cards (NICs), hubs, switches, bridges, and routers help in efficient communication over the network. These devices are called network hardware devices and they perform different jobs. First, they help in regulating the speed on the network. If the hub operates at 10 Mbps, then the other network devices would operate at the same speed. Second, these devices manage the flow of traffic by opening, closing, or directing the traffic along a specific route. For example, a router ensures the most efficient way for the traffic to reach its destination. Third, these devices help in protecting some sensitive devices as well.

Network communication takes place in layered fashion. The OSI model has seven layers, with each layer performing a specific networking function. Protocols govern these functions and manage the end-to-end communication between devices. As user data is passed from the upper layers down below, each layer adds a header (and sometimes also a trailer) to the data. These headers contain protocol information and are termed protocol data units (PDUs). The process of adding headers to the layers is known as encapsulation.

17.3.2.1 NICs

These are also known as adapters or network interfaces or simply cards. NIC is a device that connects a server, a client computer, printer, or any other component to the network. It is connected to the network via a small receptacle, called a port. For wired systems, the network cable is inserted into this port. For wireless networks, the port includes a transmitter/receiver that sends/receives the radio signals. NIC services include the connection of the computer physically to the network and, second, it converts information on the computer to and from electrical signals for the network. For example, the information on the computer is converted into electrical signals of appropriate shape and speed compatible to the network to which it is connected.

The NICs connected to the network must conform to the physical and data link level protocols for the electrical signals to be compatible and information exchange to take place effectively and successfully. If the Ethernet network runs at 10 Mbps, then the servers and printers attached to the network must have 10 Mbps compatible NICs. Each NIC connected to the network must have a unique media access control (MAC) address that helps routing information within the LAN.

17.3.2.2 Hubs

A hub is a layer 1 device and physically connects network devices together. Hubs do not have any intelligence and thus a hub forwards every frame out every port, only excluding the port from which the frame is originated. Hubs cannot differentiate between frame types and would thus forward unicast, multicasts, or broadcasts out every port, except the originating one. Hubs are unable to process layer 2 or 3 information and thus cannot make any decision based on hardware or logical addressing.

Ethernet hubs operate in half duplex mode. Ethernet employs carrier sense multiple access with collision detection to access a medium before sending a frame and transmits the same only if the link is idle. If two devices send their frames at the same time, a collision will occur and both the frames are rejected after informing the devices that sent the frames. When more than one station want to send a frame, a back-off algorithm sees to it that a single station wins the bus arbitration. Thus, this station gains sole access of the network and transmits a frame without any kind of error. All ports belonging to the same hub belong to the same collision domain. Hubs also belong to the same broadcast domain.

All devices (clients, servers, peripherals, etc.) connected to a hub share the bandwidth of the network and belong to the same collision domain.

As more devices are added to a hub, the probability of collision increases, thereby degrading performance. The probability of collision increases with increase in network traffic.

Hubs can also be divided into two types: passive and active. In passive hubs, the signal is forwarded as it is, while in active hubs, the signal is amplified before sending. Thus, an active hub behaves as a repeater. Hubs can also be cascaded to increase the network capacity.

17.3.2.3 Repeaters

The physical length of a segment should be limited to a length of approximately 1900 m due to attenuation along the length of the wires. Repeaters are used to increase the length of a network, and it also increases the number of devices. A repeater refreshes the timing of a signal and boosts its amplitude to the original level at its input. A fieldbus controller chip does the above jobs. A repeater is bidirectional in nature and galvanically isolated. A maximum of four repeaters can be employed on a single network, thereby increasing the length of the network to 9500 m—thus having five segments. Schematically, it is shown in Figure 17.8.

It makes economic sense to employ repeaters to connect intrinsically safe segments, thereby forming a larger network with a corresponding increase in device count, rather than having a single safety barrier per host communication port. Some typical field instruments consume a considerable amount of current, thereby limiting the number of such devices to two or three per segment. In a typical connection, a safe area segment can be connected to the host via a linker on one side, while several multidropped barriers with repeaters attached can be connected at the other side. Via each barrier, a hazardous area segment is connected and shown in Figure 17.9.

A segment has two terminators, placed at each end of the segment. Repeaters with built-in terminators are available such that external terminators are dispensed with. Care should be exerted when using repeaters—they should be disabled or enabled as per the requirements. The type of repeater to be used depends on the network type used. Thus, a repeater conforming to IEC 61158-2 must be used for proper system operation.

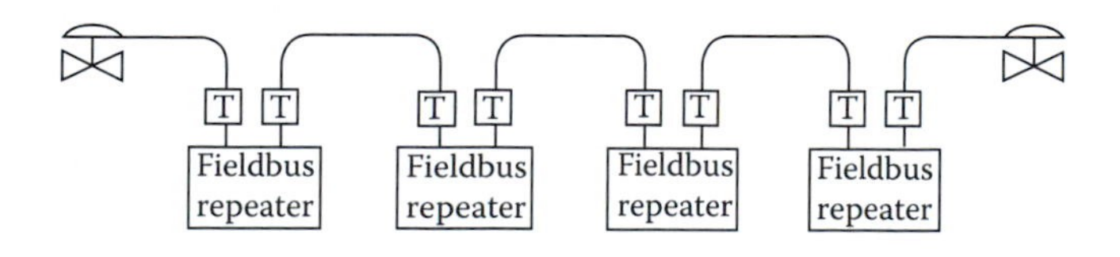

• Power supplies not shown

FIGURE 17.8 Repeaters in a fieldbus network.

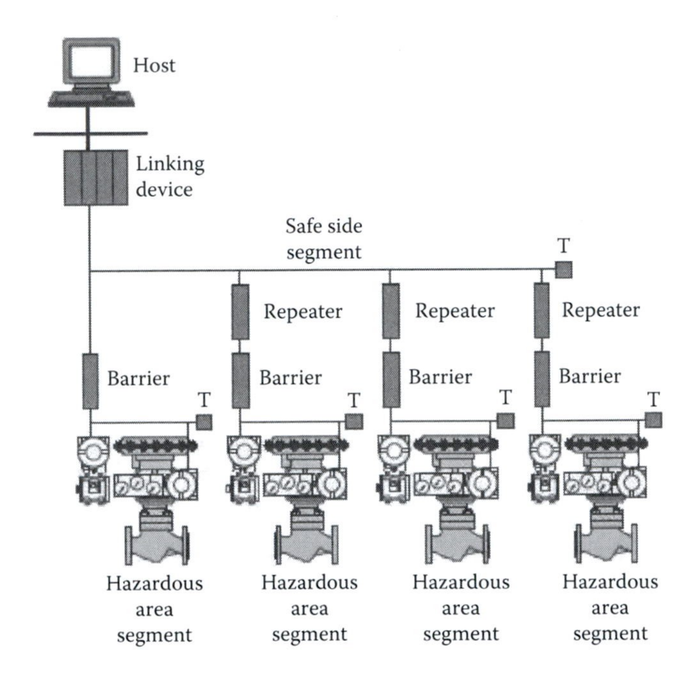

FIGURE 17.9 Large network formed with repeaters and barriers. (From B. G. Liptak. *Instrument Engineers' Handbook—Process Software and Digital Networks*, 3rd Edition. CRC Press, Boca Raton, FL, p. 166, 2002.)

Repeaters can be active bus-powered or non-bus-powered devices. Device count can be increased to 240 by using repeaters. Repeaters operate at the physical layer of OSI and extend the physical length of a LAN. A repeater is a two-port node—when it receives a frame from any port, it regenerates the original bit pattern and simply forwards it to the other port. A repeater is a regenerator of data and not an amplifier.

17.3.2.4 Switches

A switch is a layer 2 (data link layer) device and makes intelligent forwarding decisions based on the header information of this layer. It derives this intelligent decision based on special hardware circuits, known as Application Specific Integrated Circuits. Only then the frame is forwarded to the appropriate destination port and not to all ports.

A switch builds a hardware address table, which contains a hardware address for the host devices and a port address that each device is attached to. Ethernet switches build MAC address tables following a dynamic learning process methodology. When a switch is powered the first time, it behaves very much like a hub. It then floods every frame, including unicasts, out every port sans the originating one. The switch will start building the MAC address table by examining the source MAC address of each frame. As the MAC address table becomes more and more populated with time, the flooding of frames would continue to decrease and the switch becomes more and more efficient.

A drawback associated with switches, like hubs, is its susceptibility to switching loops. This generates a destructive broadcast storm. Switches avoid this by a spanning tree protocol. Thus, a switch has the following advantages that a hub does not have: loop avoidance, intelligent frame forwarding, and hardware address learning.

A switch operates in full duplex mode. Each individual port on a switch belongs to its own collision domain and hence a switch creates more collision domains resulting in fewer collisions. A layer 2 switch belongs to one broadcast domain, while a layer 3 device separates broadcast domains. A switch cannot differentiate one network from another but can discriminate one host from another. Thus, switches are not suitable for large networks, while a router, belonging to layer 3, is capable of.

A switch has the capability to momentarily connect the sending and receiving devices so that they can use the entire bandwidth of the network without interference. The use of a switch does not result in any benefit when the network traffic is poor. This is because a switch examines the information inside each signal on the network to determine the address of the sender and the receiver before sending the frame. This results in slow processing of signals and lead to latency.

A switch forwards frames using any one of the three methods: The Store and Forward, The Cut Through (Real Time), and The Fragment Free (Modified Cut Through).

17.3.2.5 Bridges

A bridge is an active bus-powered or non-bus-powered device. It connects two or more LANs or two or more segments of the same network. It enables data from one network to be transferred to another having different speeds or physical layers—like wires and optical fiber. Thus, a bridge enables different networks to talk to each other. Devices residing on different segments but on the same network must have different node addresses. However, two devices on different networks may have the same node address. It is the responsibility of the bridges to ensure that devices are not mixed up. Bridges are normally built into a linking device or an interface card. Figure 17.10 shows bridges talking to each other.

Bridges are simple and efficient traffic regulators. Bridges, like switches, filter network traffic before forwarding them. Thus, overall traffic congestion is avoided. Again, like switches, bridges learn the MAC address of the

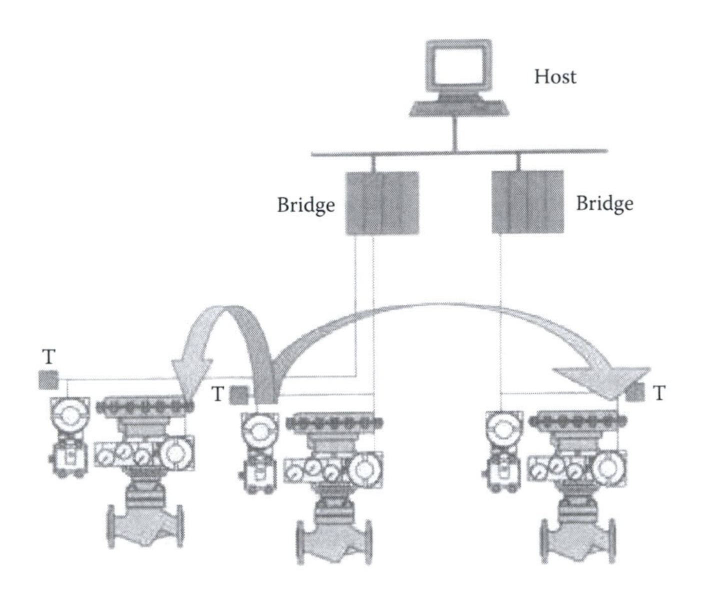

FIGURE 17.10 Different networks can talk to each other via bridges. (From J. Berge. *Fieldbuses for Process Control: Engineering, Operation and Maintenance*. ISA, Research Triangle Park, NC, p. 99, 2004.)

connected servers, peripherals, clients, and associate each address with a bridge port. Bridges are layer 2 devices.

17.3.2.6 Routers

Routers are layer 3 devices and forward a packet from one network to another, based on the contents of the network layer header. This forwarding is based on the destination network and not on the destination host. The forwarding path is based on a routing table that contains the following: routing metrics and administrative distance, the destination network and the subnet mask, and, third, the next hop the router to follow to reach the destination network.

Since each individual interface on a router belongs to its own collision domain, routers create more collision domains resulting in fewer collisions. Also, routers separate broadcast domains. A router never forwards broadcasts and usually will not forward multicasts.

A router normally copies packets in its buffers and checks the router table lookup before forwarding the same. It thus results in latency. Nowadays, improved router versions use hardware to offset the latency effect.

The routing table has two types of protocol: the routed protocol and the routing protocol. The first assigns logical addressing to devices and route packets between networks, which includes IP and IPX. The second one dynamically builds information in routing tables, which includes RIP,

EIGRP, and OSPF. A router keeps information about other routers in a routing table. This routing table is found in all routers connected to the network.

Routing, in a way, is similar to a switch, but a router is not a switch. A switch connects two computers to form a LAN, while routers connect two LANs, two wide area networks (WANs), or a LAN and its own ISP network. Routers filter network traffic so that any unauthorized entry is denied. Routers ensure the best route that a packet is to take to reach the destination by avoiding congested routes. A router ensures that (a) traffic meant for a local network does not flood the Internet and (b) a traffic existing on the Internet that is not meant for the local network stays on the Internet and does not intrude the local network. When a router receives a packet, it notices its hardware MAC address. If this address is not on a local segment, then this address is stripped by the router and it looks for the software IP address in its routing table. On the basis of this comparison, the packet is then sent to the router that contains that address.

17.3.2.7 Gateways

A gateway performs protocol conversion and, therefore, networks with different protocols can be connected together via a gateway. Gateways are protocol specific and, thus, proper gateways must be employed for any message to be transported effectively from one network to another. As an example, a data from a Foundation Fieldbus HSE network can be transported to a PROFIBUS network provided a gateway is employed, which converts a protocol from the former system to the latter.

A gateway is manually configured by mapping the device address, data type, or any other kind of information in the foreign protocol to the desired information format for every parameter that is to be sent to the foreign network.

17.3.2.8 Routers vs. Gateways

A router operates at the network layer (layer 3), while a gateway operates at the transport layer (layer 4). Both are used to regulate traffic on two or more separate networks. Routers regulate traffic between similar networks (protocol dependent), while a gateway regulate traffic between dissimilar networks (protocol independent), i.e., networks having different protocols. A local network using Windows 2000 can be connected to the Internet via the router because both use TCP/IP as their primary protocol. On the other hand, a device that uses a Windows NT network can be connected to the NetWare network via a gateway.

While routers are configured by their routing tables, gateways are configured by internal and external networks. Gateways act as a network point, which acts as an entry point to another network. Router is a device that guides the data packet to its proper destination that arrives at the gateway. A gateway provides the entry/exit point to/from a network like the door of a house.

17.4 POWERING FIELDBUS DEVICES

Fieldbus networks carry both DC power supply to power the field devices and a communication signal at 31.25 kHz from the devices. The two power supply lines carry both DC superimposed with AC. A normal power supply having both voltage regulation and noise reduction circuitry would not serve the purpose here, as such power supplies, having almost zero output impedance, would short the 31.25 kHz communication signal. The power supply must have output impedance conforming to IEC 61158-2 such that the communication signal passes through the two lines without attenuation and preventing a short circuit. On the other hand, this output impedance of the power supply would offer almost zero resistance to DC and, hence, the current drawn by the devices would cause little drop on the network. Thus, either an IEC 61158-2-compliant power supply should be used or an impedance module compliant to IEC 61158-2 should be inserted between the power supply and the network.

Figure 17.11 shows a normal power supply and an IEC 61158-2-compliant impedance module connected to the network. A terminator is attached to the impedance module and another at the far end of the cluster of devices.

Redundant power supplies should be employed to ensure higher system availability. In such cases, it should be ensured that in case of failure of primary power supply, there is bumpless switchover to the secondary power

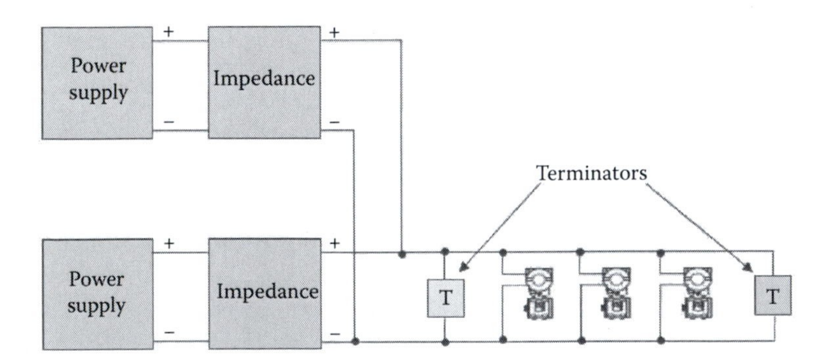

FIGURE 17.11 Conventional power supply along with an impedance module. (From J. Berge. *Fieldbuses for Process Control: Engineering, Operation and Maintenance*. ISA, Research Triangle Park, NC, p. 100, 2004.)

supply. The power supplies must be galvanically isolated, be short circuit protected, and have a failure indication.

The 250-ohm resistor employed in HART plays an identical part as the impedance module here. This resistance prevents the power supply from short circuiting the HART signal operating at 1200–2200 Hz. At the same time, this resistance acts as a shunt, which converts the modulating current into a voltage that can be picked up by the receiving devices.

Bus-powered field devices belonging to Foundation Fieldbus and PROFIBUS-PA operate at voltages between 9 and 32 V DC. Typically, the power supply voltage output is around 24 V DC. Most field devices consume around 12 mA, while some devices consume twice as much. Fieldbus power supplies provide a load of around 300 mA.

17.5 SHIELDING

Fieldbus instruments are placed considerable distances apart even for a mid-sized plant. Because of the constant operation of heavy-duty motors and other electrical appliances, noise of varying degree is always present along the communication paths. The network cable paths therefore must be shielded from such stray electromagnetic interferences. The cables may also be placed in a metal conduit, which also provides for mechanical protection against damage. The spurs and the trunks should be shielded and connected to the negative of the power supply, shown in Figure 17.12, and not at the negative of the impedance module.

Signal conductor lines should not be grounded because it may result in loss of communication. Usually, shields are connected to ground at one point. If multiple grounding is required because of existence of potential differences along the ground path, capacitive shielding is preferred. In this case, grounding is made through a capacitor. Thus, only high frequencies in the noise get grounded. It is recommended that a shield should never be used as a conductor. The communication ports of fieldbus devices are all galvanically isolated.

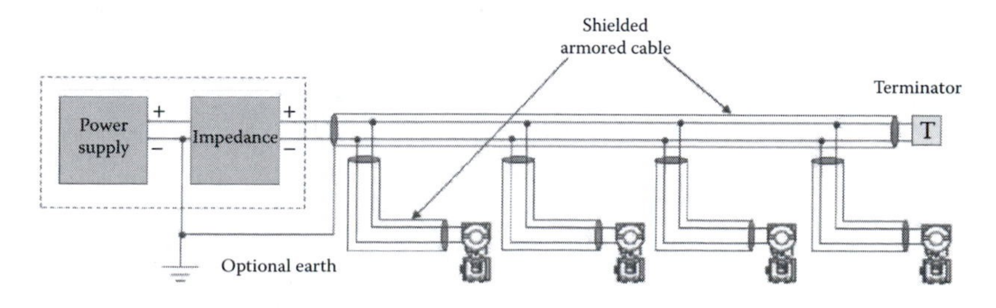

FIGURE 17.12 Shielding of spurs and trunks.

17.6 CABLES

Fieldbus signal, as it travels down via spurs and trunks, is attenuated until it becomes too weak to be recognized properly. Thus, it puts a restriction on the maximum cable run, which depends on the type of cable employed. Typical cable lengths and their types are specified in IEC 61158-2 to maximize backward compatibility of the existing cable infrastructure, and shown in Table 17.1. The specifications also support optical fiber.

Thus, type A cable supports 1900 m of total cable run, while it is only 200 m for type D. Types C and D are mainly used for retrofit applications and they have limited applications because of restricted cable length usage.

Multi-twisted-pair cables are employed in large projects, which helps reduce cable installation cost. This type of cable supports multifieldbus trunks. This is shown in Figure 17.13, where the multipair cable runs from the control room to a strategically placed junction box. From this, individual trunks run to junction boxes that are placed in the vicinity of field devices. This is facilitated by using a device coupler in the junction box that connects the spurs to the field devices.

TABLE 17.1
Typical Cable Lengths and Their Types

Types	Pair	Shield	Twisted	Size	Total Length
A	Single	Yes	Yes	0.8 mm^2 (AWG 18)	1900 m (6200 ft)
B	Multi	Yes	Yes	0.32 mm^2 (AWG 22)	1200 m (3900 ft)
C	Multi	No	Yes	0.13 mm^2 (AWG 26)	400 m (1300 ft)
D	Multi	Yes	No	1.25 mm^2 (AWG 16)	200 m (6500 ft)

Source: J. Berge. *Fieldbuses for Process Control: Engineering, Operation and Maintenance.* ISA, Research Triangle Park, NC, p. 87, 2004.

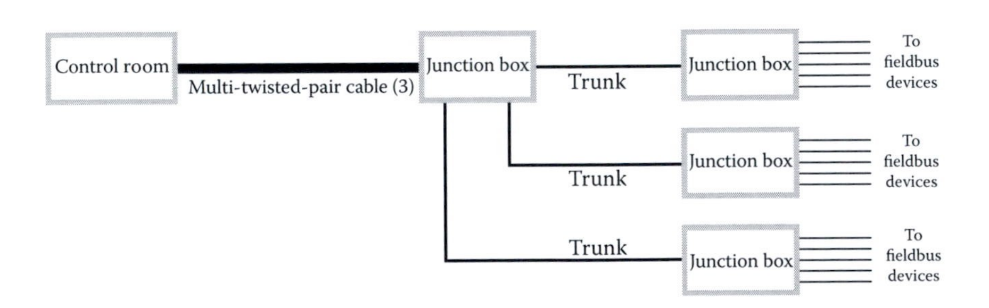

FIGURE 17.13 Use of multi-twisted-pair cables. (From Fieldbus Wiring Guide, 4th edition, Austin, TX, Doc. No. 501-123, Rev. : E.0. Available at www.relcominc.com/pdf/501-123FieldbusWiringGuide.pdf.)

Cable length restrictions are imposed by device power consumption considerations as also intrinsic safety. In such cases, the cable lengths become even shorter.

Fieldbus was originally designed such that existing cables in a plant can be utilized for the same. However, it is recommended that for new plant installations, the specifications as detailed in Table 22.1 must be adhered to. The total length of a cable (this includes the lengths of home run cable, the trunk or trunks, and the spurs) must be less than as envisaged in the table. When two or more types of cables are mixed for a segment, the total cable length is determined as per the following:

$$(L_x/L_{\max x}) + (L_y/L_{\max y}) \leq 1$$

where L_x = length of cable x, L_y = length of cable y, $L_{\max x}$ = maximum length of cable type x alone, and $L_{\max y}$ = maximum length of cable type y alone.

The cable lengths shown in Table 17.1 are for a segment only. If more than the recommended cable length is required in a particular case, then multisegments are used, with each individual segment being connected by repeaters.

In fieldbus systems, bus-powered devices are mostly used. The field devices draw current, and there is voltage drop along the wires. This limits the wire length that can be employed in any particular case because the devices cannot operate below a 9 V voltage level. For maximum cable length and the number of devices, the supply voltage should be as high as possible, device current consumption as low as possible, and wire cross section as high as possible to reduce resistance along the wires, thereby entailing lower voltage drops. The maximum cable distance is calculated on the basis of Ohm's law.

Cable length reduces with lower output supply voltage or higher device current consumption. Devices that are separately powered still draw some current from the fieldbus network power supply.

17.7 NUMBER OF SPURS AND DEVICES PER SEGMENT

The short cable length that branches out from the main trunk and connects to a device is called a spur. Cable lengths lesser than 1 m are termed a splice. Spur length is independent of the type (i.e., quality of cable) of cable, but depends on the number of devices connected to it and the topology employed. The total spur length is determined by the number of spurs and the number of devices per spur. Table 17.2 shows the spur network design involving total devices per network, number of devices per spur, and the maximum spur length.

TABLE 17.2
Relationship between Number of Devices per Spur and Maximum Spur Length

Total Devices per Network	One Device per Spur, m (ft)	Two Devices per Spur, m (ft)	Three Devices per Spur, m (ft)	Four Devices per Spur, m (ft)	Total Max. Spur Length, m (ft)
1–12	120 (394)	90 (295)	60 (197)	30 (98)	1440 (439)
13–14	90 (295)	60 (197)	30 (98)	1 (3)	1260 (384)
15–18	60 (197)	30 (98)	1 (3)	1 (3)	1080 (329)
19–24	30 (98)	1 (3)	1 (3)	1 (3)	720 (220)
25–32	1 (3)	1 (3)	1 (3)	1 (3)	32 (10)

A spur length can vary between 1 and 120 m, although normal spur length should be limited to within 10 m. Spur lengths beyond 10 m, if needed, should be kept outside high-risk areas. Special attention should be placed for tree topology, because in this case each branch is a spur. If certain spur lengths are appreciably high, the junction box should judiciously be placed so that any spur length is limited to within 120 m. For intrinsically safe installations, the spur length is limited to 30 m.

For tree topology, each device connection from the junction box is a spur. In such a case, the junction box is placed in such a manner that the recommended maximum spur length is not violated. In Figure 17.14, the network employs a trunk length of 700 m and total spur length of 52 m, making the segment equal to 752 m in length.

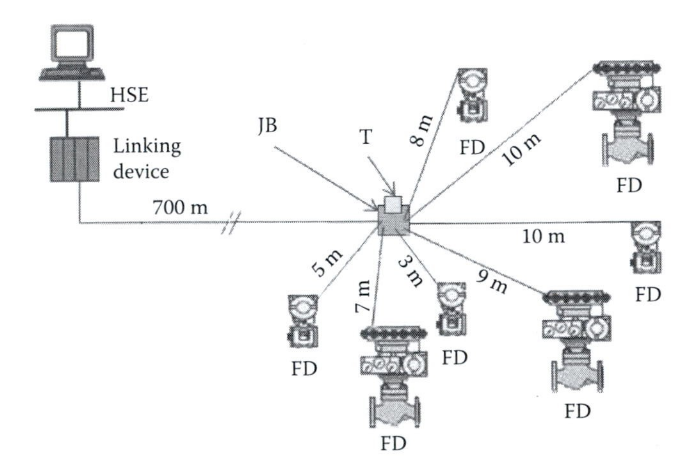

FIGURE 17.14 Tree-connected network showing spurs of different lengths.

17.8 POLARITY

In powered fieldbus networks, the alternating voltage from the field devices is superimposed on the DC voltage, while if the network is not powered only the alternating voltage from the field devices exists. The Manchester coding used by the system ensures that there will be a polarity change in each bit period. The fieldbus signal is thus polarized and all the field devices must be so connected that they all see the signal in correct polarity.

There are nonpolarized network-powered field devices that can be connected in either polarity across the network. These devices sense which terminal is positive and which is negative, and have automatic polarity detection and correction capability. Such a device can receive a message of either polarity.

If a network consists of different types of devices, the signal polarity must be taken into account with all positive terminals connected to each other, and likewise for the negative terminals. Wires are color coded to make such connections easy. It is advisable to develop a network with polarity considered so that nonpolarized devices can be blindly connected. It helps to add polarized devices later when expansion of plant necessitates addition of devices.

Signal polarity and power polarity must be the same for bus-powered devices, while nonpolar bus-powered devices can accept power and signal of either polarity.

17.9 SEGMENT VOLTAGE AND CURRENT CALCULATIONS

The fieldbus power supply is especially designed for the network to operate properly. Bus-powered field devices draw their operating power in the same way as a two-wire analog device. The farthest device in the segment of the network must a get the minimum voltage of 9 V. While designing a network with many field devices having trunks and associated spurs, the following must be taken into consideration:

1. Supply voltage
2. Current consumption of each field device
3. Resistance of every part of the network
4. Location of the power supply

DC circuit analysis is used to determine the voltage at each field device. Since a network consists of so many branches at different points in the network, calculation of field device voltages often becomes cumbersome. This becomes more so when some field devices are added to the network later at different places as per future requirements. To obviate such difficulty,

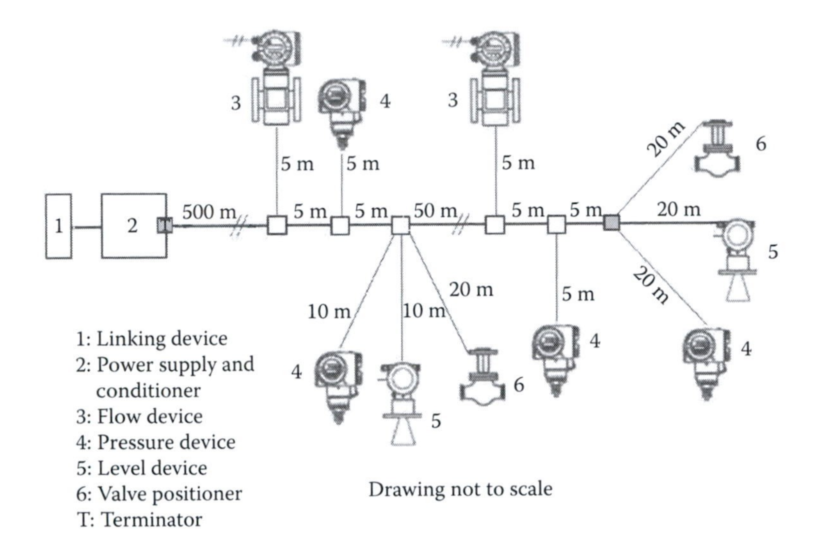

FIGURE 17.15 Voltage and current calculations for a fieldbus segment. (From B. G. Liptak. *Instrument Engineers' Handbook—Process Software and Digital Networks*, 3rd Edition. CRC Press, Boca Raton, FL, p. 595, 2002.)

fieldbus vendors have designed software that makes such calculations just a click away. Again, increased resistance due to higher temperature must be kept in mind.

Figure 17.15 shows a nonhazardous area Foundation Fieldbus bus segment, conforming to IEC 61158-2, with several trunks and spurs connected to the field devices. Type A cable is assumed to be used with a loop resistance of 44 Ω/km. In Figure 17.15, 1 is an I/O card or linking device, 2 is power supply with conditioner, 3 is a flow device with external power supply, 4 is a pressure device, 5 is a level device, 6 is a valve positioner, and T is a terminator. Typical current consumption for device types 3, 4, 5, and 6 are 12, 10.5, 11, and 13 mA, respectively.

Segment length is the sum of all trunk and spur lengths, which in this case is 690 m. The segment current I_{seg} is given by

$$I_{seg} = \sum I_B + \max I_{FDE} + I_{MOD} + \sum I_{startup}$$

where I_B is the basic current drawn by a device; max I_{FDE} is the largest disconnect electronics current in the segment; I_{MOD} is the modulation current of the segment, which is 9 mA for Foundation Fieldbus and zero for PROFIBUS-PA; and $\sum I_{startup}$ is the extra current a device may draw above the basic current on start-up.

Thus,

$$I_{seg} = ([2 \times 12] + [4 \times 10.5] + [2 \times 11] + [2 \times 13])\ \text{mA} + (4 + 4)\ \text{mA} + 9\ \text{mA} = 131\ \text{mA}$$

Again,

$$\begin{aligned} L_{segment} &= L_{trunks} + L_{spurs} \\ &= 570\ \text{m} + 120\ \text{m} = 690\ \text{m} \end{aligned}$$

$$\begin{aligned} R_{cable} &= R_{loop} \times I_{seg} \\ &= 44\ \Omega/\text{km} \times 0.690\ \text{km} = 30.36\ \Omega \end{aligned}$$

Now, the maximum supply current $I_s)_{max}$ is calculated on the basis of a minimum device supply voltage of 9 V.

$$I_s)_{max} = (V_s - 9\ \text{V})/(R_0 + R_{cable}),$$

where V_s is the supply voltage and R_0 is the internal resistance of power supply.

$$\begin{aligned} &= (19\ \text{V} - 9\ \text{V})/(2\ \Omega + 30.36\ \Omega) \\ &= 309.02\ \text{mA} \end{aligned}$$

Thus, $I_s)_{max} > I_{seg}$.

Again, the voltage at the last field device is given by

$$\begin{aligned} V_{lfd} &= V_s - I_{seg} \times (R_0 + R_{cable}) \\ &= 19\ \text{V} - 131\ \text{mA} \times 32.36\ \Omega \\ &= 14.76\ \text{V} \end{aligned}$$

Thus,

$$V_{lfd} > 9\ \text{V}$$

Hence, the segment in question will function properly because the segment current I_{seg} is much less the maximum supply current $I_s)_{max}$ and also the last field device is provided with a voltage of 14.76 V, which is much more than the minimum 9 V that a field device should be provided with.

17.10 LINKING DEVICE

A linking device connects field-mounted devices to the host by means of a higher-speed host-level network. It connects the Foundation Fieldbus H1 to the higher-speed HSE and the PROFIBUS-PA network to PROFIBUS-DP. They have buffers within to take care of speed differences and act as interface.

A linking device for Foundation Fieldbus is shown in Figure 17.16. Redundant field devices are installed in parallel, and two data paths are maintained to the host where very high data availability is a must. In the event of a linking device failing, the other takes over. The situation is shown in Figure 17.17.

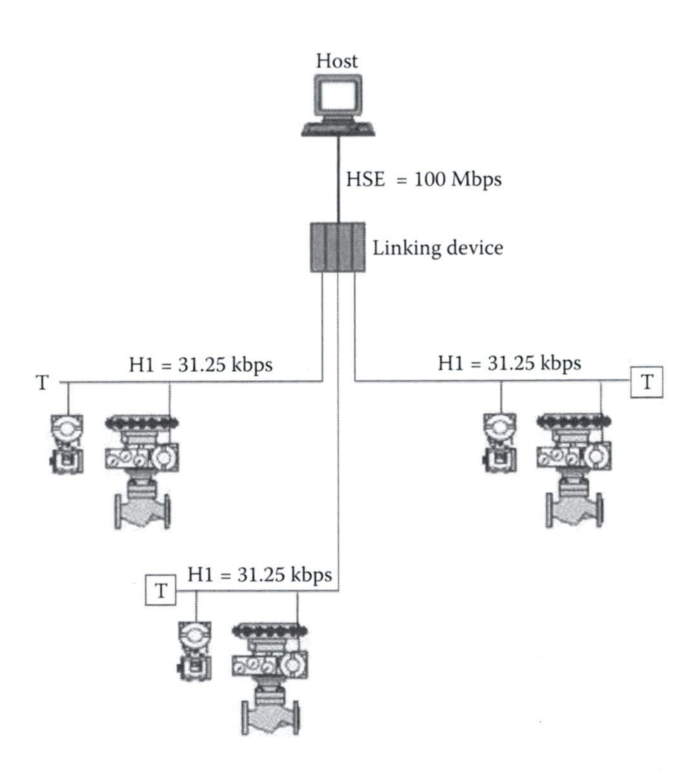

FIGURE 17.16 Linking device connects field devices to host. (From J. Berge. *Fieldbuses for Process Control: Engineering, Operation and Maintenance*. ISA, Research Triangle Park, NC, p. 97, 2004.)

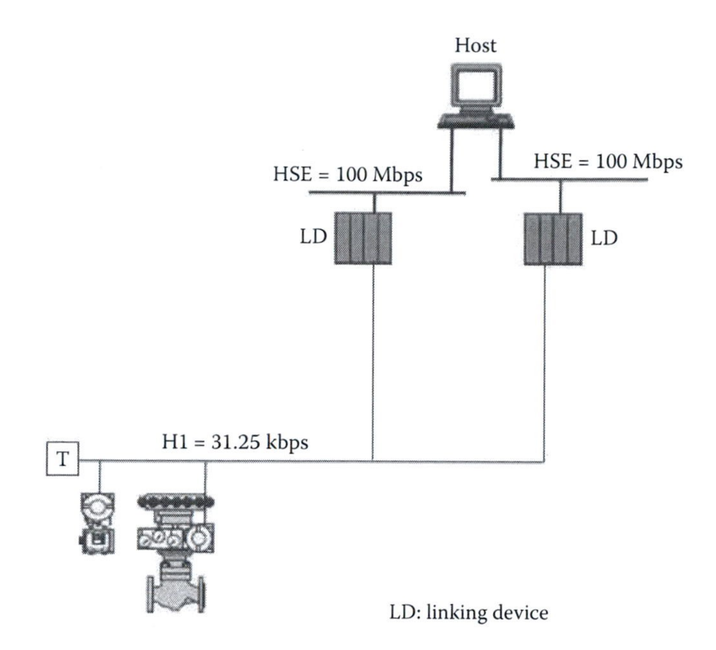

FIGURE 17.17 Redundant linking devices enhance data availability. (From J. Berge. *Fieldbuses for Process Control: Engineering, Operation and Maintenance*. ISA, Research Triangle Park, NC, p. 98, 2004.)

17.11 DEVICE COUPLER

A device coupler connects the spur cables to their corresponding trunks. FF-846 is the standard for device couplers for Foundation Fieldbus. Compliance to this standard ensures better device coupler performance that includes input impedance, voltage drop, and reaction to short circuit.

There are two types of device couplers, viz., nonisolated and isolated. Termination into the device couplers are via screw terminals, pluggable

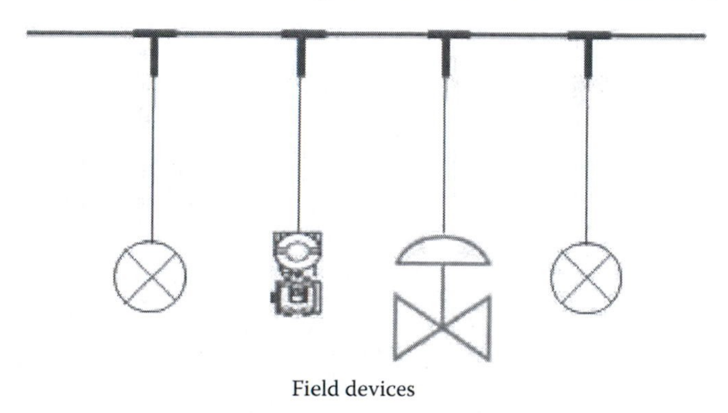

FIGURE 17.18 Simple "T" connection.

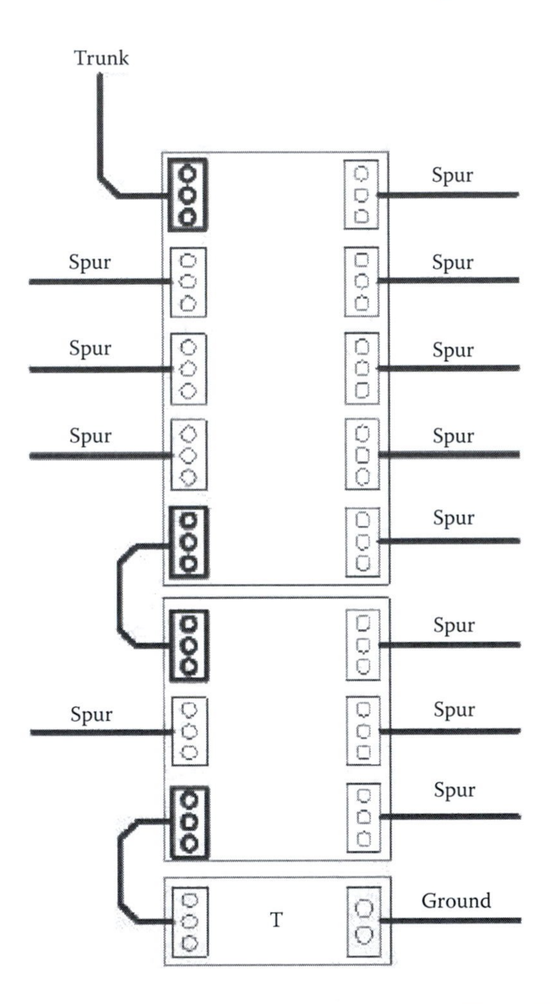

FIGURE 17.19 Schematic diagram of device coupler. (From Fieldbus Wiring Guide, 4th edition, Austin, TX, Doc. No. 501-123, Rev. : E.0. Available at www.relcominc.com/pdf/501-123FieldbusWiringGuide.pdf.)

screw terminals, and pluggable spring clamps. Device couplers route the field transmitter outputs to the bus and to the host controllers, and then route the commands back to the control devices. Device couplers provide short circuit protection against the whole segment failing if either one single device fails or the spur cable is shorted.

The simplest fieldbus device connections to the segments are via the spurs, and its simplest version is "T." However, its problem is that if a device shorts itself, it would bring down the entire segment along with it. Such a connection is shown in Figure 17.18.

Figure 17.19 shows the schematic diagram of a device coupler. It shows two device couplers—one an 8-spur and another a 4-spur, connected via

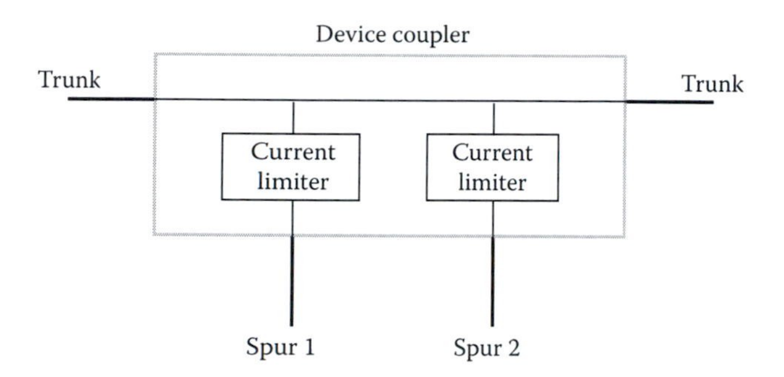

FIGURE 17.20 Device coupler with current limiters. (From Fieldbus Wiring Guide, 4th edition, Austin, TX, Doc. No. 501-123, Rev. : E.0. Available at www.relcominc.com/pdf/501-123FieldbusWiringGuide.pdf.)

a short jumper cable. The other end of the 4-spur coupler is connected to the terminator (T). The other end of the terminator is grounded for surge suppression. Thus, 12 field devices are connected to such a device coupler and housed in a junction box.

Device couplers have current limiters built into them. This is shown in Figure 17.20. This prevents a "shorted" spur from short circuiting its corresponding segment.

Second-generation device couplers that have been introduced of late have features such as autotermination, short circuit protection, and visual circuit checking. They have quick up time and low maintenance costs.

Current limiters in older device couplers limit the "lock-in" current to as much as 60 mA, which may deprive others from receiving sufficient current and such field devices may ultimately drop off the network. TRUNKGUARD device couplers from MooreHawke employ a fold-back technique that limits the current to as low as 2 mA. This is just sufficient to light a LED and removes the device from the network. TRUNKGUARDs employ an "end-of-the-line" sensing circuit that automatically provide segment termination. This ensures that local portions of a segment continue to operate even if the remote parts are accidentally disconnected.

Second-generation device couplers help accelerate device commissioning and routine maintenance by providing easy access points for handheld communicators for ease of trouble shooting.

17.12 COMMUNICATION SIGNALS

As the fieldbus signal travels along the cable, it is both attenuated and distorted. A fieldbus signal is attenuated as it travels through the cable and also at the spur cable where it branches off the trunk cable. The latter

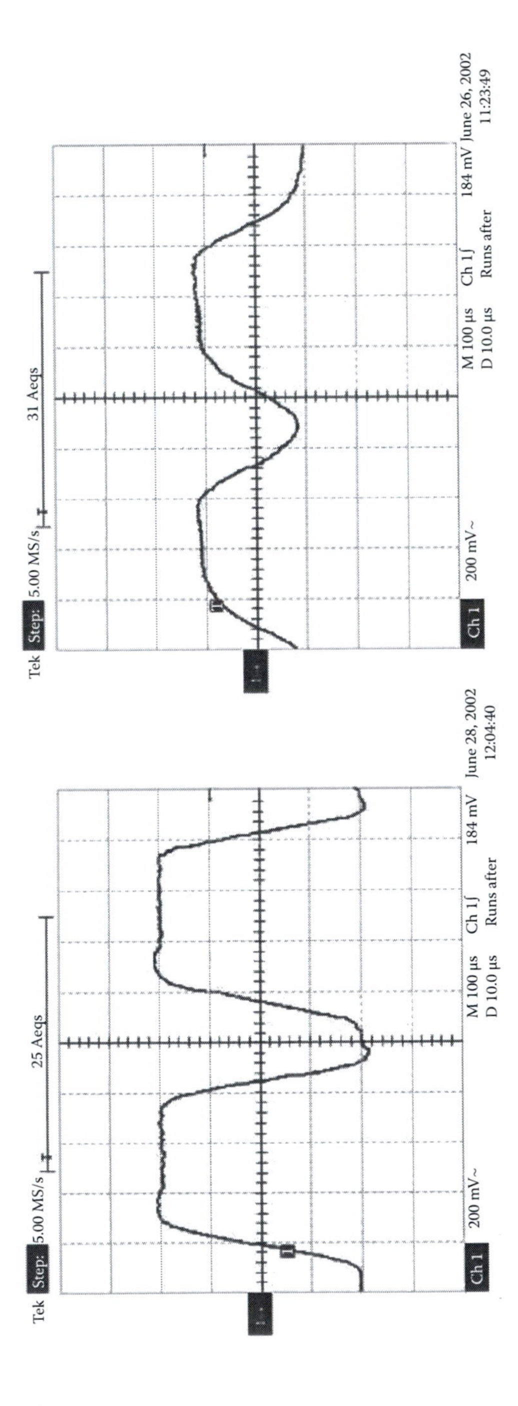

FIGURE 17.21 Transmitted and received fieldbus signals. (From Fieldbus Wiring Guide, 4th edition, Austin, TX, Doc. No. 501-123, Rev.: E.0. Available at www.relcominc.com/pdf/501-123FieldbusWiringGuide.pdf.)

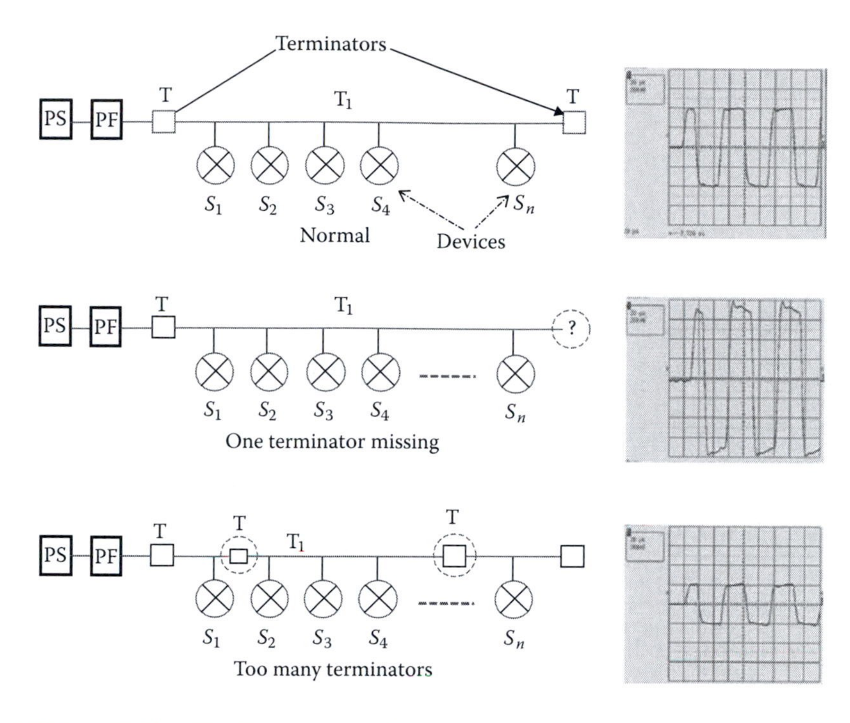

FIGURE 17.22 Fieldbus communication signals under different conditions.

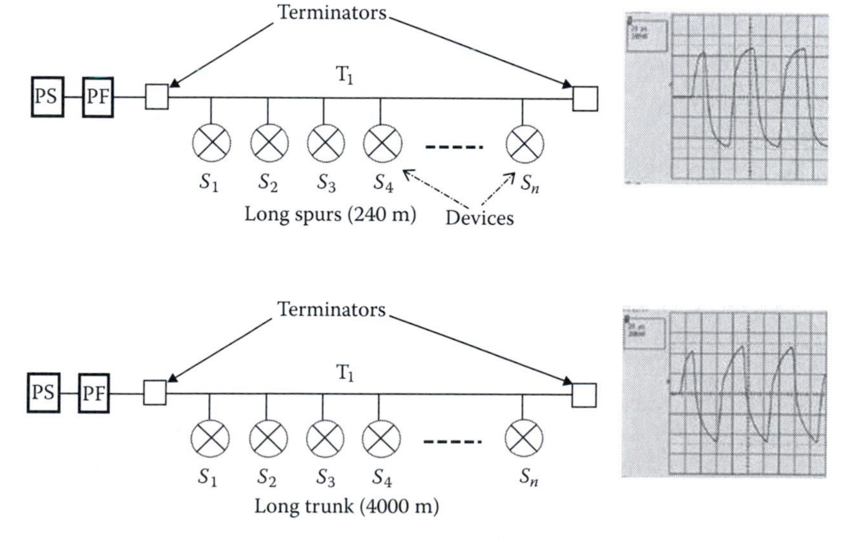

FIGURE 17.23 Fieldbus communication signals for long spurs and trunks.

attenuation is caused mainly by cable capacitance. The transmitted and received fieldbus signals are shown in Figure 17.21.

Fieldbus signal quality is dependent on characteristics of cables, reflections at spurs, etc. Received signals are degraded if trunk length or total spur lengths are not adhered to as per recommendations. Figure 17.22 shows the quality of communication signals under normal condition, one terminator missing, and too many terminators.

Again, Figure 17.23 shows the communication signals when using long spurs or long trunks in excess of recommended limits.

17.13 DEVICE COMMISSIONING

Device address assignment for Foundation Fieldbus is done automatically by the linking device or some other interface, while the same for PROFIBUS is configured manually before the device is installed on the network. Once a device is installed, it will show up in the live list, like all other devices, in the host application.

Device configurations can be started with the help of the engineering tool even before the arrival of devices at the site, and after the detailed engineering is over. Once the devices have been received at the site, the devices can be connected to the network and the device configuration files downloaded immediately. Such an approach cuts down the start-up time appreciably.

There are two possibilities when a device does not show up in the live list when connected to the network. The first is that the device is not connected properly and needs troubleshooting. The second possibility is that the device may appear in the live list of another network.

Before field device commissioning, several checks must be undertaken for proper working of the system. These are (a) continuity of cables, proper grounding, and insulation testing; (b) field device couplers; (c) power supply; (d) proper voltage and current availability at the field device at the farthest end; (e) device configuration checking and their signal analysis; (f) terminators at either end of the segment; and (g) cable insulation and capacitance.

17.13.1 Foundation Fieldbus Device Commissioning

A Foundation Fieldbus device is identified by (a) device ID, (b) device name (TAG), and (c) device address. The device ID is a 32-character device identifier. It is a hardware address that is totally unique and unambiguously distinguishes one device from the other. The manufacturer sets this 32-character identifier. A device ID identifies the manufacturer, serial number of the device, model, etc. The device name (TAG) is set by the user. It identifies the function of the device and its physical location in the process. The device address is unique and is automatically set by the system. A device is associated to its particular configuration by correlating the device configuration tag with its device ID.

Common users interact with fieldbus devices based on tags, although the node address is shown in the live list. A device is detected within seconds of its connection to the network. It is possible to connect or disconnect a device to the network without disturbing the same.

A device, once connected, is identified automatically by the network and is assigned a default node address in the range of 248–251. The process of this assignment may take around a minute or so. After that, the device is assigned its operational address in the range of 16–247. The linking device or the interface continuously probes the node addresses for new devices. The device configuration database is developed on the basis of device tags that are assigned by the user.

17.13.2 PROFIBUS-PA Fieldbus Device Commissioning

For PROFIBUS-PA, the device address range is 0–126 and is unique in nature. Care must be taken to avoid address duplication; otherwise, the system would lose control over the network. Thus, tag and address cross references are done to ensure avoidance of address duplication.

Address for each device must be set correctly so that each device is detected correctly; otherwise, disruption of communication would take place. The address may be set locally for each device or remotely by the device configuration tool. For local setting, internal hardware DIP switches are used, or externally by means of a local digital display. A device that is not on the operational network can be set remotely. This ensures nondisturbance of the network that is in operation.

17.14 HOST COMMISSIONING

An interface or a linking device connects the host to field level networks. The linking device or the interface may have multiple ports, allowing several networks to be connected to such an interface. Device support files from different manufacturers have to be loaded in the host so that it can interact with the devices placed in the field. The configuration tool has a dedicated folder for support files with subfolders for each type of device and manufacturers. The support files can be downloaded from the manufacturer's website so that configuration work can be completed even before the arrival of field devices at the site. There are certain standard support files that are required to be installed. For Foundation Fieldbus, these are device description files (FFO and SYM) and capabilities file (CFF). For PROFIBUS-PA, these are a one device data sheet file (GSD) for class I master type and additional files related to class II master types. A field device tool (FDT) has already been designed for PROFIBUS that resides in the host. The FDT configures the devices from the host. For each device type, a device type manager software is supplied by the manufacturer, which is installed on the FDT.

17.15 WIRING AND ADDRESSING VIA ETHERNET AND IP

Data transfer between two nodes in a network take place via Transmission Control Protocol (TCP) and User Datagram Protocol, while Internet Protocol (IP) is used for networking purposes. Ethernet is used for wiring in LANs. Wiring and hardware access in Foundation Fieldbus HSE, PROFInet, and MODBUS/TCP are based on Ethernet IEEE 802.3/ISO 8022.

On a network, Ethernet devices with different protocols can coexist without conflict. This is because they are built on the same platform; however, in such cases, they cannot talk to each other due to differing user layers.

Addressing in Ethernet is done via an Ethernet MAC address. This is a unique address set in hardware by the manufacturer, which ensures that no two identical MAC addresses exist and eliminates any chance of conflict. This unique MAC address has a centrally administered code along with a unique identifier. On the other hand, the IP network address can be set either by the user or else assigned automatically.

17.16 ETHERNET

A LAN is a conglomeration of several computers connected in a network in a localized geographical area. LANs can work in isolation catering to the needs of an organization. However, today's LANs are mostly linked to a WAN or the Internet.

Ethernet, Token Ring, Token Bus, FDDI, and ATM LAN technologies are used for LAN. Of these, Ethernet technology has become the most dominant.

17.16.1 IEEE Ethernet Standards

The IEEE 802.3 committee has developed standards for the different Ethernet versions available, the more common among them are shown in Table 17.3.

There is also another version, 10 Gbps Ethernet. Thus, Ethernet has travelled through four generations with the main concept remaining unchanged but has adapted it to new technologies and the changing market needs.

The IEEE standard (called Project 802) for LAN is shown in Figure 17.24, along with the traditional OSI model. In this standard, the

TABLE 17.3
Common IEEE Ethernet Standards

Transmission Rate	Ethernet System	Transmission Medium	Maximum Segment Length
10 Mbps (standard)	10Base-5	Coaxial cable (RG-8 or RG-11)	500 m
	10Base-2	Coaxial cable (RG-58)	185 m
	10Base-T	UTP/STP category 3 or better	100 m
	10Base-36	Coaxial cable (75 ohm)	Varies
	10Base-FL	Optical fiber	2000 m
	10Base-FB	Optical fiber	2000 m
	10Base-FP	Optical fiber	2000 m
100 Mbps	100Base-T	UTP/STP category 5 or better	100 m
	100Base-TX	UTP/STP category 5 or better	100 m
	100Base-FX	Optical fiber	400–2000 m
	100Base-T4	UTP/STP category 3 or better	100 m
1000 Mbps	1000Base-LX	Long wave optical fiber	Varies
	1000Base-SX	Short wave optical fiber	Varies
	1000Base-CX	Short copper jumper	Varies
	1000Base-T	UTP/STP category 3 or better	Varies

Source: W. Tomasi. *Advanced Electronic Communications Systems*. Pearson Inc., p. 263, 2004.

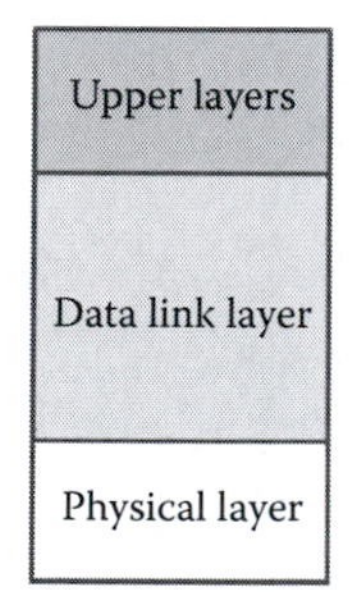

FIGURE 17.24 IEEE standard for LAN and its comparison with OSI. (From B. A. Forouzan. *Data Communications and Networking*, 4th Edition. Tata McGraw-Hill, New Delhi, India, p. 396, 2006.)

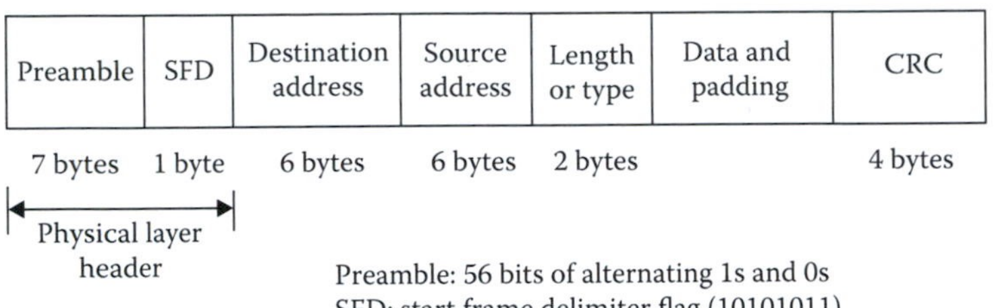

FIGURE 17.25 IEEE 802.3 MAC frame. (From B. A. Forouzan. *Data Communications and Networking*, 4th Edition. Tata McGraw-Hill, New Delhi, India, p. 398, 2006.)

Data link layer is divided into two sublayers—logical link control (LLC) and MAC, with the former residing over the latter. LLC provides flow and error control for the upper layer protocols. It also supports a part of the framing actions. Framing is jointly supported by both LLC and MAC. It may, however, be mentioned that most upper layer protocols, including IP, do not use LLC in their OSI formatting. The standard Ethernet sublayer is represented by 802.3 MAC frame consisting of seven fields. It is shown in Figure 17.25, and the minimum and maximum data lengths are 46 and 1500 bytes, respectively.

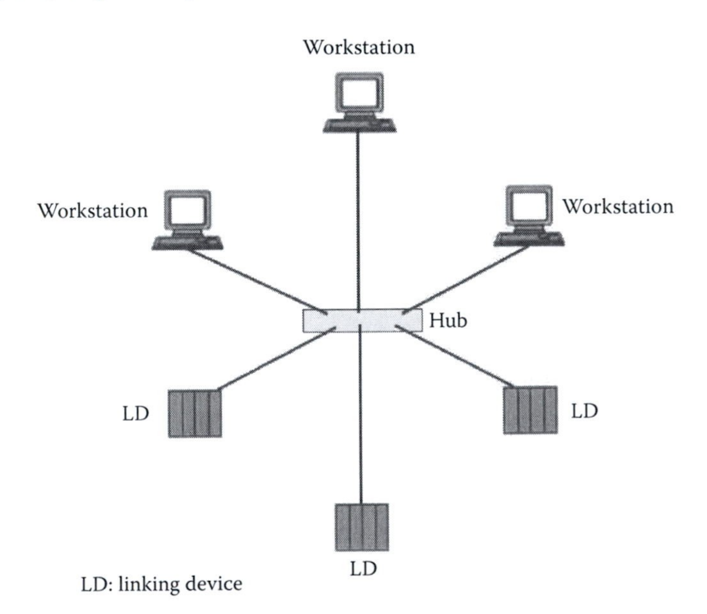

FIGURE 17.26 Star topology. (From J. Berge. *Fieldbuses for Process Control: Engineering, Operation and Maintenance*. ISA, Research Triangle Park, NC, p. 116, 2004.)

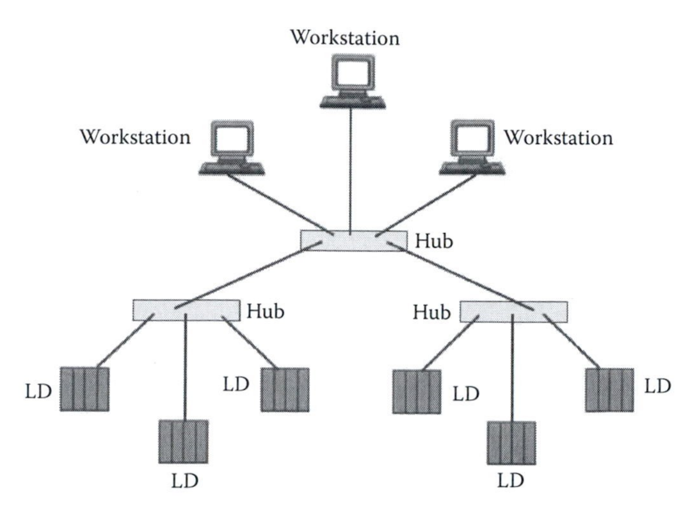

FIGURE 17.27 Tree topology resulting from several star connections. (From J. Berge. *Fieldbuses for Process Control: Engineering, Operation and Maintenance*. ISA, Research Triangle Park, NC, p. 116, 2004.)

17.16.2 Topologies

Ethernet uses mostly the star topology in which each node is connected to its own port on the central hub. Each segment is totally separate and is connected to the hub like spokes in cycle. Even if a segment fails, normal operation is not hampered and segments can be connected or disconnected under power. Such a star connection topology is shown in Figure 17.26. There is a limit to the segment length due to signal attenuation along the wires.

Again, when several star-connected hubs are connected by means of a hub, a tree formation results, shown in Figure 17.27. In tree, branch, or star topology, disconnection of a device does not affect other devices, while it is so in the case of the daisy chain topology.

17.17 IP BASICS

In fieldbus-based process control systems, the linking devices, servers, workstations, etc., belong to the host level, while Foundation Fieldbus HSE and PROFInet belong to the Internet protocol. The IP addresses in FF HSE and PROFInet devices are usually configured by network administrator.

The total IP network address consists of two parts: the first part represents which network node is on and the second part represents the network node address. Each of the host networks has an IP address. In IP version 4,

TABLE 17.4
IANA Allocated Three Private IP Address Blocks

Start Address	End Address
10.0.0.0	10.255.255.255
172.16.0.0	172.31.255.255
192.168.0.0	192.168.255.255

Source: J. Berge. *Fieldbuses for Process Control: Engineering, Operation and Maintenance.* ISA, Research Trianle Park, NC, p. 133, 2004.

the address is 4 bytes, each separated by a dot. These numbers are assigned by Internet Assigned Number Authority (IANA). It has set aside three blocks of IP addresses for private use, exactly like the ISM band. The process control system can be thought of as a private network that does not require it to be permanently connected to the Internet. In a way, it is a private network and uses the three free IP network addresses for the host level networks. The three private IP addresses are given in Table 17.4.

Routers forward packets of information based on the IP network addresses. Locally, at the host level, the switches and bridges take care of Ethernet MAC address. Because both Foundation Fieldbus HSE PROFInet use the Internet, they both use IP routers for transporting information across the Internet, i.e., inbetween networks.

17.18 IP COMMISSIONING

IP network addresses can be configured manually even before a device is connected to the network. It can also be assigned automatically by Dynamic Host Configuration Tool (DHCP) server.

A Ping command checks the proper operation of a device after it is connected to the network. If the Ping times out, several possibilities arise: the device is not wired or operating properly, it is connected to the wrong network, or the network parameters are configured wrongly. The Ping command can be used to confirm whether an IP address is already in use or not. Again the activity LED associated with each device helps confirm whether the correct device is being accessed with the help of Ping command.

It should be noted at this point that IP routers are used by Foundation Fieldbus HSE and PROFInet to reach out to the destination network node, while the Ethernet MAC address is used by switches and bridges to identify the particular device on a local network.

17.18.1 Subnet

Help of IP routers is taken to divide a large network into smaller ones to counter the effect of huge broadcast network traffic. Likewise, subnet (sub-network) is used to divide the local network into smaller segments having fewer devices in each. In the subnet mask, a portion is dedicated for identifying the IP part of the network, while the rest portion is meant for device address identification. A device is always identified with a particular subnet mask.

A private IP address, given in Table 17.4, shows the starting address at 10.0.0.0 and the end address at 10.255.255.255. If such a private space is used for total plant network, it would be too many. Thus, such a space is judiciously segregated into subnet segments so that connection establishment with a particular device belonging to a particular subnet becomes much faster.

17.19 MANUAL IP CONFIGURATION

Configuration involves setting the subnet mask and also the router IP address in the devices and workstations. This IP address is called the Default Gateway. Manual configuration involves simply keying in the address information. The addresses of either the devices or that of the network nodes must be unique. Address duplication must be avoided. Inadvertent address duplication would result in failed communication and disrupt supervision and control. To avoid the above, it is suggested to have an address and its corresponding tag cross reference.

A manual addressing mode is static in nature. A device that does not have a local interface must be connected to computer on a separate network that is running a special configuration tool. The host part of the IP address cannot have all binary zeroes or ones for the purpose of configuration.

17.20 AUTOMATIC IP CONFIGURATION

When configuring a large network, it is advisable to configure the system automatically because it is the best option. A DHCP server helps in automatic address assignments by simply connecting the devices to the network. The DHCP server assigns unique IP addresses to the nodes in the network. This addressing is dynamic because they may change from time to time.

18 Wireless Communication

18.1 INTRODUCTION

A trend we have been witnessing for quite some time now is the emergence of wireless communication from the office world to the industrial world. Wireless communication in the field of mobile technology has curved out a niche for itself for a couple of decades now. Today's mobile communication is offering a host of services that could not be thought of even a couple of years ago. Advancements in wireless networking technology, particularly in short-haul communication, offer a tremendous opportunity for networking field devices with the host stations. The process measurement, control, and communication field is steadfastly moving toward the wireless strategy as the initial shortcomings are being removed and as the system is becoming more and more rugged and dependable. For any process control system to be networked wirelessly, factors that must be taken into consideration are reliability, scalability, and real-time mixed data transmission, availability, security, and coexistence with other networks.

18.2 WIRELESS COMMUNICATION

In wireless communication, there is no physical wire link between transmitter and receiver. Unguided media transports electromagnetic waves between the transmitter and the receiver with the help of antenna. At the transmitting end, the antenna radiates electromagnetic waves in the medium, which is usually air. At the receiving end, the receiving antenna picks up the electromagnetic waves from the medium. It is a multicast system in which any receiver can pick up the transmitted electromagnetic waves. Wireless transmission uses the whole electromagnetic spectrum, which has a range from 3 kHz to 900 THz.

In wireless communication, two types of transmissions are used: directional and omnidirectional. In the first, the transmitting antenna puts out a focused electromagnetic beam. Thus, in this case, both the transmitting and the receiving antennas must be carefully aligned so that the receiver may get the maximum transmitted power. In the omnidirectional case, the transmitting antenna spreads the electromagnetic spectrum in all directions and thus any receiving antenna can receive the same.

Unguided signals can reach the receiver via several methods: ground propagation, sky propagation, and line-of-sight propagation. In the first, the signals that come out of the transmitting antenna are of low frequency and travel in all directions following the lowest part of the atmosphere. The transmission path follows the curvature of the earth, and the distance they can travel depends on the power that emanates from the transmitting antenna. The second, i.e., sky propagation, enables relatively longer path travel with less power of the transmitting antenna. In this case, relatively higher radio waves are radiated into the ionosphere where they are reflected back into the receiving antenna. In line-of-sight propagation, the antennas face each other (directional) and very high frequency signals are transmitted in straight lines. The antennas must either be very tall or else close enough to each other such that the propagation is not affected by the curvature of the earth.

The part of the electromagnetic spectrum that includes radio waves and microwaves is segregated into eight range bands and shown in Table 18.1.

TABLE 18.1
Bands in Electromagnetic Spectrum

Bands	Ranges	Propagation Method	Application Areas
VLF (very low frequency)	3–30 kHz	Ground	Long-range radio navigation
LF (low frequency)	30–300 kHz	Ground	Radio beacons and navigational locators
MF (middle frequency)	300 kHz–3 MHz	Sky	AM radio
HF (high frequency)	3–30 MHz	Sky	Citizens band (CB), ship/aircraft communication
VHF (very high frequency)	30–300 MHz	Sky and line of sight	VHF TV, FM radio
UHF (ultra high frequency)	300 MHz–3 GHz	Line of sight	UHF TV, cellular phones, paging, satellite
SHF (super high frequency)	3–30 GHz	Line of sight	Satellite communication
EHF (extremely high frequency)	30–300 GHz	Line of sight	Radar, satellite

Source: B. A. Forouzan. *Data Communications and Networking*, 4th Edition. Tata McGraw-Hill, New Delhi, India, p. 204, 2006.

The radio waves have a frequency range between 3 kHz to 1 GHz, while microwaves have frequencies 1 to 900 GHz. Thus, the band of radio waves is just less than 1 GHz, while for microwaves it is 299 GHz. Therefore, subbands within the radio waves are relatively narrow leading to low data rate for digital communication, while the same for microwaves is relatively wider leading to higher data rate for digital communication. Radio waves are mostly omnidirectional, while microwave is unidirectional. Wireless communication takes place with the help of radio waves, microwaves, and infrared waves. Infrared waves have frequencies between 300 GHz and 400 THz.

The technique of wireless message transmission is shown in Figure 18.1. The data to be transmitted is so encoded that the receiver can decode it without error. To decrease latency, pipelined architecture for data transmission is adopted. Transmission may begin as soon as the first byte is encoded. The successive bytes are serially encoded with the preceding bytes lined up for transmission. Actual transmission begins with the execution of the media access control (MAC) protocol, which allows multiple transmitters to share a single communication channel. Start symbol and synchronization pattern precede the actual encoded data stream.

The receiver is tuned to receive the start symbol first. To eliminate noise and correct reception, the receiving clock is made at least twice the transmitted clock. Once the start symbol is recognized by the system, it checks for the synchronization pattern and the receiver is synchronized with the incoming data bits. The received bits are then decoded and processed as per the need.

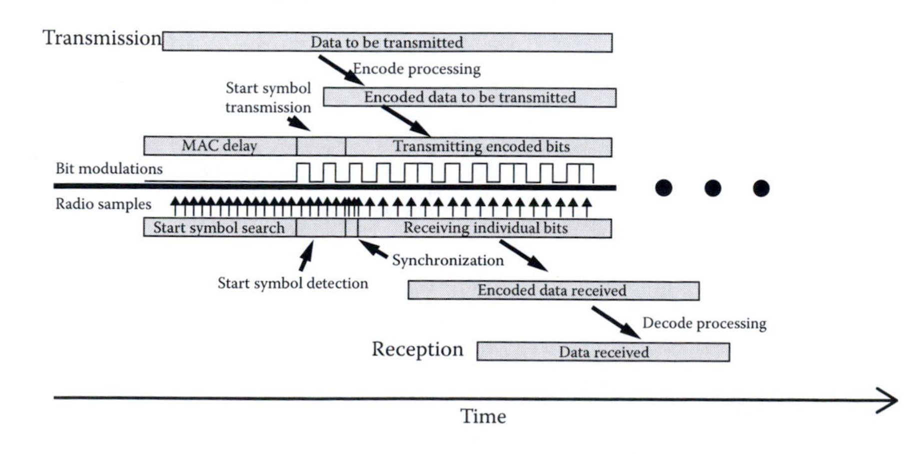

FIGURE 18.1 **(See color insert.)** Wireless communication methodology. (Hill J. L. System architecture for wireless sensor networks. PhD Dissertation, Department of Computer Science, University of California, Berkeley, CA, Spring, 2003.)

18.2.1 Wired vs. Wireless

As wireless communication in process automation is gaining momentum, some key benefits with regard to employing wireless technology are fast emerging compared with its wired counterpart. Wireless technology is becoming more attractive by the day as it eliminates the problems inherent with wired networks. Some of the important points are now discussed.

Installation and maintenance costs are less in wireless networks compared with the wired ones. Sometimes wired systems face difficulty in troubleshooting, particularly in cases of loose connections. For wired networks, a higher inventory is needed for maintenance, keeping an eye on future plant expansions. Maintaining a wired system in a hostile and geographically huge network is troublesome, time consuming, requires more manpower, and susceptible to disconnections. It is less flexible compared with its wireless counterpart. Wireless systems reduce blind spots in process visibility, leading to higher yields at a lesser cost. Some process tasks require field personnel to be physically present at plant locations, necessitating deployment of manpower. Employing wireless networks help in collecting such data for better plant surveillance.

Of late, very rapid progress has been made in the field of mobile phone technology by using digital techniques where the receiver can have a high speed. In the case of industrial process measurement and control applications, the installation is fixed and even if the receiver is moving, it would be very restricted and also the range of movement is very moderate. There are certain concerns, however, in case of industrial wireless communication, which must be addressed for reliability, security, and ultimate acceptance by the users. Some of the problem areas that need to be taken care of are interference, cross-talk from neighboring channels, and jamming the transmission by using the same frequency in the vicinity for malicious purposes such as inception and eavesdropping.

Lightning surges flow through both power supply cable and communication cable. A wireless system is thus not affected by such surges, unlike its counterpart of wired communication. However, only the on-sight smart transmitters are to be protected from such surges.

18.2.2 ISM Band

Industrial, scientific, and medical (ISM) band is an unlicensed band having three bands in its fold and operate in the ranges of 902–928 MHz, 2.4–4.835 GHz, and 5.725–5.850 GHz. Initially, the application areas in these ranges were medical diathermy, microwave ovens, radio frequency (RF) heating, etc. Electromagnetic radiations emanating from operations of these devices can interfere with communications taking place in the same band in the vicinity of the former. Thus, these device operations are contained in certain ranges so that their electromagnetic radiations do not disrupt communications of the others. In recent years, the growing applications in these bands pertain to short-range low-power communications such as Bluetooth devices, near-field devices, cordless phones, and wireless computer networks using local area networks (LANs). The most important application is, however, in wireless sensor networks that use these license-free bands for error-tolerant communications.

Figure 18.2 shows the license-free electromagnetic spectrum in red colors. Out of several bands shown, the 2400–2483.5 MHz band is used for industrial purposes. The ISM bands can be used without taking any prior approval from the regulating authorities, but the flip side is that many users can use the same band at the same time at the same place leading to interference. As shown in Figure 18.2, the license-free bands are at 433.05, 2400, and 5150 MHz, apart from another one that starts at 863 MHz and is used in Europe.

The different frequency bands used in Europe for license-free operations and otherwise are shown in Table 18.2. The different frequency bands are characterized by different output powers, bandwidths, and duty cycles. In Europe, the band 1880–1900 MHz is reserved for DECT standard, while the same for United States is a 2400 MHz ISM band.

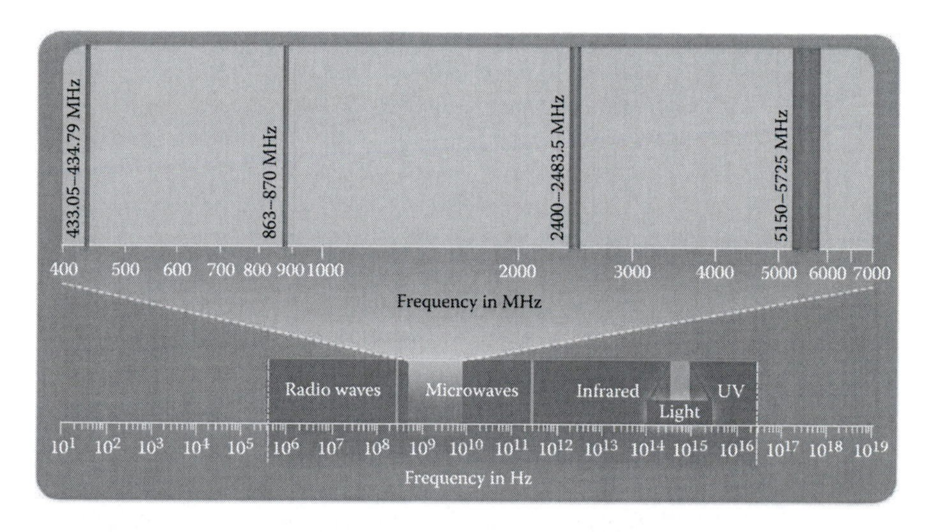

FIGURE 18.2 **(See color insert.)** License-free frequency bands (in red). (Bentje H. et al. "Wireless in Automation" Working Group. *Coexistence of Wireless Systems in Automation Technology: Explanations on Reliable Parallel Operation of Wireless Radio Solutions*, 1st Edition. ZVEI-German Electrical Manufacturers' Association, Automation Division, Frankfurt, Germany, April 2009.)

TABLE 18.2
Frequency Bands, Both Licensed and License Free, as Used in Wireless Automation

Frequency in MHz	Type of Use	Utilization Conditions/ Output Power	Properties
433–434	License free (ISM)	Output power max. 10 mW ERP[b], max. 10% duty cycle[a]	Good penetration, reduced data rate
448 and 459	License required	Output power max. 6 mW, time synchronized, limited duty cycle[a]	Good penetration, low data rate, wide ranges
410–470	License required	Output power depends on frequency assignment, typically 6 W/12 W for mobile devices, channel spacing typically 12.5 kHz/25 kHz	Good penetration, wide ranges
863–870 (in US: 902–928 as ISM band)	License free	Output power 5–500 mW ERP[b], channels partially with 25-kHz bandwidth, duty cycle[a] partially only 0.1%	Wide ranges
1880–1900	License free according to DECT standard	Output power max. 250 mW peak ERP[b], time and frequency division multiple access (TDMA/FDMA)	Good availability, high output power
2400–2483.5	License free (ISM)	Output power 10 mW (100 mW when using spread spectrum techniques, used within buildings without restrictions), no limitations regarding duty cycle[a]	Available almost worldwide, broad bandwidth, already widely used
5150–5350, 5470–5725	License free (partially ISM)	Output power partially up to 1 W, power control and dynamic frequency selection sometimes required	Low penetration of walls, quasi-optical propagation, high data rate

Source: Bentje H. et al. "Wireless in Automation" Working Group. *Coexistence of Wireless Systems in Automation Technology: Explanations on Reliable Parallel Operation of Wireless Radio Solutions*, 1st Edition. ZVEI-German Electrical Manufacturers' Association, Automation Division, Frankfurt, Germany, April 2009.

[a] Duty cycle refers to relative duration of medium utilization. Some frequency bands put restrictions on such relative usage.

[b] ERP: "effective radiated power" from an antenna.

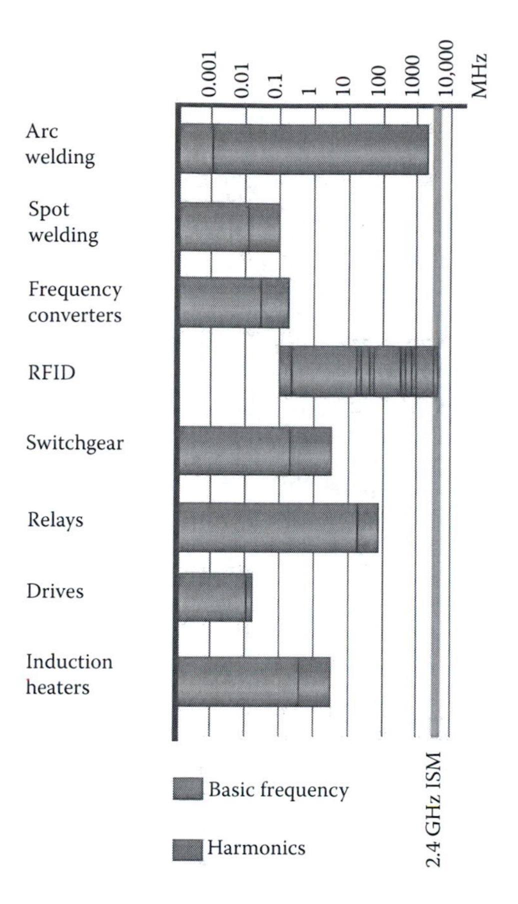

FIGURE 18.3 **(See color insert.)** Interference spectra of some devices used in industries. (Bentje H. et al. "Wireless in Automation" Working Group. *Coexistence of Wireless Systems in Automation Technology: Explanations on Reliable Parallel Operation of Wireless Radio Solutions*, 1st Edition. ZVEI-German Electrical Manufacturers' Association, Automation Division, Frankfurt, Germany, April 2009.)

In industrial automation and manufacturing sectors, a host of devices are used to facilitate an increase in production. Some of these are motor drives, relays, arc welding machines, RFIDs, and switchgears, frequency converters. The interference spectra of such devices are shown in Figure 18.3. As shown in the figure, the frequency spectra for most of these devices do not extend to 2400 MHz. Only RFIDs have their spectra extending up to the same.

18.2.3 Wireless Standards

Wireless LAN is governed by a set of standards (IEEE 802.11) in the 2.4-, 3.6-, and 5-MHz frequency bands. They are developed by the IEEE LAN/MAN committee.

Apart from IEEE 802.11, there are other standards catering to other wireless communication needs. IEEE 802.11 is for WiFi, while ZigBee, WHART, and ISA100.11a fall under the purview of IEEE 802.15.4. WiMax is based on IEEE 802.16 initially, while its latter version is IEEE 802.16a.

Within the IEEE 802.11 Working Group, some standard and amendments exist based on the IEEE Standards Association. Some of these standards are given below:

- IEEE 802.11k: radio resource measurement enhancements, 2008
- IEEE 802.11n: higher throughput improvements using MIMO (multiple input, multiple output antennas), 2009
- IEEE 802.11s: mesh networking, Extended Service Set (ESS), 2011
- IEEE 802.11v: wireless network management, 2011

Some of the standards that are in the pipeline and would be standardized in the near future are as follows:

- IEEE 802.11ai: fast initial link setup, 2014
- IEEE 802.11ah: sub-1-GHz sensor network, smart metering, 2015

18.2.3.1 WiFi

WiFi, also known by the names Wi-Fi or Wi-fi or wifi, was originally licensed by Wi-Fi Alliance. It covers the technology of wireless LAN based on IEEE 802.11 specifications, sometimes called wireless Ethernet and introduced in the year 1997. It is a local area technology that adds mobility to private wired LANs. WiFi supports a maximum range of a few hundred meters.

WiFi uses a MAC protocol called carrier sense multiple access with collision avoidance (CSMA-CA), which is different from carrier sense multiple access with collision detection (CSMA-CD) used in Ethernets. Half-duplex shared media configurations are used in WiFi, where transmitting and receiving stations exchange information on the same radio channel. Thus, a station cannot transmit and receive at the same time, but it clearly supports avoiding collisions. The collision avoidance mechanism adopted in IEEE 802.11 is called distributed control function.

18.2.3.2 WiMax

WiMax stands for worldwide interoperability for microwave access and is based on IEEE 802.16, first released in 2001. It is an emerging standard for Broadband Wireless Access in licensed frequency bands under 6 GHz and is promoted by WiMax Forum. WiMax is applied in accessing wireless sensor networks in a better way than others in terms of greater coverage and cell capacity. WiMax is positioned on the gateway between wireless sensor networks (WSNs) and the Internet. VoIPs and video, which are time-sensitive traffic, get priority in WiMax. WiMax operates in non-line-of-sight (NLOS) conditions. WiMax uses a request/grant mechanism that assumes separate channels for inbound and outbound transmissions. These channels are either separated by Time Division Duplex (TDD) or Frequency Division Duplex. Thus, in WiMax, message transmission is full duplex in nature, while it is half duplex in case of WiFi.

The IEEE 802.16a was released in 2003 and has a frequency range of 2–11 GHz. Within this range, it operates in four subranges.

WiMax traffic architectures can be both point to point and point to multipoint. Data traffic for the former can be from data sources—such as data center and central office—to the broadband subscriber; for the latter, it is distributive type with a single base station handing out data to many dissimilar subscribers having different bandwidth and services offered by them. WiMax is a much more complex technology that uses both licensed and unlicensed frequencies. Security is of much concern in WiFi, while security is taken care of by an encryption mechanism in WiMax. A comparison between WiFi and WiMax technologies is shown in Table 18.3.

18.2.3.3 Bluetooth

Bluetooth is an open wireless communication protocol, sometimes called small wireless LANs, targeted at personal area networks (PANs) and used for low-power, short-range office applications. Bluetooth uses IEEE 802.15.1 and uses channel-hopping scheme that supports low latency and high throughput that increases the reliability of the system. The hopping frequency of transreceivers is 1600 hops/s to overcome interference and fading. However, it does not guarantee any end-to-end communication delay. Bluetooth uses a quasi-static reduced star topology that makes it unsuitable for wireless sensor networks where scalability is of prime importance. Bluetooth cannot be applied in industrial environments because of scalability, security, and power consumption problems. Bluetooth uses Time Division Duplex–Time Division Multiple Access (TDD–TDMA) technology for transmission purposes in which transmission and reception do not take place at the same time (half duplex). The to-and-fro communication uses different hops.

The maximum data rate with Enhanced Data Rate (EDR) is 2–3 Mbps. Antennas used in Bluetooth are mostly omnidirectional, although in some cases it works in NLOS situations. Communication range is 10 m for low-power devices and 100 m for the high-power ones. There are three power classes for Bluetooth devices: classes 1, 2, and 3. Maximum output power is 20 dBm for class 1, 4 dBm for class 2, and 0 dBm for class 3 devices.

TABLE 18.3
Comparison between WiFi and WiMax Technologies

Attribute	WiMax (802.16a)	WiFi (802.11b)	WiFi (802.11a/g)
Primary application	Broadband wireless access	Wireless LAN	Wireless LAN
Frequency band	Licensed/unlicensed 2 GHz to 11 GHz	2.4 GHz ISM	2.4 GHz ISM (g) 5 GHz U-NII(a)
Channel bandwidth	Adjustable 1.25 MHz to 20 MHz	25 MHz	20 MHz
Half/full duplex	Full	Half	Half
Radio technology	OFDM (256 channels)	Direct sequence spread spectrum	OFDM (64 channels)
Bandwidth efficiency	Less than 5 bps/Hz	Less than 0.44 bps/Hz	Less than 2.7 bps/Hz
Modulation	BPSK, QPSK 16-, 64-, 256-QAM	QPSK	BPSK, QPSK 16-, 64-QAM
FEC	Convolution code Reed-Solomon	None	Convolution code
Encryption	Mandatory-3DES Optional-AES	Optional-RC4 (AES in 802.11i)	Optional-RC4 (AES in 802.11i)
Access protocol	Request/grant	CSMA-CA	CSMA-CA
Best effort	Yes	Yes	Yes
Data priority	Yes	802.11e WME	802.11e WME
Consistent delay	Yes	802.11e WMS	802.11e WMS
Mobility	Mobile WiMax (802.16e)	Under development	Under development
Mesh	Yes	Vendor proprietary	Vendor proprietary

Source: M. F. Finneran. *WiMax Versus Wi-Fi: A Comparison of Technologies, Markets and Business Plans.* Copyright dBrn Associates, Howlett Neck, NY, 2004.

Bluetooth is a wireless LAN technology that is designed to connect laptops, desktops, telephones, mobiles, printers, cameras, computer mice, etc. Because of its failure to provide guarantee in end-to-end communication delay and very low distance coverage, it cannot be used in industrial applications.

Bluetooth defines two types of network architectures: piconet and scatternet. A piconet is a Bluetooth network having a maximum of eight stations, of which one is primary (master) and the remaining seven are secondaries (slaves). A piconet is a wide personal area network (WPAN). The individual clocks and the hopping sequences of the secondaries are in synchronism with the primary. The communication between the primary and the secondaries can be one to one or one to many (broadcast). A piconet system can be expanded, beyond the recommended seven secondaries, by an additional eight secondaries. In such a case, these additional eight secondaries are in a parked state.

Bluetooth piconets can be interconnected to form what is called a scatternet. A time division multiplexing scheme is applied to synchronize the different piconets that form the scatternet. A secondary station in a piconet can act as the primary for its succeeding piconet, i.e., it acts as a gateway. This particular secondary accepts message from its primary and delivers the same to the secondaries connected to it to which it itself acts as a primary.

18.2.3.4 ZigBee

Like Bluetooth, the ZigBee protocol is also a PAN, designed for ultra-low-power wireless applications in areas like monitoring and control. It is a low-power, low-data-rate, and low-cost communication protocol based on IEEE 802.15.4 physical and MAC layers.

ZigBee is a secure communication with a 128-bit Advanced Encryption Standard (AES) cipher algorithm and a user-defined security application layer. It provides interoperability and conformance testing specifications. ZigBee, like Bluetooth, does not guarantee end-to-end communication delay. Apart from the above, it cannot provide reliability against interferences and obstacles so often encountered in industries. For these reasons, ZigBee cannot be applied in industries, although it is a secure communication.

ZigBee uses mesh topology and provides for a fast communication system. It uses the direct sequence spread spectrum (DSSS) technique. It is surely advantageous over channel hopping because synchronization does not have to occur before initiating communication. On the other hand, it does not provide frequency and path diversity.

18.2.3.5 WHART

Wireless Highway Addressable Remote Transmission (WHART) is an IEEE 802.15.4-based centrally managed mesh protocol network. It is an extension of the widely used HART communication protocol and is backward compatible with existing HART devices and applications. WHART is considered to be the first open standard protocol for WSNs in the area of process automation and control.

WHART uses the same 2.4-GHz ISM nonlicensed frequency band as several other wireless technologies. Its data rate is 250 kbits/s and uses TDMA technology. It is interoperable and supports channel hopping with a 5-MHz gap between any two adjacent channels. Industrial security services are provided by MAC layer and network layer through a 128-bit AES algorithm. Reliability of WHART systems is achieved by using frequency, path diversity, and message delivery methods. Power consumption is optimized by properly managing the communication schedule. Chapter 19 discusses WHART in an elaborate manner.

18.2.3.6 ISA100.11a

ISA100.11a is based on IEEE 802.15.4 in the nonlicensed 2.4-GHz ISM band. It provides reliable and secure wireless communication for noncritical monitoring and control of industrial automation systems. Some of its features are low cost, low power consumption, robustness to RF interferences, low complexity, scalability, and interoperability. However, unlike WHART, it is not backward compatible. It operates in a time-synchronized channel-hopping scheme to ward off RF interference and also to reduce power consumption. It can operate in both star and mesh topology. The former one is used for quick response—necessary for some time-critical industrial applications. Mesh topology is used for increased robustness of the system, greater tolerance to RF interferences, and higher reliability. Messages are protected by way of AES 128 block cipher, which includes asymmetric cryptography and object-based application layer security. ISA100.11a provides protocol translation, and its low power consumption is because of TDMA, which allows for sleeping routers. Chapter 20 discusses ISA100.11a in an elaborate manner.

18.2.4 Media Access

For proper communication to take place, the transmission medium, i.e., air, has to be accessed. If the transmitters try to access the medium at the same time, there would be total chaos and no reception would be possible at the corresponding receivers. Various techniques are used for media access and each one has its own characteristics.

The simplest among them is that data is modulated by a fixed frequency. Although it requires less bandwidth, it is more susceptible to interferences. In TDMA, all the stations participating in transmission of messages are allotted different and precise predefined time slots to avoid collisions. In CSMA, collision is avoided by introducing random delay times after a free channel is recognized. Until the transmission through this free channel is finished, other channels wait for their turns, thus avoiding collision.

In frequency hopping spread spectrum, the transmitter frequency hops from one frequency to another as per some predefined schedule that is known to the receiver. The pattern received at the receiver is affected only in part by interference, which can be retransmitted. In DSSS, a chipping sequence spreads the transmitted spectra, making the transmission less sensitive to narrow band noise. In chirp spread spectrum, the energy of the transmitted data is spread over a wider frequency range. Lastly, in orthogonal frequency division multiplexing (OFDM), properties of narrow band modulation are utilized to operate several channels very close to each other. The process involves a very high data rate since a narrow band may be affected by interference.

18.2.5 Topology

Three topologies are normally employed for networking of wireless sensors in industrial automation. These are star, mesh, and star–mesh. The star topology is shown if Figure 18.4.

In this topology, the message or data is directly sent to the gateway and then sent to other systems via the gateway. The star topology is applied when the power consumption is at a premium and the data sources must be confined in a very short geographical range. This mode is fastest for data transmission purposes. No router is employed in this case.

Data sources or wireless sensors act as routers for a mesh network, which is shown in Figure 18.5. This topology is suited for a network where the sensors are spread over a wide geographic area with high redundancy. Self-configuring networks automatically determine the best path for the messages from the sensors to be taken to the gateway and the system automatically bypasses a failed sensor. However, the flip side is that enough power must be available to all the data sources so that the same can be routed to the gateway.

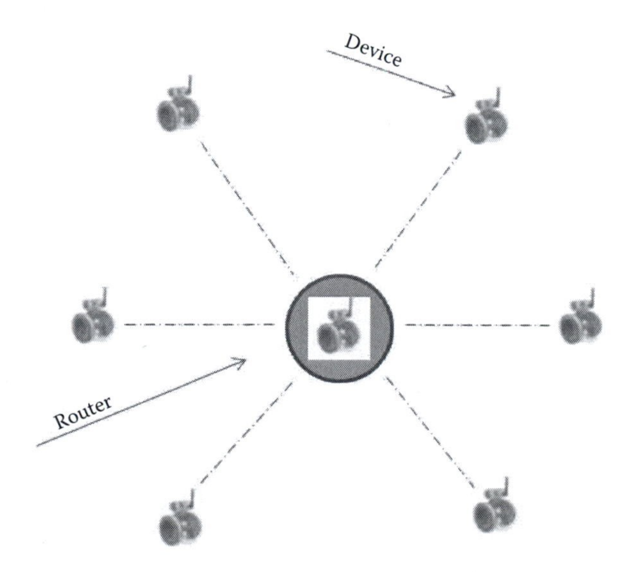

FIGURE 18.4 Star topology.

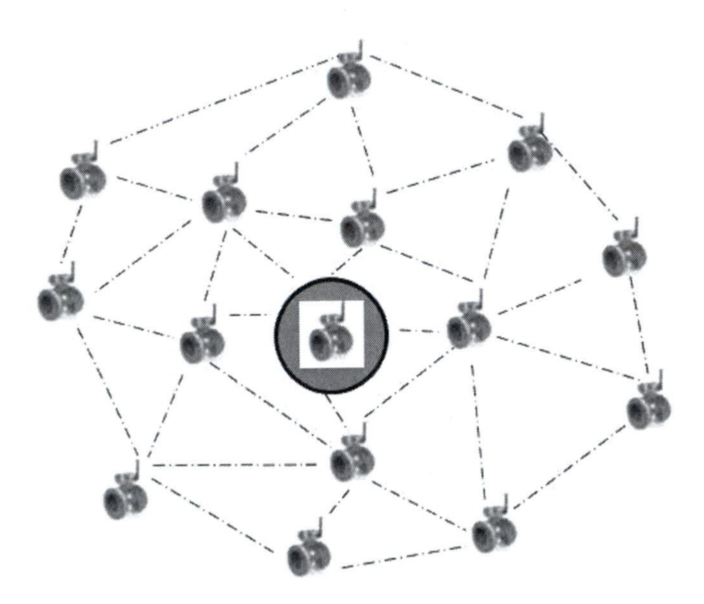

FIGURE 18.5 Mesh topology.

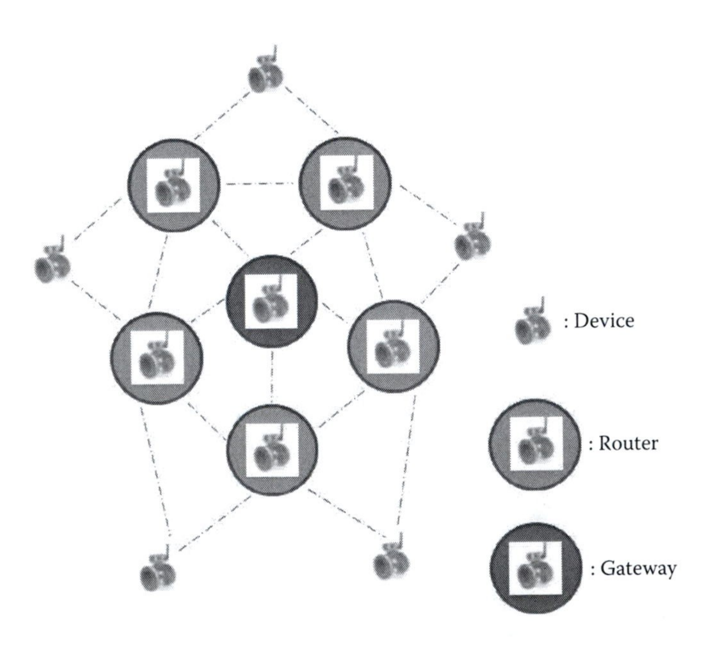

FIGURE 18.6 Star–mesh topology.

Star and mesh topologies are combined together in a star–mesh topology to derive the best of them. Such a topology is shown in Figure 18.6. Thus, this combined topology has the high speed of star topology and the self-repairing capability of mesh networks.

18.3 WIRELESS SENSOR NETWORKS

Wireless technology applied to process industries is gaining increasing market acceptance because of the numerous advantages they offer. Such industries employ sensors that collect process variable information and can be used to effect control actions, data analysis and logging, health monitoring of structures, environmental condition monitoring, etc. The networking is particularly suited for low-frequency applications. Lower installation cost and time and much faster deployment than traditional wired networks are some of the advantages of the wireless network. WSNs can dynamically adapt themselves to changing environments, such as change in topology, or can switch from one mode of operation to another. It helps in enhancing process visibility, i.e., it reduces blind spots, increases productivity and product quality at the same time, while employing fewer manpower. The smart sensors in WSNs combine sensing, computation, and communication into a single tiny device.

18.3.1 Coexistence Issues

Although numerous advantages accrue by deploying wireless network technologies in process plant operations, some serious concerns do arise because of RF interference between various wireless technologies. For example, if IEEE 802.11b and IEEE 802.15.4 protocols are employed in the vicinity of each other, RF interferences may seriously impede the quality of reception of either. Both these protocols use the 2.4-GHz unlicensed ISM band for message transmission, and one transmission may affect the other if the issue of interference is not addressed properly. This is known as coexistence issue. The problem may occur when two messages with sufficient energy collide or overlap with each other in time or frequency.

Coexistence is defined as "the ability of one system to perform a task in a given shared environment where other systems have an ability to perform their tasks and may or may not be using the same set of rules." Successful coexistence of different wireless protocols implies reliable delivery of message belonging to each protocol in the presence of RF interference. Several techniques are adopted to combat and minimize the coexistence issue, which include the following: frequency diversity, time diversity, power diversity, coding diversity, space diversity, blacklisting, and channel assessment.

Protocols 802.15.4 and 802.11b/g both operate in the 2.4-GHz ISM frequency band in which the former is divided into 16 channels and the latter into 3. In this band, channel numbers 15, 20, 25, and 26 (for North America) or 15, 16, 21, and 22 (for Europe) of 802.15.4 will be influenced by a lesser degree from the side slopes of 802.11 b/g. This is shown in Figure 18.7. WHART uses a pseudo-random channel-hopping sequence to reduce this interference by using these nonoverlapping channels. Thus, frequency decoupling involves proper channel selection and channel blacklisting.

WHART uses TDMA technology to transmit one message per frequency channel at a time to avoid collision. It is possible that via two separate channels, two different messages are sent—they would not interfere with each other because they use different frequency bands. Since WHART uses only a part of the ISM band at any given instant of time, chances of RF interference with other wireless networks are minimized.

IEEE 802.15.4 protocol–based instruments have low power and, thus, suitable for industrial wireless process control applications. The power of this protocol reduces the chances of interference with WiFi (IEEE 802.11 b/g). They can communicate up to a distance of 150–200 m to the next instrument, which, in turn, can serve as a router to transmit the same onward ultimately to the gateway. For lower communicating distances, the associated instrument power is also reduced accordingly, thus reducing even more the chances of RF interference. Thus, spatial decoupling

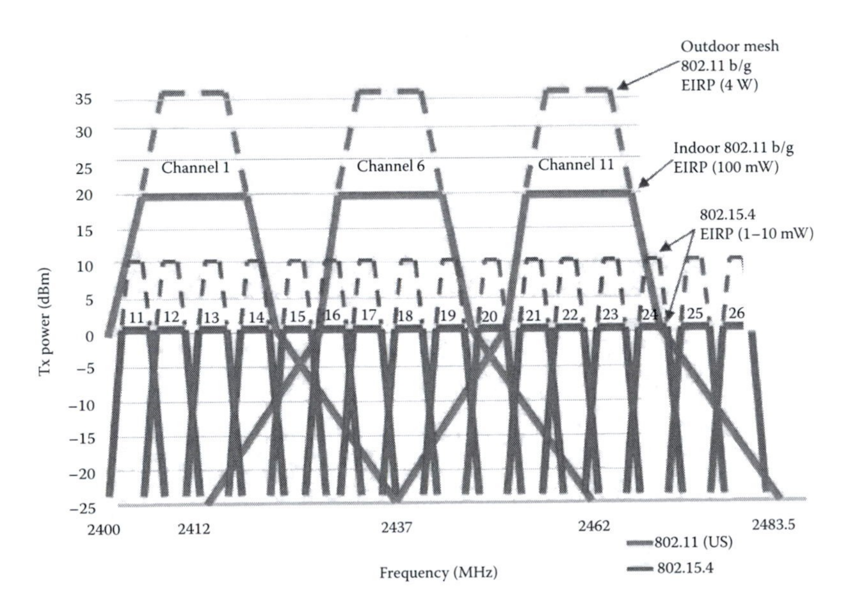

FIGURE 18.7 **(See color insert.)** Response of 802.15.4 and 802.11 b/g in 2.4-GHz ISM band.

involves proper adaptation of transmitted power, antenna selection, positioning, and orientation.

In coding diversity, the message is spread over the entire bandwidth of the selected channel using DSSS. This is decoded by the receiving device specially meant for this.

Mesh networking supports physical space diversity in the implementation of the low-power IEEE 802.15.4 protocol. In mesh networking, the message from the original instrument reaches the gateway via different instruments along its path—the intervening instruments guide and route the original message ultimately to the gateway. Mesh networking supports path redundancy and, thus, are more reliable than if each instrument has a direct line-of-sight path to the gateway. Mesh networks are also adaptive to changing communication needs and also other environmental conditions.

WHART networks can be configured to avoid certain channels that are prone to be heavily used by other networks, thus minimizing chances of conflict and interference. It also has the capability to listen to a particular channel before initiating transmission. If the said channel is busy, the attempted message transmission is aborted and sent in another time slot.

Time-based decoupling involves minimizing the utilization of a particular channel. Thus, occupation of channels on the time scale is reduced, leading to less chances of interference.

18.3.2 WSNs in Industrial Networks

In process automation, sensors are employed at different locations within the plant for process parameter measurements and their subsequent control. Data from the sensors are processed and control actions taken such that the process variables remain within their set values. Closed-loop process control is synonymous with reduced cost, increased productivity, less manpower deployment, increased safety, etc. Today's automation industries include a whole gamut of process plants such as chemical, pharmaceutical, steel, and oil and gas.

The hierarchical structure of an industrial plant automation system comprises stages where specific decisions are taken pertaining to that level. It starts from the field sensor level to the enterprise level, where decisions are taken. Working conditions existing on the plant shop floor level is sometimes very harsh, such as blast furnaces and steel-melting shop in steel plants, compared with the other upper layers.

Figure 18.8 shows a comparison between the network architectures of a conventional, fieldbus, and wireless system. The major advantage of a wireless system (as apparent from Figure 18.8) is that since it is devoid of

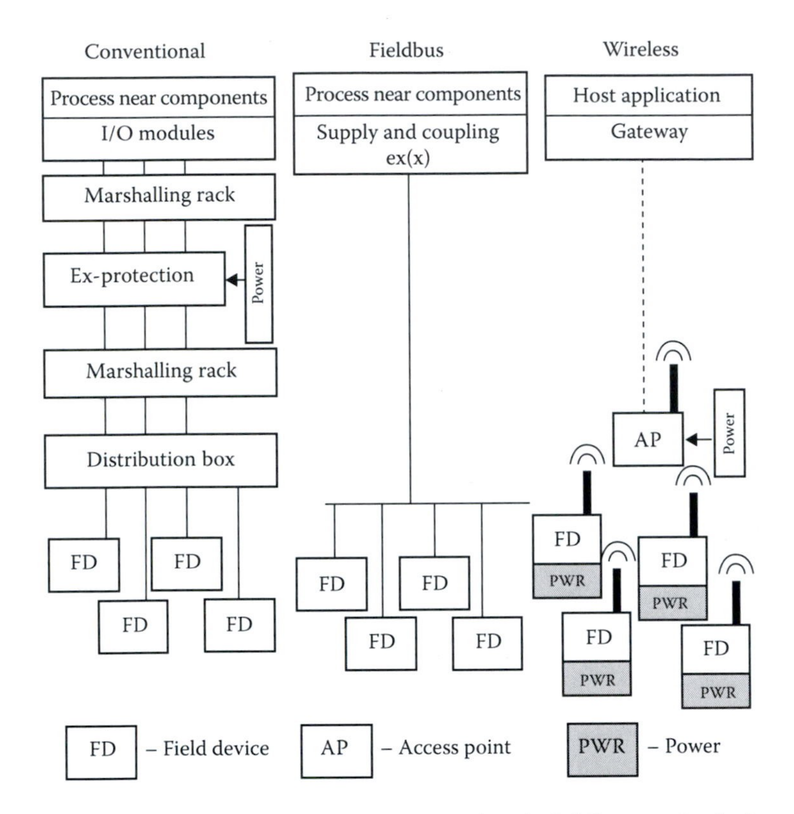

FIGURE 18.8 Comparison between conventional, fieldbus, and wireless network architecture. (W. Ikram and N. F. Thornhill, *Wireless Communication in Process Automation: A Survey of Opportunities, Requirements, Concerns and Challenges*, Control 2010, Coventry, UK 2010.)

any wired connections, it enjoys all the associated advantages of a wireless system compared with its wired counterparts, such as maintenance of costly cables, conduits, etc.

While designing an industrial wireless network, focus must be given to the following: data sent from the stations or nodes must be encrypted and protected against eavesdropping, the network should stay connected irrespective of whether the nodes are static or mobile, energy consumption in each station must be kept as low as possible, transmission should be interference free from the neighboring RF sources/noises, there should be data retransmission in case of packet loss, etc.

18.3.3 Benefits of Industrial WSNs

There are numerous advantages that accrue in applying wireless sensor networks in industrial automation. Although there are certain coexistence issues that must be addressed while implementing WSNs, the benefits far outdo the drawbacks.

In WSNs, no signal and power wiring are needed. In remote and harsh locations where power source is not available and the site is a "hard-to-wire" one, the use of wireless sensors comes to the rescue of industries. For such locations where power is available, the wiring cost becomes prohibitive. Modernization of an existing plant is quite easy with wireless technology. Reduced maintenance and construction cost along with enhanced productivity and safety and fewer on-site personnel requirements are the other major benefits in employing wireless technology in process and automation plants.

19 WirelessHART

19.1 INTRODUCTION

Wireless Highway Addressable Remote Transmission (WirelessHART or WHART), like ISA100.11a, is fast becoming an emerging standard in industrial wireless communication arena that promises to revolutionize the process control and automation industry. Existing wireless technologies such as Bluetooth, ZigBee, or WiFi cannot be used in industrial plants because of their inherent drawbacks. Bluetooth is primarily targeted at personal area networks (PAN) whose range is limited to within 10 m. Furthermore, Bluetooth supports only the star-type network topology with a master having a maximum of seven slaves. This obviously limits its application in process industries. Although ZigBee supports direct sequence spread spectrum (DSSS), its performance is severely degraded and affected in the presence of noise normally encountered in industries. WiFi is not at all a good choice for industrial communication because of its inability to support channel hopping. Power consumption is a matter of concern in WiFi, which also goes against its possible use in industries.

WHART is the first open wireless communication standard specifically designed for process industries. It was officially released in September 2007 as part of the HART specification and was a part of IEC 61158. The WHART specification was approved by the International Electrotechnical Commission (IEC) as an international wireless standard (IEC 62591 Ed. 1.0) for wireless communication and process automation in March 2010.

WHART was developed keeping in mind industrial applications such as device status and diagnostics monitoring, calibration, critical data monitoring, troubleshooting, commissioning, and supervisory process control. WHART provides robust and reliable communication using a DSSS channel-hopping scheme based on IEEE 802.15.4, mesh (redundant data paths), and retry mechanisms.

Wireless mesh networks (WMNs) are gaining more and more market acceptance because of advancements in wireless technology with multi-input/multi-output systems and smart antennas. Some of the key features of WMNs are as follows: self-configured and self-organized, a mix of wired and wireless devices, every network device acts as a router so that the nodes communicate with each other through multihopping, uses gateway to communicate between devices, etc.

19.2 KEY FEATURES

WHART is a highly reliable, self-healing, self-organizing, time-synchronized redundant path wireless mesh network communications protocol designed to meet process industry requirements and applications. It is backward compatible with existing HART devices, i.e., it supports HART command structure and Device Description Language (DLL). WHART operates in the 2.4-GHz license-free ISM radio band utilizing IEEE 802.15.4-compatible DSSS radios with a channel-hopping facility on a packet-to-packet basis.

It uses TDMA technology to arbitrate and coordinate communications between various devices connected to the network. It supports multimessaging modes that includes one-way publishing of process, autosegmented block transfers of large data sets, and has facilities like spontaneous notification of exceptions and also *ad hoc* request–response.

It provides highly secure centrally managed communications using AES-128 block ciphers with individual join and session keys and a data link level network key. The security services are provided by the MAC layer and the network layer.

WHART is a hybrid network consisting of wireless and wired devices and instruments. It has a channel "blacklisting" facility that disallows use of determined channels. Use of redundant paths and redundant communications lead to increased reliability.

Using TDMA for medium access reduces probability of collisions. Also, power consumption is drastically reduced by ensuring radio transmitters remain awake in the allotted slots by prescheduling the network manager.

WHART employs 15 channels in the ISM 2.4-GHz spectrum. Channel selection is based on absolute slot number (ASN). The slot numbers are unique and increase monotonically. Slots and superframes are assigned by the network manager. Slots can be keep-alive, join request–response, *ad hoc* request–response, or special purpose types. Pseudo-random channel usage by employing the channel-hopping technique ensures that interference on one or several channels does not prevent reliable communication. It frequently changes channel for transfer of packets for reliable data delivery. Data traffic can be periodic or sporadic. The field devices of WHART must have mandatory routing capability, while the adapter connects to a wired HART network. The access points provide access to the wireless network.

There is a single network manager that is responsible for configuring the network as well as scheduling and routing of data packets. The security manager is responsible for management and distribution of keys. Sometimes the gateway, network manager, and security manager are packed into a single device.

19.3 WHART NETWORK ARCHITECTURE

The architecture of a WHART network is shown in Figure 19.1. It is a hybrid network consisting of wired and wireless devices and components. It consists of WHART field devices, HART-enabled field devices, handheld field devices, access points, gateway, adapters, network manager, security manager, host applications, etc.

Communication between the wireless field devices and the wired host system is enabled by the gateway. Manufacturers of WHART are trying

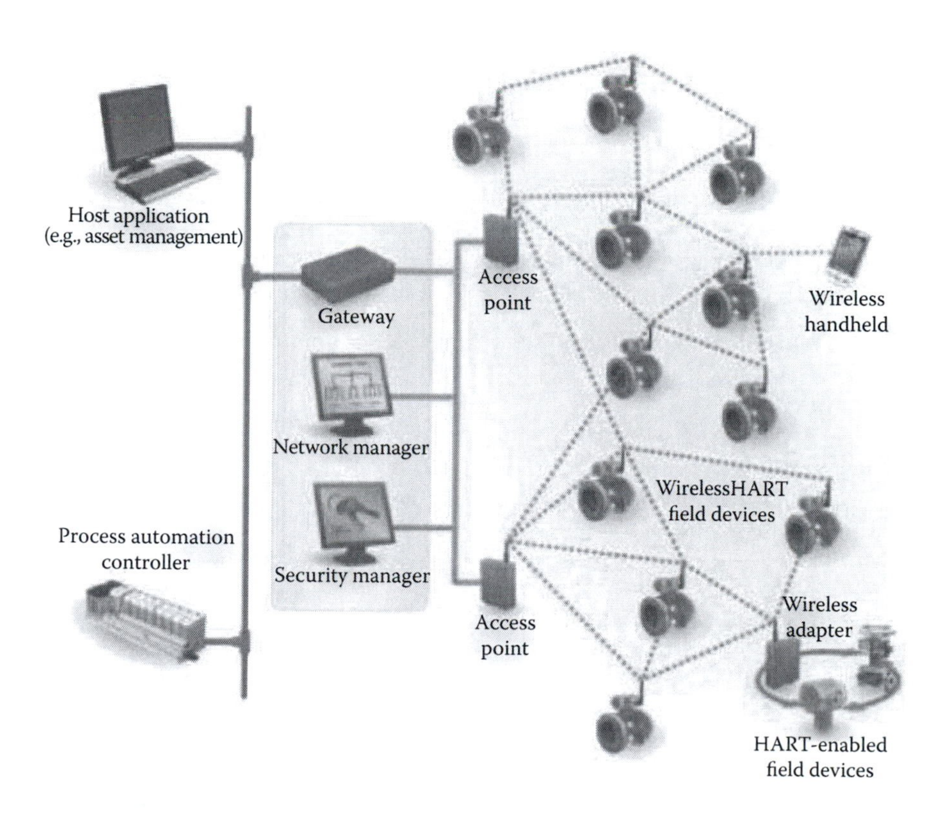

FIGURE 19.1 **(See color insert.)** WHART network architecture. (Available at www2.emersonprocess.com/siteadmincenter/PM%20Central%20Web%20Documents/EMR_WirelessHART_SysEngGuide.pdf.)

to integrate the functionalities of the network manager, security manager, and access points into one product.

Point-to-point topology and mesh topology are used to connect the WHART field devices. The network manager provides new schedules and routing information as new devices continue to join the network. For any wireless device, there should be at least two connected neighbors that can route the traffic using graph routing. Use of mesh topology along with frequency hopping and TDMA for communication makes the system robust and reliable in industrial process control applications.

19.4 PROTOCOL STACK

The protocol stack of WHART is shown in Figure 19.2, along with the seven-layer OSI protocol. The WHART protocol stack consists of five layers: physical layer, data link layer, network layer, transport layer, and application layer. In addition, a central network manager manages the routing and arbitration of the communication schedules.

19.4.1 Physical Layer

The physical layer is based on IEEE 802.15.4-2006 with a data rate of 250 kbps (62.5 kbaud), having an operating frequency of 2400–2483.5 MHz. Its channels are numbered from 11 to 25, with a 5-MHz gap between two adjacent channels. The modulation used is O-QPSK with DSSS. The nominal transmit power is 10 dBm, which is adjustable in discrete steps. The physical layer protocol data unit (PDU) is IEEE compliant with a

Application layer	Command-oriented, predefined data types, and applications
Presentation layer	
Session layer	
Transport layer	Autosegmented transfer of large data sets, reliable stream transport, and negotiated segment sizes
Network layer	Power-optimized, redundant path, self-healing, and mesh network
Data link layer	Secure, reliable, time-synchronized TDMA/CSMA frequency agile with ARQ
Physical layer	2.4 GHz, wireless, 802.15.4-based radios, 10 dBm Tx power

FIGURE 19.2 WHART protocol stack. (Available at www.mindteck.com/images/Wireless%20HART%20-%20Overview.pdf.)

Preamble	Delimiter	Length	PPDU payload

FIGURE 19.3 Physical layer PDU.

maximum payload of 127 bytes. Channel hopping takes place on a packet-by-packet basis. It has clear channel assessment (CCA) facility by which it first checks whether a channel is free before sending a data packet.

The physical layer PDU (PPDU) is shown in Figure 19.3. It consists of a preamble of 4 bytes, a delimiter of 1 byte, a PPDU of length 1 byte, and a variable length payload.

19.4.2 Data Link Layer

The WHART data link layer rests on the top of the 802.15.4 physical layer. The DLL establishes the mechanism for reliable error-free and secure communication between devices to occur in distinct time slots. To ensure error-free communication, WHART DLL introduces channel hopping with a channel blacklisting facility using TDMA technology. TDMA uses precise time slots between which secure and deterministic communication takes place. The data link layer PDU (DLPDU) is ciphered by using a 128-bit Advanced Encryption Standard (AES) cipher algorithm for achieving secure communication at this layer. The DLPDU is shown in Figure 19.4.

The DLPDU consists of several fields. The first contains a 1-byte 0×41, followed by a 1-byte address specifier, a 1-byte sequence number, a 2-byte network ID, a 2- or 8-byte destination and source address, a 1-byte DLPDU, a 1-byte keyed message integrity code (MIC), a 2-byte cyclic redundancy

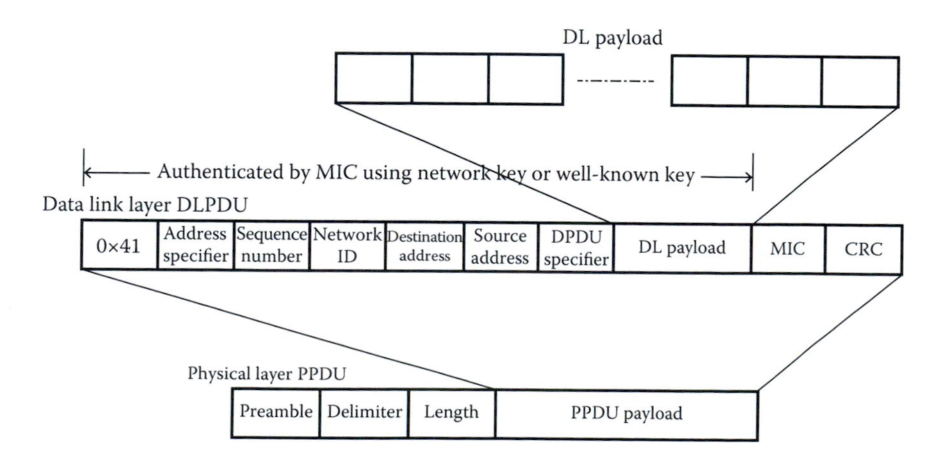

FIGURE 19.4 Data link layer PDU.

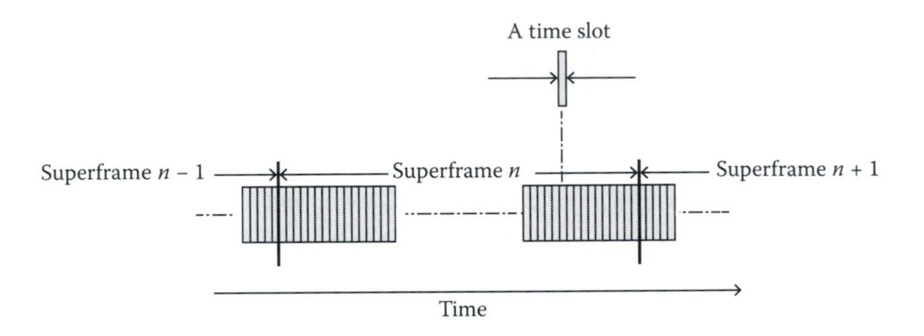

FIGURE 19.5 WHART slot time and superframes.

code (CRC), and a variable-length DLL payload. The DLPDU packet type defines the contents of the DLPDU payload.

A series of time slots (100 time slots per second) form a superframe, and WHART consists of multiple superframes. Figure 19.5 shows such superframes with time slots. One superframe is always enabled at a time, while the other ones can be enabled or disabled throughout the network lifetime. To ensure contention-free access to the wireless medium, a time slot is allotted to two devices at the same time—one as the source and the other as the destination. The length of a superframe is fixed when it is active, but can be varied or modified when inactive. Thus, different superframes may have different lengths.

A management superframe has 6400 slots. The slots in such a superframe are used for join request–response, *ad hoc* request–response, keep-alive request, or special purpose slots used for block transfers or handling requests from handhelds.

In a time slot, a DLPDU is transmitted from the source, followed by an ACK DLPDU from the destination. If the destination receives the DLPDU from the source successfully, only then will the ACK DLPDU be transmitted by the destination. Again, if the source does not successfully receive and validate the ACK DLPDU from the destination, the transmission would be regarded as invalid and the DLPDU will be retransmitted by the source in the next available time slot. If a particular link fails to deliver data to a destination repetitively, the link is discarded and data would be sent to the destination via some alternate route based on the routing table of the source device.

When a source device is transmitting a DLPDU, the destination (RX) is in the listening mode, while when the destination sends back the acknowledgment ACK DLPDU, the source (TX) would be in the listening mode. This is shown in Figure 19.6. The acknowledgment packet includes the

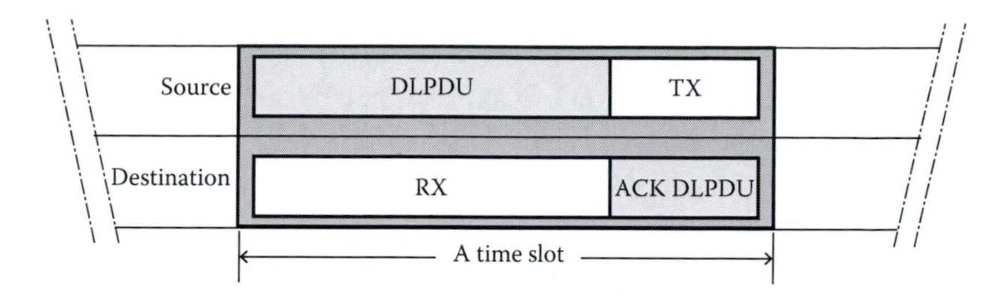

FIGURE 19.6 Message transmission and reception in a time slot.

relevant timing information to continuously synchronize TDMA operation across the whole WHART network.

In the data link layer, message prioritization is done to achieve latency management and flow control. There are four priority levels:

1. Command or highest priority: This corresponds to any packet that contains network management payloads.
2. Process data: It contains process data.
3. Normal: This corresponds to any packet not belonging to "command," "process data," or "alarm."
4. Alarm or lowest priority: Any packet that contains only alarm and event payload.

WHART defines five frame types: acknowledgment DLPDU, advertise DLPDU, keep-alive DLPDU, disconnect DLPDU, and data DLPDU. Data DLPDU contains the higher layer's information in its payload. The other four frames are meant for data link information only. Advertise DLPDU is very important because it contains information about joining superframes and links for the joining devices.

DLL addressing is based on the IEEE 802.15.4 physical layer with a unique 64-bit IEEE extended address (IEEE EUI 64), a short 2-byte unique address within the network, and a 2-byte network ID. Services available are data transfer, data receive, event service (connect, disconnect), management service (*Set*, *Get* parameters), security in the form of authentication but not encryption, and QoS (priority).

Time synchronization is maintained by the gateway, which acts as the ultimate time source. The gateway synchronizes with the access points. The time synchronization is based on a tree-based clock adjustment procedure.

Data transfer on a data link layer can be periodic or sporadic, i.e., occasional in nature. Again, data transfer may require one slot for the first transmission, one slot for a possible retry on a separate channel, or one slot for a second retry on another channel yet again.

19.4.3 Network Layer

The network layer performs many functions: packet routing, encapsulating the transport layer message exchanged across the network, ensuring secure end-to-end communication, blocking data transfer, end-to-end security, acknowledging the broadcasts, etc.

The network manager is responsible for configuring the routing tables of every device in the network, scheduling, and managing communications between different WHART devices. Every device in the network should be able to forward the incoming packets on behalf of the other devices to support mesh technology. There are three routing protocols defined in WHART: graph routing, source routing, and proxy routing.

1. Graph routing: It is a collection of paths that connect the network nodes. The paths in each graph is created by the network manager and downloaded by each network device. Each device maintains a graph table with a list of graph IDs. That is, each network device is preconfigured with graph table (information). When a packet is to be sent, the source device writes a specific graph ID (this is determined by the destination address) in the network header. When the device receives a packet containing the graph ID, the information stored in the graph table is utilized to select the next path (hop).
2. Source routing: In this, the source device includes in its header the entire route in the packet. Thus, as the packet progresses through the nodes, each time the current node just forwards the packet to the next hop as per the information that is already embedded in the header of the packet. This way, it goes on until the destination node is reached.
3. Proxy routing: It is used when the device is yet to join the network.

The network layer PDU (NPDU) is shown in Figure 19.7. The frame begins with a control header that contains an addressing scheme employed, and it also shows whether any special routes are used in the remainder of this byte. TTL stands for time-to-live and it is a counter that continues to decrement as the message passes through the hops, one after the other. Thus, the counter value at any instant indicates how many more hops the message is to pass through before it reaches its destination. The ASN Snippet field provides performance information about the network. It specifies the time passed since a packet was formed. Graph ID routes the packet about the path the packet takes to reach the destination. The next two fields are destination and source addresses. The expanded routing information field indicates any additional routing information that may be needed to guide the packet to the destination.

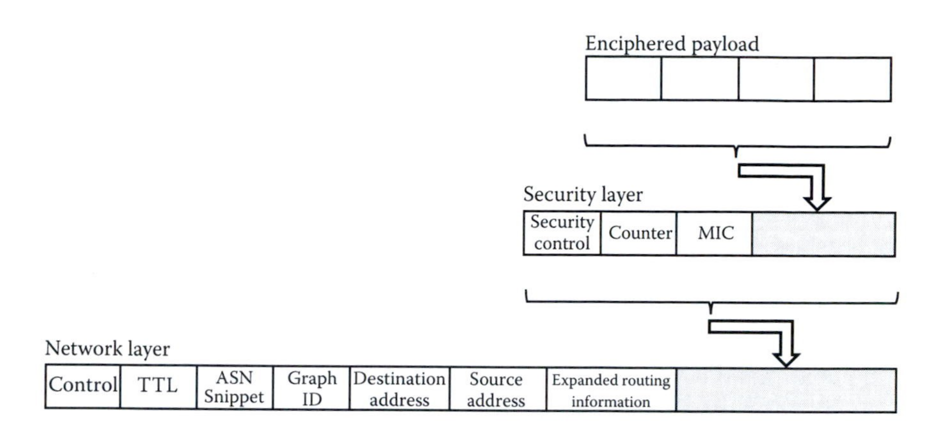

FIGURE 19.7 WHART network layer PDU.

The next field of the network layer is a security sublayer that has several fields and is responsible for data encryption and NPDU authentication. The security field indicates the type of security employed: join key, unicast session key, or broadcast session key. Data integrity is checked by the message integrity code (MIC) field. The total length of the header frame depends on the source and destination address, any special route(s) taken, and the length of the counter. The minimum length of the NPDU header is 21 bytes.

19.4.4 Transport Layer

Figure 19.8 shows the transport layer PDU. It ensures end-to-end packet delivery for communications that require acknowledgment, such as request–response traffic. Data sets are automatically segmented at the source device and reassembled at the destination device.

Either the join key or any one of the session keys is used to encipher the TPDU. This layer ensures reliable communication. Devices that communicate with each other are provisioned with identical join keys. The transport layer encapsulates the application layer data. It also acts as the convergence point between HART and WHART.

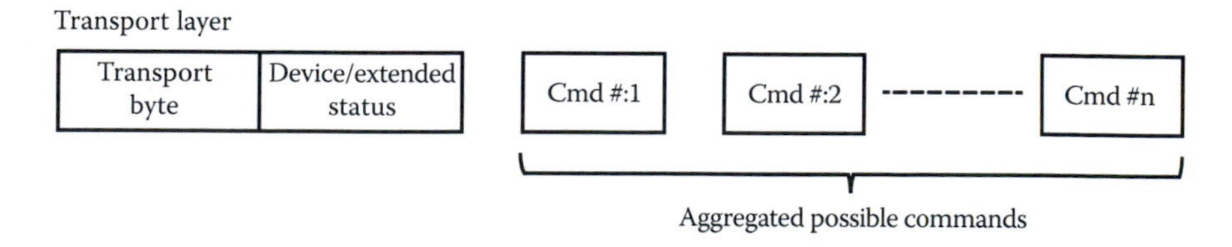

FIGURE 19.8 WHART transport layer PDU.

16-Bit command number	Length	Data

FIGURE 19.9 WHART application layer format.

19.4.5 Application Layer

WHART uses the standard HART application layer, which is command based and resides at the topmost layer in WHART. The commands are universal commands, common practice commands, device family commands, and wireless commands. The universal commands are a set of commands that are supported by all WHART devices. The common practice commands are optional in nature and apply to a wide range of devices. The device-specific commands are manufacturer specific according to field device needs, and the implementation of such commands is also optional in nature. The wireless commands support WHART products. All devices supporting WHART must implement these commands for proper network operation.

The application layer in WHART is responsible for parsing the message content, extracting the command number, executing the specific command, and generating responses. The network manager utilizes the application layer commands to configure and manage the whole WHART network. In WHART, the communication between the gateway and the devices is based on commands and responses. Figure 19.9 shows the general format of the WHART application layer.

19.5 NETWORK COMPONENTS

There are several network components in a WHART network. These are network manager, security manager, virtual gateway, access points, host interface, adapters, routers, and field devices. Field devices are connected to the process or to plant equipments and instruments. These are the sensors distributed throughout the plant that route and forward the data or information packets.

The concept of a distributed architecture with more than one access point, which is normally the case, is shown in Figure 19.10.

The access points shown in Figure 19.10 receive and send the wireless messages. An access point receives the data or messages from the virtual gateway and passes the same to the sensors via the host interface, taking the help of the security and network managers. On the other hand, device data is put on the WHART network via the access point.

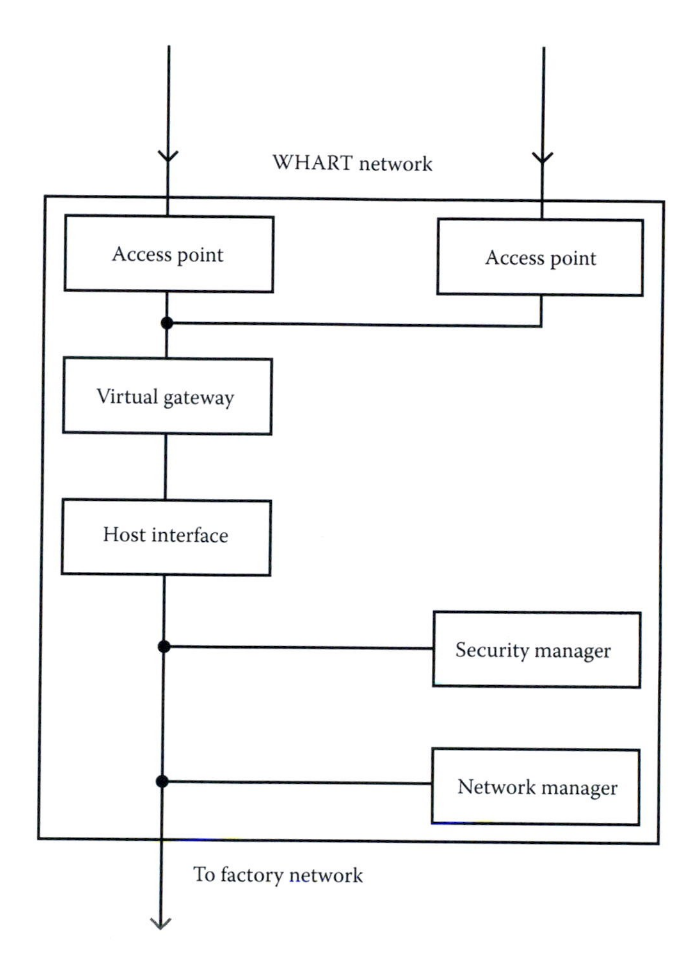

FIGURE 19.10 Distributed architecture of network components. (From WirelessHART, Device Types—Gateways, HCF_LIT-119 Rev. 1.0, June 23, 2010, Gerrit Lohmann–Pepperl+Fuchs, HART Communication Foundation, Austin, TX 78759, USA.)

19.5.1 Network Manager

A network manager can be thought of as the receiver and distributor of HART commands. It is the control center of the whole network. The network manager is not a physical device but simply a software. It deals with forming, organizing, and monitoring the HART network. It configures the network, schedules the required communication among the devices, and manages the routing in the network, as well as reports the health of the network to the host application. In a network, there is only one active network manager at any given instant. The network manager receives the status of all the devices in the network. On the basis of this, it schedules the communication between two devices. The network manager is responsible for managing dedicated and shared resources. When a device joins or leaves a network, it is the responsibility of the network manager to manage the same.

A measure of the performance of the WHART network manager can be gauged by the following: maximum number of devices connected to the network, time required to initialize the network, time required for a device to join the network, and the overall throughput of the system.

19.5.2 Security Manager

The security manager works in close association with the network manager. It allows only authorized devices to join the network. It provides authorization and encryption keys to ensure proper encryption of messages. It manages the security resources and also monitors the network security.

Some important points with regard to the security manager are as follows:

- There is only a single security manager for a WHART network.
- A security manager can serve more than one WHART network.
- The security manager and the network manager work in a client–server fashion.
- The interface between the security manager and the network manager is not defined in the WHART standard.

19.5.3 Gateway

A gateway enables communication and acts as a link between host applications and field devices in the network. It is functionally divided into a virtual gateway and one or more access points. There is one gateway per network. It is responsible for buffering, protocol conversion, and clock source. Conceptually, a gateway can be thought of as the wireless version of marshaling panels and junction boxes.

A gateway should have the following attributes:

- It should provide for network and security management functionalities.
- It should have multiple output protocols that would ensure proper integration to a range of different host applications such as DCSs, PLCs, and data historians.
- It should support multiple connections; i.e., it acts like a server. Thus, data can be sent to different types of end users without any difficulty.
- A gateway should be interoperable; i.e., it can support field devices from different vendors.

- A gateway should support secure transfer of all protocols over an Ethernet connection through a robust encryption process.
- A gateway catering to different users must have different security access for each of them ensuring a secure network administration.

19.5.4 Adapter

An adapter allows existing HART field devices to be integrated into a WHART network. It provides a parallel communication path to the existing 4–20 mA current loop for the WHART network. The uses of an adapter are as follows:

- Provides a wireless communication path to HART devices.
- Diagnostics not available with a legacy HART network can easily be analyzed with WHART.

An adapter should have the following:

- An adapter should have a HART tag.
- An adapter should employ identical security functions as available in a WHART network.
- An adapter must not affect the normal 4–20 mA current signal either under normal condition or in a failed condition.
- An adapter should behave like any other WHART device.

19.6 ADDRESSING CONTROL

WHART is envisaged to be used in industrial process measurement and control applications. For proper feedback control to take place in process automation field, the simple rule of thumb is applied. It mentions that control should be executed 5–10 times faster than the process response time (process time constant) plus the associated process dead time.

19.6.1 Sample Interval

Figure 19.11 shows that a change in the process input results in process output where the output starts changing only after the process dead time. In most of the cases involving process measurements and control, measurements are unsynchronized, as is the case shown in Figure 19.11. Thus, measurements and corresponding controls must be much faster than the process response time. Fortunately, in process control, demand on measurement and control is not fast enough so that even with unsynchronized measurement, no difficulty is encountered for effective control to take place.

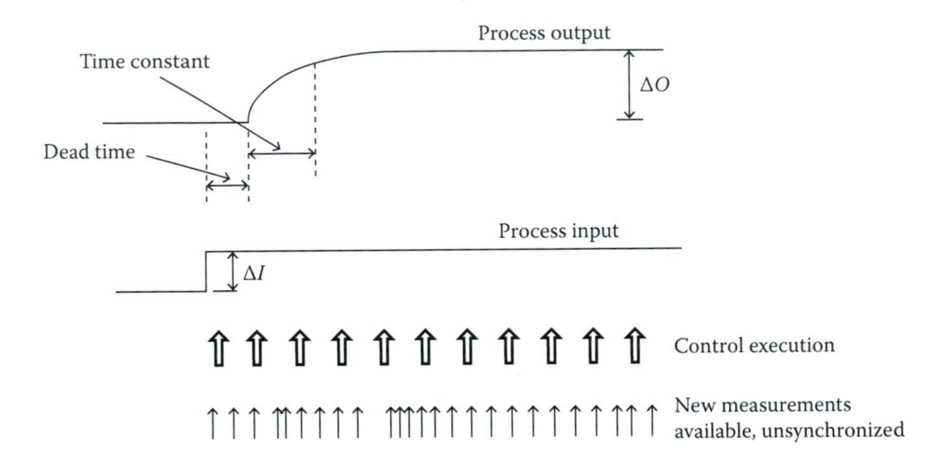

FIGURE 19.11 Unsynchronized process measurement and control system. (From Control with WirelessHART, HART Communication Foundation, Austin, TX 78759, USA.)

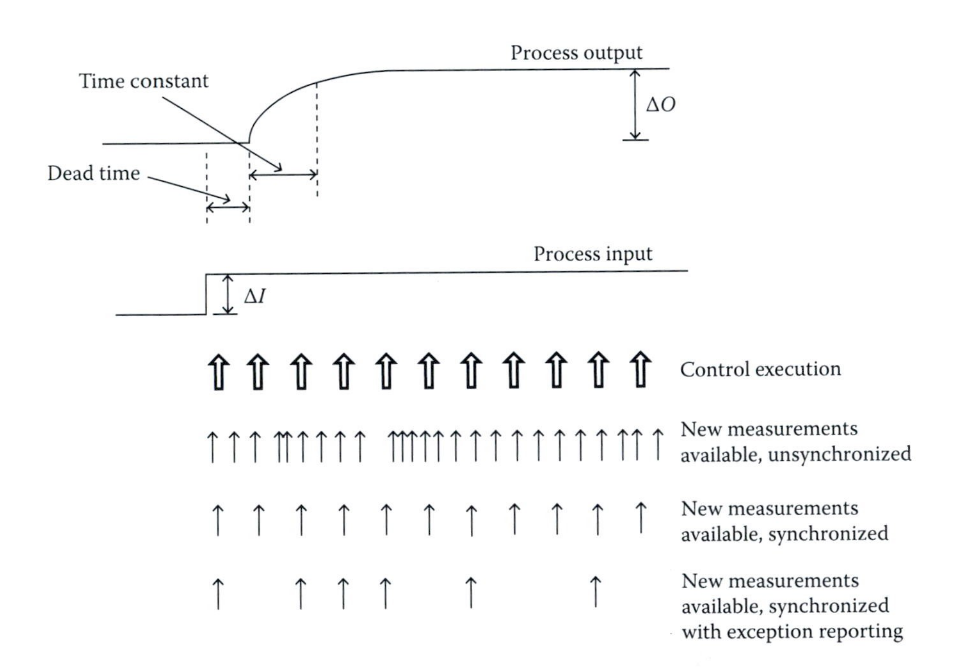

FIGURE 19.12 Synchronized process measurement and control system. (From Control with WirelessHART, HART Communication Foundation, Austin, TX 78759, USA.)

To extend the life of battery of the measurement device, it is desirable to reduce the measurement frequency to the extent that effective control still can take place that does not compromise control reliability. This can be achieved with proper scheduling of communications. There are two methods to achieve this: synchronized and synchronized with execution reporting.

In the former, measurements are taken and transmitted only when they are needed for control execution. In the latter, measurements are taken at scheduled intervals, but transmitted only if the measured value exceeds a certain specified limit or if the time since the last communication exceeds a certain time limit. This is shown in Figure 19.12.

19.6.2 Latency and Jitter

Time is required for a measurement data to reach the controller input. This latency (time delay) and jitter (variation in latency) may affect proper control of the process.

Fortunately, WHART is a time-synchronized control protocol with every field device having a common sense of time accurate to 1 ms across the entire network. This synchronization is not available with most of the other protocols. This time synchronization is used to schedule measurements, almost eliminating latency and jitter.

Compared with some wired fieldbus systems, WHART is faster in nature. For example, in Foundation Fieldbus, the transmission rate is 31.25 kbits/s, which involves a communication delay of 32 μs/bit. This figure is only 4 μs/bit in WHART for a data rate of 250 kbits/s.

Again, a time slot involves 10 ms for transmission and its acknowledgment to take place. A typical WHART message consisting of 128 bytes needs only 4 ms for the message to reach its destination. Thus, the latency problem can be minimized.

Communication latency does not pose any problem as long as the latency time is less than the process response time. For all practical purposes, this is so. Appropriate communication schedules, along with blacklisting of channels and channel hopping and proper networking with appropriately placed access points can reduce the latency to such an extent that latency always remains much less than the response times of most of the physical processes encountered in practice.

19.7 COEXISTENCE TECHNIQUES

The process industry has seen the application of wireless signals for quite some time, and WHART has been in the picture for a few years. It operates in the license-free ISM band, which is also being used by other wireless systems such as Bluetooth, WiFi, and ZigBee. Thus, any wireless industrial application has to confront signals coming from such systems. Efforts have been made so that WHART can coexist with these signals and can reliably communicate with other wireless field devices as well as the wired host systems and applications. Several techniques have been adopted to achieve the above, which are discussed below.

19.7.1 Channel Hopping

As already mentioned, WHART is a highly reliable, self-healing, self-organizing, time-synchronized redundant path wireless mesh network. This self-organizing property of WHART ensures that the network can hop from one path to another in its passage to the receiver. The different paths are subject to varied amount of radio frequency (RF) interferences. Their effects are minimized by the self-organizing property of WHART networks.

WHART uses 15 channels having different frequencies in the ISM band. Hopping from one frequency to another will help in ensuring interference-free communication. WHART uses a pseudo-random channel-hopping sequence to reduce the chances of interference from other possible interfering systems such as Bluetooth, WiFi, and ZigBee. A pseudo-random method ensures that any particular channel is not in use for any prolonged length of time, thus reducing any prolonged interference.

WHART utilizes a channel-hopping scheme called *slotted hopping*, in which pattern variation is done for different channels, and shown in Figure

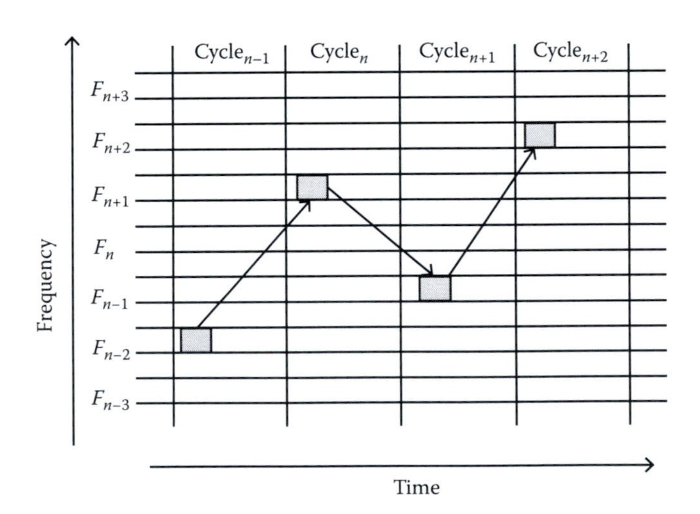

FIGURE 19.13 Channel hopping in WHART. (From G. Wang. *Comparison and Evaluation of Industrial Wireless Sensor Network Standards ISA100.11a and WirelessHART.* Master of Science Thesis, Communication Engineering. Department of Signals and Systems, Chalmers University of Technology, Gothenburg, Sweden, Report No. EX036, 2011.)

19.13. The channel-hopping order is dependent on channel offset, ASN, and also on the number of channels that are currently active. The active channel is represented by

Active channel = (channel offset + ASN)% number of active channels

19.7.2 DSSS

DSSS provides about 8 dB additional gain employing a unique coding scheme. DSSS thus provides an improved signal-to-noise level. It extends receiver sensitivity via digital processing. The coding ensures that the transmission is spread over the entire frequency band allocated to a WHART network. WHART-enabled field devices can correctly decode this coded information, while others see the transmission as a white noise and consequently reject it. In consequence, multiple overlapping radio signals are received, understood, and properly decoded only by the WHART network devices.

DSSS features include less sending power, signal hidden in the background noise and, hence, the signal cannot be tapped or jammed. Figure 19.14 shows how the DSSS signal spreads over the entire frequency range. Figure 19.14 also shows the original message signal.

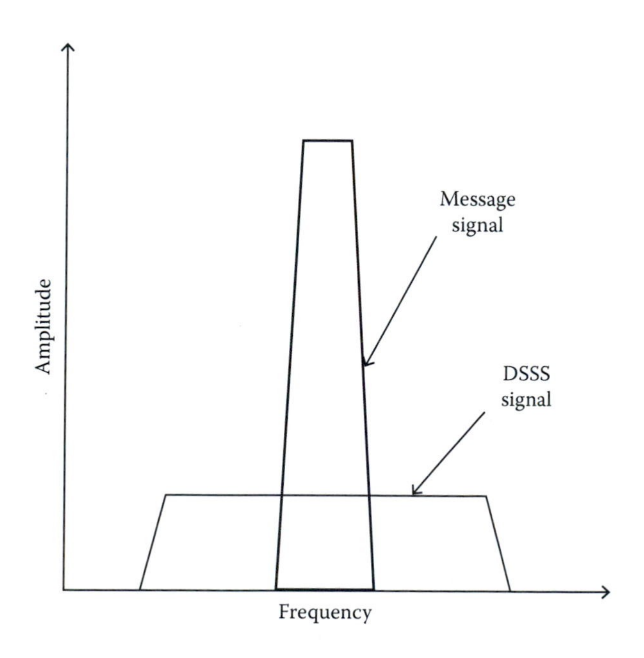

FIGURE 19.14 Spread of DSSS signal.

19.7.3 Low Power Transmission

WHART devices are provided with low power compared with other wireless applications such as RFID readers and Bluetooth. Thus, WHART devices are subject to less interference from such applications. For process control applications, IEEE 802.15.4 has been chosen for its relatively low power. To cover a distance up to 200 m to the next device, 10-dB amplifiers are used. They act as routers to pass the message along en route to the access point. In case lesser distance is involved, even lower-power amplifiers are used for hopping from one device to the next, thus reducing chances of interference still further with IEEE 802.11b/g networks. This limits the RF pollution for users using the same spectrum. It must be kept in mind that the radius of interference increases faster with transmit power than the radius of reception. Thus, better spatial density and throughput over a given area would be obtained with low-power RF control.

19.7.4 Blacklisting and Channel Assessment

WHART can be configured to avoid channels that are relatively used more than others. This would result in lesser interference. However, in reality, most networks are not loaded continuously and, hence, chances of such interference is not that high.

Before transmitting, a WHART device can listen to the frequency channel it intends to use for the purpose of transmission. If another transmission is detected, the device will back off for the present and would attempt transmission at a later time slot on a different frequency.

19.7.5 Spatial Diversity

It involves placing the wireless devices in such a manner that coexistence issues are minimized with nearby wireless devices. It requires careful planning during the installation phase and also during plant expansion.

One example of spatial diversity is antenna diversity in which the receiver receives the best signal from multiple antennas. Again, directional antennas transmit in a specific direction, which helps in mitigating RF interference.

19.8 TIME-SYNCHRONIZED MESH PROTOCOL (TSMP)

Time synchronization is critical in WHART. Communications on WHART are precisely scheduled and timed. It is based on the Time

Division Multiple Access (TDMA) mechanism and employs a channel-hopping scheme for system security. Communication schedules are structured by graph routes that create communication paths between one or more paths (hops). All network devices keep track of time in terms of ASN, which helps them communicate with each other using the correct scheduled time slots. Gateway is the device that provides the time clock to the whole network.

Time is required for measurement data to reach the controller via different paths incorporating latency in the system. In WHART, each field device is time stamped with time accurate to 1 ms across the entire network. This frequent synchronism of the devices is not available with most of the other protocols. With a data rate of 250 kbits/s, WHART is much faster than most of the wired fieldbus systems, like a transmission data rate of 31.25 kbits/s in Foundation Fieldbus systems.

19.9 SECURITY

WHART is an IEC-approved standard mesh networking technology and is used for comparatively secure and reliable message transmission in process control and automation environment. A field device can have one or more sensors that collect information about the process and send them to other field devices. This routing information, security keys, and the timing information are sent to the devices in a very secure manner. Data travel along the WHART network in the form of commands, and confidentiality, integrity, and authenticity of the commands are ensured.

Security requirements for both wireless and wired portions should be addressed properly, although WHART neither enforces nor specifies any means to provide security and reliability in the wired part of the whole network.

19.9.1 OSI Layer-Based Security in HART and WHART

Figure 19.15 shows the security and reliability aspects as applied in HART and WHART at different layers of the OSI protocol. In WHART, data from the application layer down to the transport layer does not provide any cryptographic protective cover, although from the network layer downwards, WHART data, i.e., WHART commands, are protected.

It is apparent from Figure 19.15 that network along with the data link layer provide the security services, i.e., confidentiality, data integrity, source integrity (authenticity), and availability.

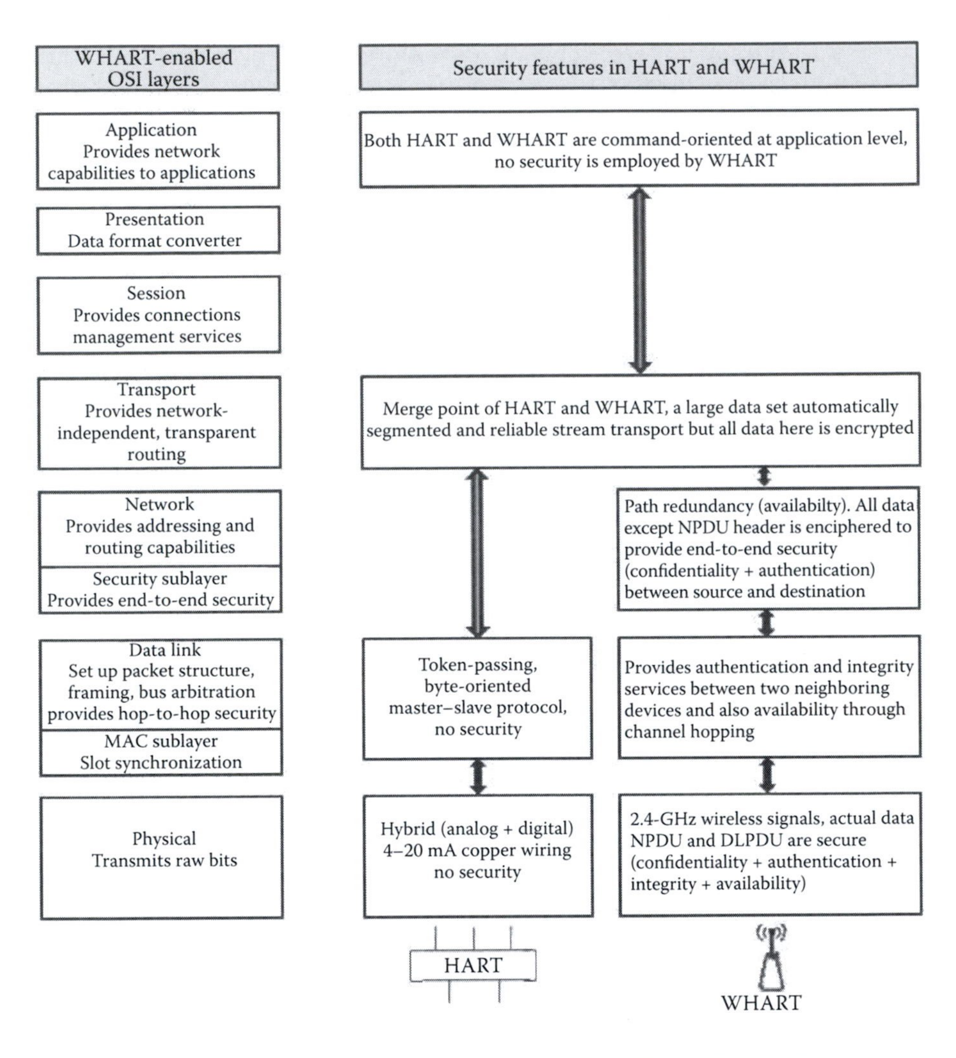

FIGURE 19.15 Security at different layers of HART and WHART based on OSI protocol. (From S. Raza. *Secure Communication in WirelessHART and Its Integration with Legacy HART*. Swedish Institute of Computer Science, Technical Report T2010:01, ISSN: 1100-3154. SICS Networked Embedded Systems (NES) Laboratory, Kista, Sweden, p. 22, 2010.)

19.9.2 End-to-End Security

This is all about the security between the source and the destination devices when data or message travels through the network. The network layer is responsible for providing this security.

A field device cannot create a session with another one. Thus, data from a source field device to a destination field device has to travel through a gateway. A session can be created between a field device and a gateway, or between a field device and the network manager. Thus, data from a source field device is first encrypted with a unique symmetric session key and sent to the destination device via the gateway. The gateway decrypts the data from the source and encrypts it again with the session key of the destination device. The gateway finally sends it to the destination device.

19.9.3 NPDU

The NPDU is already shown in Figure 19.7 and consists of several fields, which are shown as a header field in Figure 19.16. The security sublayer of NPDU consists of three fields: a security control byte, a counter, and the MIC.

The NPDU payload shown is the transport layer PDU. It is encrypted using an AES with a 128-bit key. The other fields present in the NPDU are responsible for routing of data. Network layer provides confidentiality, integrity, and authentication.

19.9.3.1 Security Control Byte

The security control byte (SCB) is part of the NPDU. It is used for defining the type of security employed. The upper nibble of the SCB is reserved for future security requirements, while the lower nibble defines the security type. It is shown in Figure 19.17. Three types of security are defined:

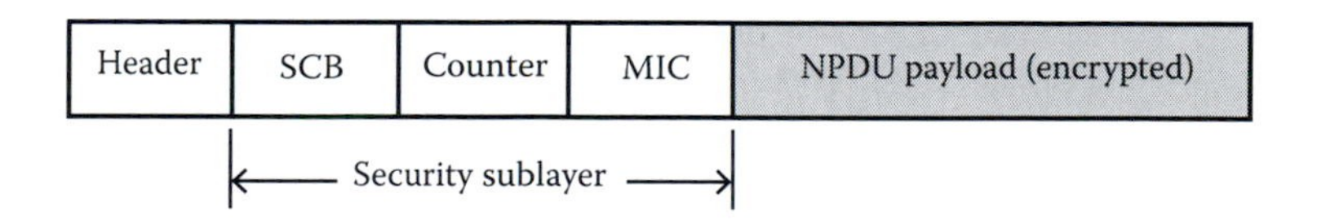

FIGURE 19.16 WHART NPDU.

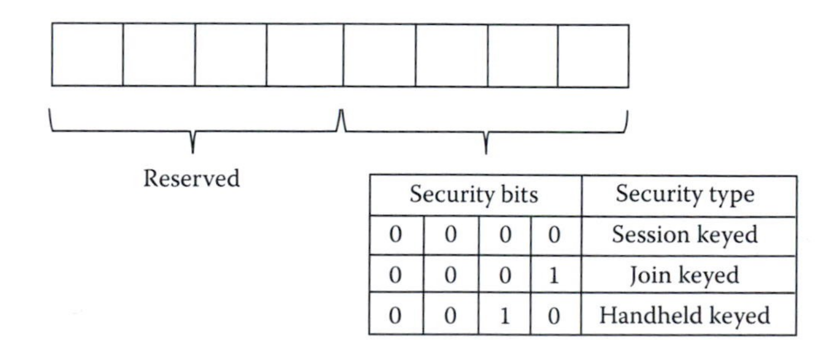

Security bits				Security type
0	0	0	0	Session keyed
0	0	0	1	Join keyed
0	0	1	0	Handheld keyed

FIGURE 19.17 Security control byte. (From S. Raza. *Secure Communication in WirelessHART and Its Integration with Legacy HART.* Swedish Institute of Computer Science, Technical Report T2010:01, ISSN: 1100-3154. SICS Networked Embedded Systems (NES) Laboratory, Kista, Sweden, p. 24, 2010.)

session key, join key, and handheld key. These three key types are defined till HART 7.1.

19.9.3.2 Message Integrity Code (MIC)

The MIC provides source integrity (authenticity) and data integrity. It is calculated on the NPDU after setting the TTL, counter, and MIC to zero. Four byte strings are needed for MIC calculation: NPDU header, NPDU payload, AES key, and the nonce. The AES key is used for encryption of the NPDU payload.

The nonce is 13 bytes in length and is used to defend reply attacks. The first byte is either all ones (for join response message only) or all zeroes. The nonce counter consists of the next 4 bytes. The last 8 bytes make up the source address.

19.9.3.3 AES-CCM

The NPDU can be encrypted or decrypted with the help of AES-CCM in counter mode. Here the message blocks are created in the same manner as in AES-CCM–CBC-MAC mode. However, in this case, no padding is inserted and the blocks can be manipulated in parallel. The process of enciphering the NPDU payload by AES-CCM in counter mode is shown in Figure 19.18.

In this mode, an initialization value (vector) initializes the first counter block, and subsequently the counter blocks go on increasing with each

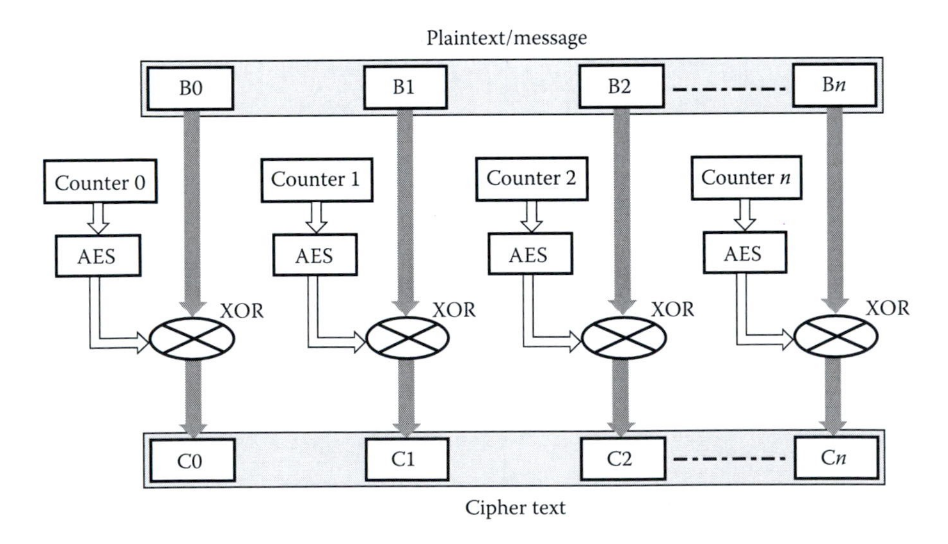

FIGURE 19.18 Enciphering NPDU payload with the help of AES-CCM in counter mode. (From S. Raza. *Secure Communication in WirelessHART and Its Integration with Legacy HART.* Swedish Institute of Computer Science, Technical Report T2010:01, ISSN: 1100-3154. SICS Networked Embedded Systems (NES) Laboratory, Kista, Sweden, p. 26, 2010.)

block of plaintext. The algorithm uses the same AES-128 key to encrypt the counter blocks, one after the other. This way the final enciphered text C0, C1, C2 … C*n* is formed as shown in Figure 19.18.

The figure shows that the plaintext message is divided into B0, B1, B2 … B*n* as per the length of the counters—counter 1 to counter *n* to get the final ciphered text after encryption.

19.9.3.4 AES-CCM–CBC-MAC

The CCM mode is the combination of counter mode and cipher block chaining message authentication code (CBC-MAC) mode. This is used to calculate the MIC and is shown in Figure 19.19.

The CBC-MAC is implemented to obtain data integrity and authentication. Both plaintext and cipher text can be used to generate MIC. This can be verified with the help of a shared secret key. The mode requires an exact number of blocks with padding in the last block if the situation so demands. The same session key is used for both encryption and MIC calculation.

The plaintext is divided equally into several blocks according to the length of IV, wherein zero padding may be required in the last block to fill in the total block size. Actually, a formatting function is applied on the unencrypted NPDU header, encrypted NPDU payload, and on nonce to produce B0, B1 … B*n*. XORing of B0 and IV is encrypted by AES 128. This is then XORed with B1 followed by a second AES operation. The final MIC is obtained as shown in Figure 19.19.

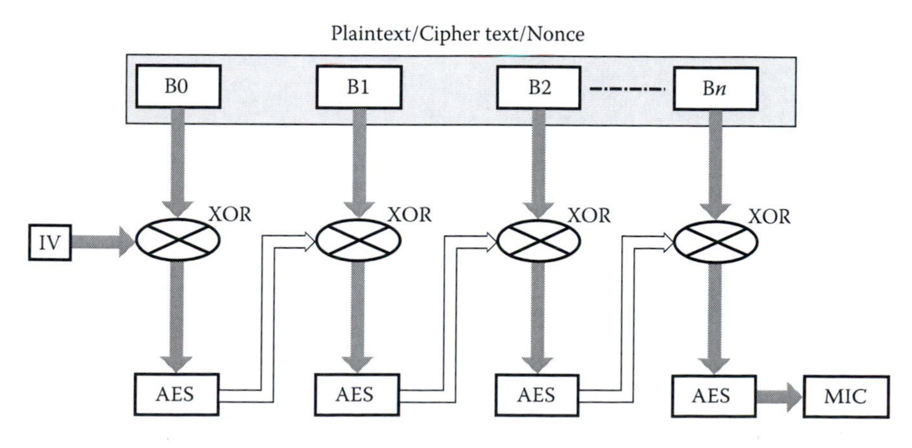

FIGURE 19.19 AES-CCM–CBC-MAC mode for MIC calculation. (From S. Raza. *Secure Communication in WirelessHART and Its Integration with Legacy HART.* Swedish Institute of Computer Science, Technical Report T2010:01, ISSN: 1100-3154. SICS Networked Embedded Systems (NES) Laboratory, Kista, Sweden, p. 25, 2010.)

19.10 SECURITY THREATS

Security in an industrial process control and automation system depends on both its wired and wireless portions. Since WHART shares the same ISM band along with other wireless communication systems such as WiFi, Bluetooth, and ZigBee, there is always a security threat from these. A threat is the presence of some unwanted signal other than that expected to be present. The presence of such unwanted signals may lead to security breach in the form of an information leak, disruption in network traffic, or a change in the information expected at the receiving end. A security threat is also encountered at the point where the wireless part of the network meets its wired counterpart. Figure 19.20 shows the different kinds of security threats and the ways to combat them.

19.10.1 INTERFERENCE

Interference is an unforeseen and unintentional disruption to the intended communication signal that is meant for the receiver. The possibility of such an interruption overriding and dwarfing the original signal occurs when the interfering signal has the same frequency and modulation as that of the original signal. WHART operates in the ISM band of the 2400–2483.5 MHz band with a bandwidth of 5 MHz for each of its 16 channels. This

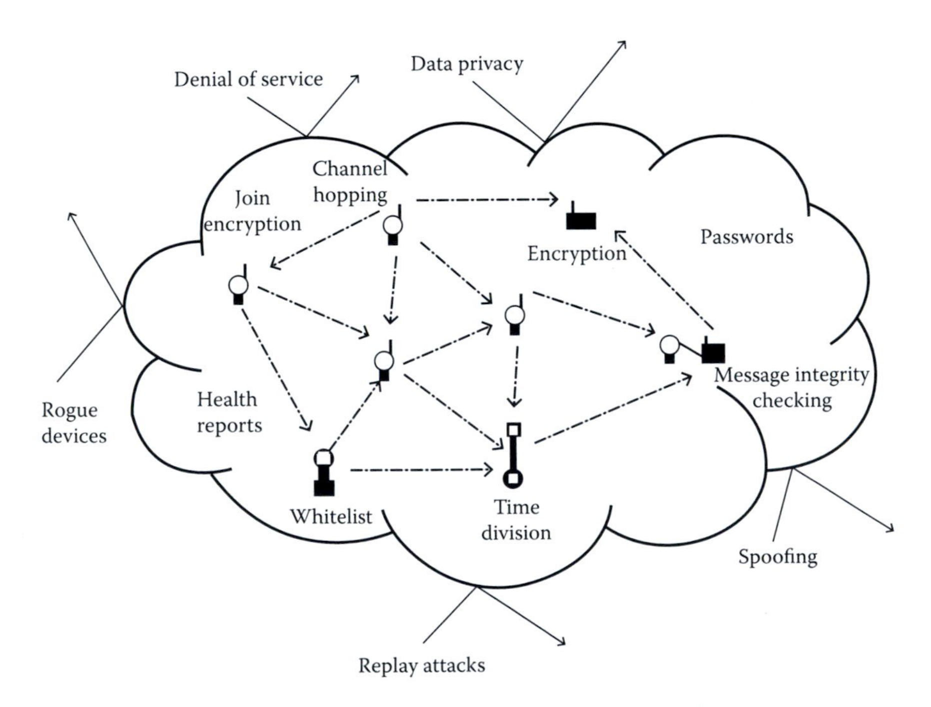

FIGURE 19.20 Different security threats and ways to combat them.

range is shared by Bluetooth, ZigBee, and WiFi of IEEE 802.11b/g. Thus, operating any of these three in the vicinity of a WHART system may cause interference. Extensive use of such IEEE 802.11b/g may ultimately lead to failure of the WHART-based communication. The problem may be mitigated by using frequency hopping spread spectrum (FHSS—frequency diversity), time diversity, and path diversity.

19.10.2 Jamming

In jamming, an intentional disruption to the network signal is introduced in the form of some noise having the same frequency and modulation technique as used in the network communication signal. The effect of such intentional jamming on the network can be much more severe and damaging to the network performance, and it depends on the intruder's ability to inject a noise signal to jam the network.

To overcome the problem arising from channel jamming, WHART has an FHSS feature that blacklists the channel(s) that is/are a constant source(s) of noise. However, such blacklisting of channels results in fewer available channels to carry WHART signals. Blacklisting of a channel is done manually by the network administrator.

19.10.3 Sybil Attack

In this, an attacker may hold multiple identities in the form of a node or a software introduced in the system. Such a Sybil attack on the system may hamper the performance of the network badly. The attack can be taken care of effectively in the following manner.

Each device used in the network has a globally unique ID, which is a combination of device type and device ID. These unique IDs are maintained by a WHART gateway, while the network manager assigns individual nicknames to each of the devices. Each device in the network uses these two (unique ID and nickname) to establish sessions with the gateway and network manager, respectively.

19.10.4 Collusion

Collusion occurs when more than one device try to access the same channel at the same time. It can be either intentional or otherwise, and can be reduced by channel hopping (frequency diversity) and time diversity (TDMA).

A collusion problem is detected by CRC-16, and its effect on the system performance is reduced by proper coordination and implementation of physical and data link layers.

19.10.5 Tampering

If an intruder to the network has knowledge of the network key and unencrypted DLPDU, the latter can be very easily tampered with, and a new MIC can be formed to make it appear authentic. The attack can be serious and damaging since the DLPDU is unencrypted and the packet can be sent to another destination. The attack becomes more serious when the packet is sent back to the source. This can have adverse consequences on network performance.

19.10.6 Spoofing

A device intending to join the network does so by advertisement time slots earmarked for the same. This is done with the help of the well-known network key. An attacker can spoof a new device intending to join the network by sending a fake advertisement. On receipt of the join request, it can simply reject the new device and debar it from joining the network. The proximity of the spoofing device to the device intending to join the network can permanently discard the latter from joining the network. If the spoofing device has access to the network key, it would then result in more malicious attack leading to network traffic blockage.

19.10.7 Exhaustion

An attacker can use a device to send messages to other devices in the network by using the well-known network key. This is possible when the attacking device has knowledge of network parameters and conforms to WHART protocol stack. Using the network key, the MIC of the DLPDU can be calculated and the attacking device would then use the join key to encrypt and authenticate the NPDU. Such fake messages, however, would be discarded by the network manager finally; however, it would consume both network resources and time, thereby seriously affecting network performance and blocking network traffic. Flooding a network with such repeated join attempts would lead to severe traffic bottlenecks affecting network performance and reliability.

19.10.8 DOS

DOS, or denial of service, can be effected in any one of the following ways:

- Sending fake time advertisements.
- Flooding the network with fake join requests.
- Jamming the network signal.

- Replacing the unencrypted DLPDU and recalculating the CRC. In such cases, message integrity can be calculated by MIC, which can be found out with the help of AES in CCM. However, verification of MIC involves a lot of network time, which would eventually reject the unverified packet.

19.10.9 Traffic Analysis

Traffic analysis can be done on NPDU and DLPDU, which are unencrypted. The NPDU header fields involve source and destination addresses, security control byte, ASN Snippet, and nonce counter, which are sent in clear. Again, DLPDU header fields involve the address specifier and DLPDU specifier, which are also sent in clear. These fields in the two layers can easily be analyzed by the attacker. The attacker would then be able to know the device usage rate, peak usage period, join requests, etc., to launch attacks much more effectively to bring down network performance.

19.10.10 Wormhole

In this, the attacker creates a tunnel between two legitimate devices by connecting them through a stronger wired or wireless link. The wormhole attack is launched through a HART device that is connected to the WHART via adapters. The tunnel is established between these two devices by using their maintenance ports, although WHART denies establishing such a tunnel by restricting network access in this mode. Such a tunnel can be created by the wireless connection if the network and session keys are known to the attacker.

If WHART uses graph routing (it supports redundant paths), then the system is prone to wormhole attack. Source routing uses a device-by-device route from source to destination that can repulse a wormhole attack. However, such a routing scheme is unreliable since if any one intermediate link fails, it would lead to a packet failing to reach the destination. A packet-leaching method is used to ward off a wormhole attack.

19.10.11 Selective Forwarding Attack

Selective forwarding attack can occur only if there is a Sybil attack on the network. In this attack mode, the compromised node selectively drops packets, instead of forwarding all. A black hole is created when no packet is forwarded. Normally the attacker selectively drops packets so that it appears to be genuine and cannot be retrieved by the recovering mechanism. A proper traffic analysis would lead to such an attack being more effective.

19.10.12 Desynchronization

A WHART network has its operation based on a very strict time synchronism. A timer module maintains this time synchronism of 10 ms. The MAC sublayer takes care of this. When a node receives an ACK, it adjusts its clock accordingly. The source of timing can be a sender and if the sender is compromised, it can disrupt the timing between the two

TABLE 19.1
Security Threats at Different Protocol Layers

OSI Layers	Security Threats	General/WirelessHART Defense Mechanism
Physical	Interference	Channel hopping and blacklisting
	Jamming	Channel hopping and blacklisting
	Sybil	Physical protection of WirelessHART devices
	Tampering	Protection and changing of network key
Data link	Collusion	CRC and time diversity
	Exhaustion	Protection of network ID and other information that is required to join the device
	Spoofing	Use different path for resending the message
	Sybil	Regularly changing the network key
	Desynchronization	Using different neighbors for time synchronization
	Traffic analysis	Sending dummy packets in quite hours and regular monitoring of WirelessHART network using handhelds, etc.
	Eavesdropping	Network Key protects DLPDU from eavesdroppers
Network	Wormhole	Physical monitoring of field devices and regular monitoring of network using source routing. The monitoring system may use packet leash techniques.
	Selective forwarding attack	Regular network monitoring using source routing
	DOS	Protection of network-specific data such as network ID. Physical protection and inspection of network
	Sybil	Resetting of devices and changing of session keys
	Traffic analysis	Sending dummy packet in quite hours and regular monitoring of WirelessHART network using handhelds, etc.
	Eavesdropping	Session keys protect NPDU from eavesdroppers.

Source: S. Raza. *Secure Communication in WirelessHART and Its Integration with Legacy HART*. Swedish Institute of Computer Science, Technical Report T2010:01, ISSN: 1100-3154. SICS Networked Embedded Systems (NES) Laboratory, Kista, Sweden, p. 35, 2010.

nodes. Thus, the two nodes would try to adjust their time synchronism and precious resources are simply wasted.

19.10.13 Security Threats at Different Protocol Layers

Security threats at different protocol layers for a WHART-based network are shown in Table 19.1.

The WHART standard is strong enough to protect wireless devices against security threats; poor implementation of the different protocol layers may put the system vulnerable to malicious attacks.

19.11 REDUNDANCY

Data and information flow at all levels in a process control and automation system is a must for proper control to be effective. Redundancy is used particularly at those positions where process data is operationally very critical or the failure of a device leads to considerable loss of data. Adding redundancy to a system increases the probability of data availability even if a device fails, or else the normal data communication path is blocked due to some reason. WHART provides redundancy at the following levels:

- At the wireless sensor network
- At the network access points
- At the gateway, network manager, or security manager

19.11.1 Redundancy in WSN

Redundancy in the wireless sensor network can be achieved in any one of the following manner:

- Spatial diversity: using multiple paths between the network components
- Frequency diversity: using the frequency-hopping technique
- Time diversity: accessing the network at different times

Referring to Figure 19.21, if data from device 1 to the gateway fails via the path AGK, it can take place via BIL. In case this second path also fails, data can reach the gateway via a third path, AEIL. A maximum of three tries are provided to device 1 for transporting its data to the gateway. The communication between the gateway and a device is bidirectional.

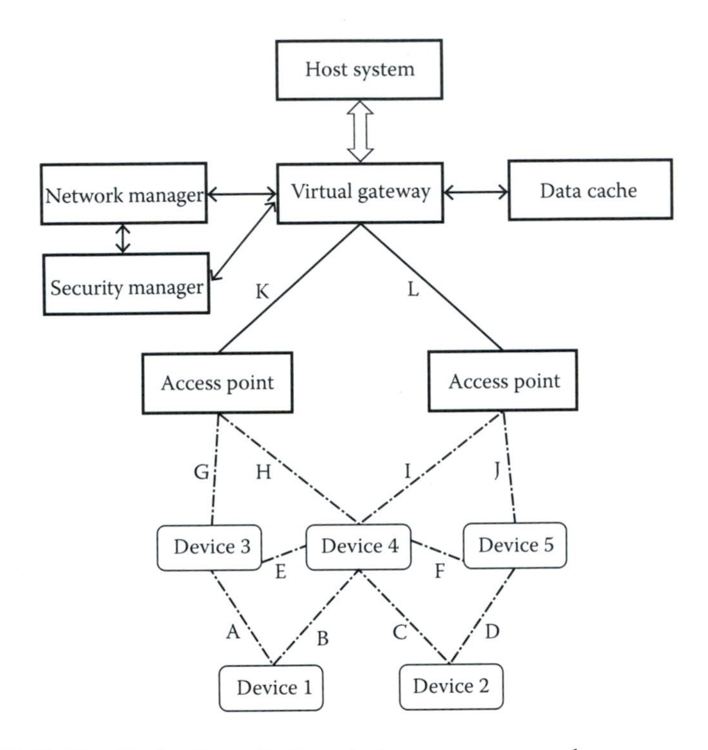

FIGURE 19.21 Redundancy in the wireless sensor network.

19.11.2 Redundancy at Network Access Points

Multiple access points allow additional path diversity and, hence, redundancy to the system. A network access point provides a communication path between a device and a gateway. An access point can be thought of as a network device having a higher bandwidth. It provides path diversity as well as redundant communication paths for the higher-level gateway and network manager.

Theoretically, a network can have any number of access points, but finally a tradeoff is made between cost, overall network bandwidth, and redundancy requirements. Also, multiple access points would lead to a low latency time.

19.11.3 Redundancy at Gateway, Network Manager, and Security Manager

Redundancy can be enforced at the gateway, network manager, and security manager by clubbing the three together in one physical gateway device and repeating this block all over again. The same is shown in Figure 19.22. Redundancy decision is taken on the basis of the designer of the network

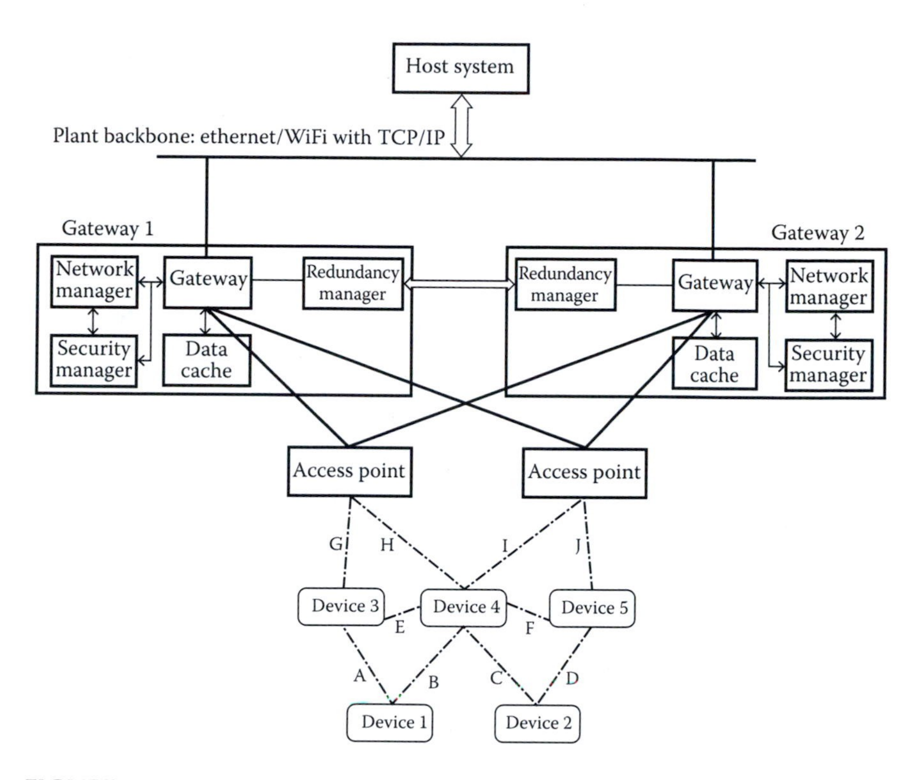

FIGURE 19.22 Redundancy at gateway, network manager, and security manager.

who decides about which component is most critical to system's proper operation. In that case, either of the three, i.e., gateway, network manager, or security manager, can be made redundant instead of making the three redundant at the same time.

In Figure 19.22, one gateway acts as the primary and the other takes over once the running gateway fails. The two gateways communicate with each other via a redundancy manager, which also keeps the system synchronized.

19.12 SECURITY KEYS IN WHART

A complete WHART network comprises both wireless and wired portions. However, a comprehensive key management scheme is not specified for either of the two portions. The WHART standard specifies the need to have a security manager; however, its design and functionalities are not explicitly mentioned.

It is mandatory to have secure communication in WHART, which provides a reasonably strong security. Keys used in WHART have the following features:

- Utilizes AES-128 block ciphers, unsigned with symmetric keys
- Supports multiple key architecture
- Separate join key per device
- Data link PDUs are authenticated by network key
- End-to-end communication is authenticated by session key

The WHART standard specifies the different keys needed and their functions, but does not mention the manner of generating, storing, revoking, renewing, or distributing the keys. The standard specifies the network manager ID responsible for sending the keys to the devices via the gateway, while the security manager is used for the generation, storage, and vetting of keys. The WHART standard does not explicitly specify the manner of distribution of keys and the interface between the network manager and the security manager. A number of keys are specified in WHART for proper functioning of the network. These are

- Join key
- Session key
- Network key
- Handheld key
- Well-known key

The security provided by WHART is transparent to the application layer. Each and every join key has to be distributed to each device before network initialization. The security keys are renewed and rotated as per the plant control automation requirements.

19.12.1 Join Key

Each and every device belonging to the WHART network will have its individual join key in order to join the network. It acts as a password to the WHART network. The security administrator manually distributes this key to each device. The handheld device is normally used to enter the join key to the device using the maintenance port of the device. Whether the join key is being added to the device initially or changed later, the device must remain isolated from the network. The join key is a "write-only" key, like all other keys in WHART.

Command 768, which stands for "Write Join Key," is used to write the join key to the device. The network manager can change the join key later using the same command. The network manager authenticates the new device using the join key and on successful authentication, it writes the network key and session key into the device.

19.12.2 Session Keys

Each device has four session keys that are allocated to them after successful joining, restart, or after power on. The four session keys are

1. Unicast session key between the gateway and the device
2. Unicast session key between the network manager and the device
3. Broadcast session key from the gateway to all the devices
4. Broadcast session key from the network manager to all the devices

Two field devices cannot communicate directly with each other, but do so via the gateway. The first (source) field device sends the message to the gateway via the unicast gateway session key of the source device. The gateway then transfers it to the second (destination) device using the unicast gateway session key of the destination device.

Command 963 (Write Session) is used to write the sessions into the devices. Only the network manager can initiate this command. A session can be deleted by the network manager using command 964 (Delete Session).

19.12.3 Network Key

There is only one network key that is shared by all the wireless devices connected to the network. It provides protection to the DLPDU against outside malicious attack. It is used to calculate the keyed MIC, which protects and secures the DLPDU. Two communicating devices authenticate each other on the basis of this MIC.

Command 961 (Write Network Key) can be initiated only by the network manager and is used to write the network key in the device.

19.12.4 Handheld Key

This key is provided by the network manager using the join key, when requested by the handheld device. It is used for peer-to-peer connection between a handheld device and a device belonging to the network. This mode is used for device maintenance by using the device's maintenance port. Such a connection does not involve a gateway. The handheld key is used to secure the NPDU.

A Handheld Superframe, residing in the handheld, is used for device maintenance. This can be checked by issuing command 806 (Read Handheld Superframe) and can be enabled by issuing command 807 (Request Handheld Superframe Mode). Command 823 (Request Session) is used to establish a peer-to-peer connection between the handheld and the device.

If the session is established, then the network manager will return with the peer device nickname, nonce counter, and handheld key.

19.12.5 Well-Known Key

All messages in the WHART network must be encrypted. During the join process, a device has only the join key, which protects the NPDU but does not possess the network key to protect the DLPDU. A network key, called the well-known key (777 7772E 6861 7274 636F 6D6D 2E6F 7267), is used to calculate the MIC for the join request–response messages. The well-known key is also used for sending join advertisements.

19.13 KEY MANAGEMENT

Automatic key management is a vital and key area without which network security would be jeopardized. Manual key adjustments would lead to occasional system operation failure. WHART employs the network manager for various key distributions, i.e., it acts as a key distribution center, while the security manager is employed for key management purposes; however, the key management issue is not very comprehensive in a WHART network. Key management involves the following:

- Key generation
- Key storage
- Key distribution
- Key renewal
- Key revocation
- Key vetting

The keys used are symmetric in nature, i.e., asymmetric cryptography is avoided.

19.13.1 Key Generation

Key generation is the responsibility of the security manager. The WHART standard does not explicitly specify ways for key generation but specifies the keys needed to secure the channels.

19.13.2 Key Storage

Network security in a WHART system is dependent on keys. Storage of keys in a secure manner is of paramount importance, although WHART does not provide means for key storage, like key generation.

TABLE 19.2
Various Key Distribution Commands in WirelessHART

WirelessHART Keys	Commands
Session keys	Command 963 (Write Session)
Network key	Command 961 (Write Network Key)
Handheld key	Command 823 (Request Session)
Join key	Command 768 (Write Join Key)

Source: S. Raza. *Secure Communication in WirelessHART and Its Integration with Legacy HART*. Swedish Institute of Computer Science, Technical Report T2010:01, ISSN: 1100-3154. SICS Networked Embedded Systems (NES) Laboratory, Kista, Sweden, p. 46, 2010.

19.13.3 Key Distribution

In a WHART system, distribution of keys to the devices is the responsibility of the network manager. How the network manager requests the keys from the security manager remains a gray area and not specified fully. The various WHART commands that distribute the keys are given in Table 19.2.

The WHART standard provides information regarding how the join key is provisioned in a device and provided to the network manager.

19.13.4 Key Renewal

Any key in WHART is breakable using brute-force attack, provided there are no time or resource constraints. Key renewal or key changing is important so that outside attacks can be repulsed. Thus, all keys should frequently and automatically be changed. WHART has mechanism to do the above, i.e., to alter or renew the keys.

While the network key can be changed by using a secure broadcast session, other keys can be changed by using a unicast session between the network manager and the devices. The security of the join key and the session key is interdependent; hence, lack of security of one would affect the security of the other.

19.13.5 Key Revocation

Key revocation means deactivation of a key when its corresponding device has left the network. If the device leaves the network permanently, the network key should also be changed. A device, initially, has only one secret key, which is the join key. When the device joins the network with its help, the device gets the network and session key details.

In a WHART network, each key belongs to a particular device. Thus, if a device is captured, its authenticity is lost and the device should self-destruct for security reasons. Else the key must be automatically deleted from the network.

19.13.6 Key Vetting

Authentication or verification of a key is called key vetting. It is mostly used in public key infrastructure (PKI). There is no specific command in WHART for key vetting purposes. However, command 779, which stands for Report Device Health, helps in key vetting.

19.13.7 Shortcomings

The shortcomings of the keys in the wireless portion of WHART are shown in Table 19.3. It is seen from the table that key generation, key distribution, key revoking, and key vetting do not have any WHART solution. Two keys—key distribution and key renewal—have partial WHART solutions enforced in them.

TABLE 19.3
Shortcomings in Wireless Portion of WirelessHART

Key Management	WirelessHART Solution	Comments
Key generation	No	Not specified
Key storage	No	Not specified
Key distribution	Limited	Key distribution from the network manager (NM) to the devices is specified but the key distribution from the security manager (SM) to the NM is not specified. Key distribution for the join key is not specified.
Key renewal	Limited	WirelessHART provides solutions for the key renewal in the wireless network but does not specify how the keys are renewed in the security manager.
Key revoking	No	Key revoking is not specified. These is a need for the network key and join key revoking.
Key vetting	No	No specific ways for key vetting

Source: S. Raza. *Secure Communication in WirelessHART and Its Integration with Legacy HART*. Swedish Institute of Computer Science, Technical Report T2010:01, ISSN: 1100-3154. SICS Networked Embedded Systems (NES) Laboratory, Kista, Sweden, p. 48, 2010.

19.14 WHART NETWORK FORMATION

Network formation in WHART has to go through several stages before it becomes ready for industrial use. The network manager sets the network ID and password, followed by join process and session security. At the shop floor, the password and network ID of the device are entered. The device is then configured as required, and the update rate is specified. The whole process is explained in Figure 19.23.

After this initial setup, the network manager sends an "advertisement" to the concerned device, and the device responds with a join request. It is shown in Figure 19.24.

In the third stage, the network manager authorizes the device with particular keys, schedules, and routing. On the basis of this information, the device publishes data, shown in Figure 19.25.

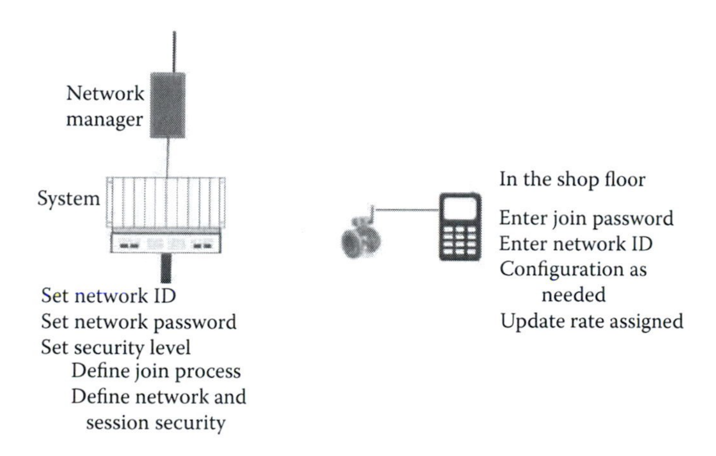

FIGURE 19.23 Network formation: stage I. (From L. Krammer. *Wireless in Automation: WirelessHART, ISA 100.* Institute of Automation, Automation Systems Group, Vienna University of Technology, Vienna, Austria, 2012.)

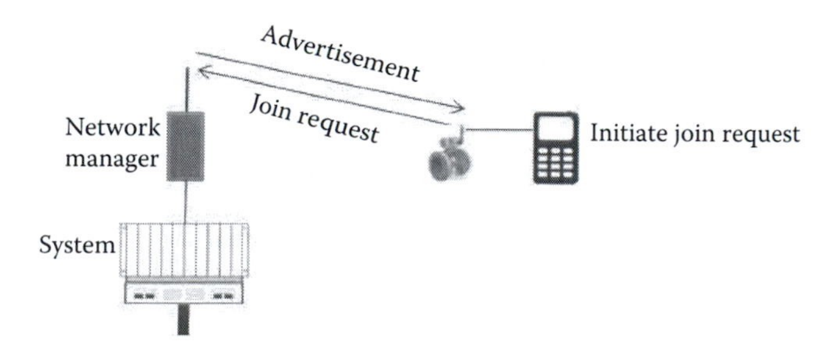

FIGURE 19.24 Network formation: stage II. (From L. Krammer. *Wireless in Automation: WirelessHART, ISA 100.* Institute of Automation, Automation Systems Group, Vienna University of Technology, Vienna, Austria, 2012.)

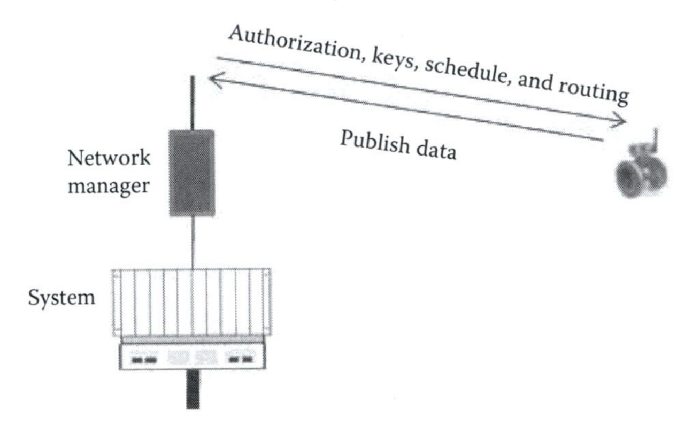

FIGURE 19.25 Network formation: stage III. (From L. Krammer. *Wireless in Automation: WirelessHART, ISA 100*. Institute of Automation, Automation Systems Group, Vienna University of Technology, Vienna, Austria, 2012.)

19.15 HART AND WHART—A COMPARISON

A comparison between HART and WHART is shown in Table 19.4. WHART is introduced because of increasing applications of wireless technology in industrial automation and control systems.

A general comparison between HART and WHART based on PDU is shown in Figure 19.26.

TABLE 19.4
Comparison between HART and WHART

HART	WHART
Not a secure protocol	A secure protocol
Legacy HART belongs to HART 6 and below	WHART belongs to HART 7 and above
Based on the OSI 7 layer protocol but loosely coupled	Based on OSI 7 layer protocol but strongly coupled
Transport and higher layers (up to application layer) identical to WHART	Transport and higher layers (up to application layer) identical to HART
Application layer consists of command number, byte count, and data	Application layer consists of command number, byte count, and data
A 16-bit command number	A 16-bit command number
Does not explicitly specify network or higher layer except application layer	Uses network layer for wireless routing and end-to-end security
Based on Token-Passing Data Link layer	Based on Time Division Multiple Access (TDMA) Data Link layer
Physical layer based on 4–20 mA analog wiring	Physical layer based on IEEE 802.15.4-2006

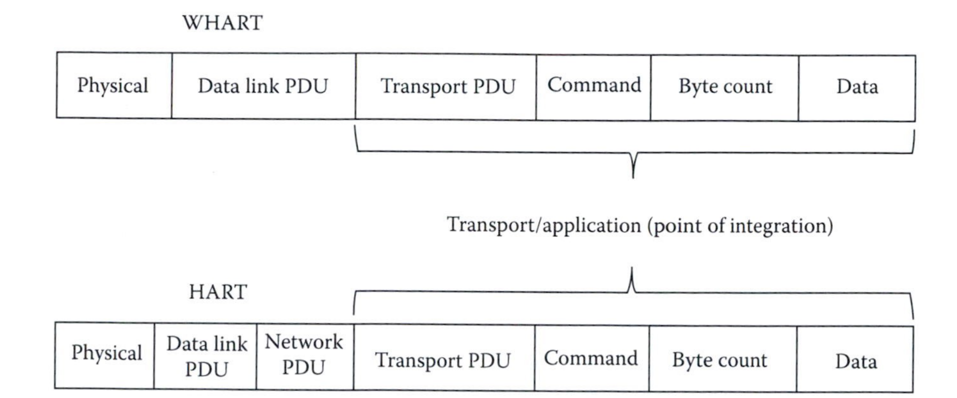

FIGURE 19.26 Comparison between HART and WHART PDUs. (From S. Raza. *Secure Communication in WirelessHART and Its Integration with Legacy HART.* Swedish Institute of Computer Science, Technical Report T2010:01, ISSN: 1100-3154. SICS Networked Embedded Systems (NES) Laboratory, Kista, Sweden, p. 51, 2010.)

19.16 HART AND WHART—INTEGRATION

The legacy HART protocol has been around for some 20 years, while its later counterpart, WHART, is less than a decade old. There is a huge industrial base for HART in the industrial control and automation sector. Thus, a need is felt that the existing HART network in a plant be integrated with WHART, introduced for either extension or modification purposes. The reasons for integration between HART and WHART can be summarized as follows:

- WHART devices and networks should be backward compatible with existing HART, so that integration becomes easy.
- Since the users are used to HART, the newly introduced WHART should support the legacy HART.
- Hundreds of HART-based devices are working properly worldwide. The newly introduced WHART should be able to communicate with the legacy HART and vice versa. Thus, network-to-network integration must be present between the two protocols.

Integration between HART and WHART was undertaken by the HART Communication Foundation.

HART devices can be integrated to WHART by employing adapters. An adapter can be used to connect one or more HART devices to a WHART network. Integration between a HART and a WHART network is done by using a gateway. HART and WHART have different protocols, and a gateway acts as a protocol converter to integrate the two. A WHART gateway can be used in a PC card, as part of an I/O subsystem or as a standalone device.

20 ISA100.11a

20.1 INTRODUCTION

ISA100.11a is a wireless standard developed by the International Society of Automation (ISA). Work on this standard started at around the same time as Wireless Highway Addressable Remote Transmission (WHART). It was endorsed as an ISA standard in September 2009. The standard provides reliable and secure wireless communication in the field of industrial automation for noncritical monitoring and control applications. ISA100.11a can integrate with HART, Foundation Fieldbus, PROFIBUS, and others through device adapters, network protocol pass through tunneling, by mapping using interface objects, etc.

20.2 SCOPE OF ISA100

ISA100 is one of the three major competing standards in wireless sensing and communication area and was formed in 2006. It is managed by the ISA. The Automation Standards Compliance Institute (ASCI) was formed by the ISA for certification, conformance testing, and compliance assessment of the activities being carried out in the ISA's automation field. It acts as a bridge between testing of the ISA standards and their effective implementation. Again, the Wireless Compliance Institute (WCI) operating under the fold of ISA100 is responsible for compliance certification of ISA100. Activities of the WCI include the following:

- Assures interoperability via testing, standards, and conformance testing
- Conducts independent testing and certification of field devices and systems to the ISA100 family of standards
- Updates users and suppliers with technical support, text materials, and tools
- Certifies devices and systems with regard to their meeting a common set of specifications and standards

WCI, in conjunction with ASCI, would provide test certificates for conformance to ISA100 standards. Devices and systems having such standard certificates would thus be interoperable in nature.

- Sensor node network connectivity
- Host node network connectivity
- Multivendor device interoperability among field devices
- Multivendor device interoperability between routers and field devices
- Multivendor device interoperability between gateways and devices
- Data flow for sensor data
- Network health metrics, metric collection, and presentation
- Field device parameterization/configuration

20.3 ISA100 WORKING GROUP

The broad objective of ISA100 is to establish standards, prepare technical reports, recommend allowed practices, and provide information for correct implementation of wireless communications in process industries. A number of working groups were formed to address specific tasks. The different groups and their corresponding specific tasks are shown in Table 20.1.

The most important among them is WG3, which corresponds to ISA100.11a. It is dedicated to "Wireless Systems for Industrial Automation: Process Control and Related Applications." WG12 is working with issues

TABLE 20.1
ISA100 Working Groups and Group Tasks

ISA100 Working Group	Working Group Focus
WG1	Integration
WG2	RFP evaluation group
WG3	ISA100.11a
WG5	RF coexistence
WG6	Interoperability
WG7	Networking
WG8	Users
WG9	User guide
WG10	Marketing
WG12	Standards convergence
WG14	Trustworthy wireless
WG15	Backhaul/backbone
WG16	Factory automation
WG17	ZigBee
WG18	Power sources
WG21	People/asset tracking

Source: H. Forbes. *ISA100 and Wireless Standards Convergence, ARC Brief*, by Harry Forbes, October 1, 2010. USA ARCweb.com.

pertaining to convergence of different standards in the wireless field, while WG5 and WG6 deal with radio frequency (RF) coexistence and interoperability issues, respectively.

20.4 FEATURES

The major features associated with ISA100.11a are given as follows:

- Topologies supported are star, mesh, and star–mesh networks.
- Adheres to the IEEE 802.15.4 standard physical layer.
- Blacklisting of channels that constantly interfere with reliable data transmission.
- Time Division Multiple Access (TDMA) method for real-time data transfer; i.e., it supports deterministic communication.
- Variable slot time.
- Uses channels 11–25 as defined by IEEE 802.15.4 with optional channel 26.
- Field devices may or may not have router capability.
- Number of devices may be typically of the order of 50–100.
- Does not have the presentation and session layers of the OSI model. However, the application layer and the data link layer (DLL) have two and three sublayers respectively.
- A combination of the frequency hopping spread spectrum (FHSS) and the direct sequence spread spectrum (DSSS) is used for the purpose of modulation.
- Channel hopping takes place on a packet-by-packet basis.
- Channel hopping supports slotted, slow, and a combination of these two.
- It supports five preprogrammed hopping patterns.
- Duocasting is supported for transmitting data to two backbone routers.
- Time synchronization is maintained by an international time clock.
- High reliability.
- Network scalability.
- Redundancy.
- Interoperability.
- Clear channel assessment (CCA), i.e., checks if the channel is "free," before sending data.
- DLL routing by graph routing and source routing.
- Addressing modes: 128-bit long address (IPv6) and 16-bit short address.
- Very low battery power: 2–5 years battery life on end devices.

TABLE 20.2
ISA100.11a Sensor Classes and Their Respective Roles

ISA100.11a Sensor Classes	Definition	Role
Class 0	Emergency action (always critical)	This class includes safety-related actions that are critical to personnel and plant, such as safety interlock, emergency shutdown, and fire control.
Class 1	Closed-loop regulatory control (often critical)	This class includes motor and axis control as well as primary pressure and flow control.
Class 2	Closed-loop supervisory control (usually noncritical)	This class of closed-loop control usually has long time constants, with timeliness of communications measured in seconds to minutes.
Class 3	Open-loop control (human in the loop)	This class includes actions where an operator other than a machine "closes the loop" between input and output. Such actions could include taking a unit office when conditions so indicate. Timeliness for this class of action is human scale, measured in seconds to minutes.
Class 4	Monitoring with short-term operational consequences (event-based maintenance)	This class includes high-limit and low-limit alarms and other information that might instigate further checking or dispatch of a maintenance technician.
Class 5	Monitoring without immediate operational consequences (preventive maintenance)	This class includes items without strong timeliness requirements. Some, like sequence-of-events logs, require high reliability; others, like reports of slowly changing information of low economic value, need to be so reliable since loss of a few consecutive samples may be important.

Source: CISCO Systems Inc. *Integrating an Industrial Wireless Sensor Network with Your Plant's Switched Ethernet and IP Network*, White Paper. New York, 2009, pp. 1–18. Available at http://www.cisco.com/web/strategy/docs//manufacturing/swpIIWS_Emerson_wp.pdf.

20.5 SENSOR CLASSES

There are six different sensor classes defined by the ISA100 committee and shown in Table 20.2. These are class 0 to class 5, with class 0 being the most sensitive class. Since class 0 is the most sensitive, it is normally not recommended to transport data over this class. Classes 0 to 4 correspond to *control*, while classes 5 and 6 are used for *monitoring*. The importance of message timeliness increases from class 5 to class 0.

20.6 SYSTEM CONFIGURATION

Figure 20.1 shows a typical system configuration based on ISA100.11a. It consists of field devices, backbone routers, a backbone network, gateway, security manager, system manager, control system, and applications.

It is seen from Figure 20.1 that redundant paths exist between the backbone routers and the field devices on the one hand, and between the controller and field devices via backbone routers on the other. The gateway acts as an interface between the wireless network and the control applications such as a distributed control system (DCS). The security manager manages the security aspects at every level of the network, whereas the system manager is responsible for controlling and managing the activities of the total network in the desired way.

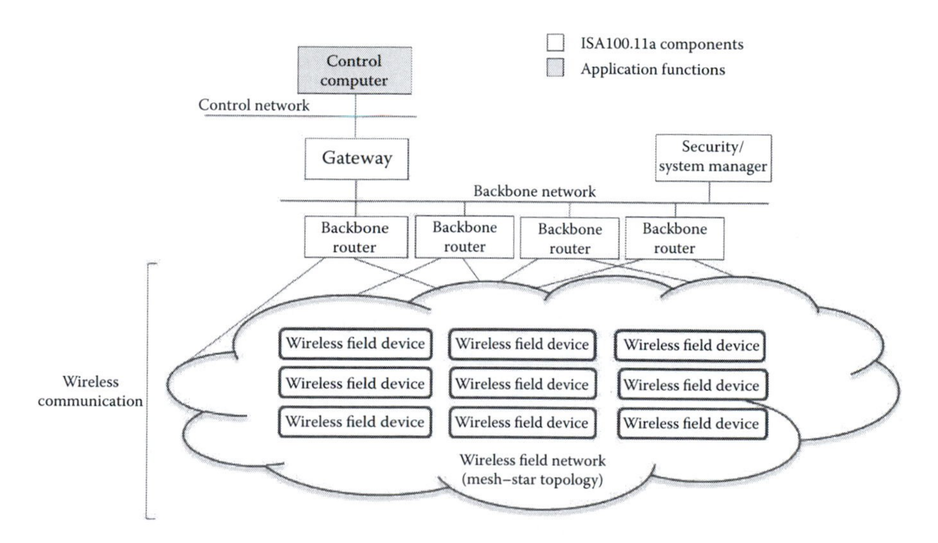

FIGURE 20.1 System configuration based on ISA100.11a.

20.7 CONVERGENCE BETWEEN ISA100.11A AND WHART

To achieve convergence between ISA100.11a and WHART, ISA100 formed a convergence subcommittee, designated as ISA100.12. This subcommittee solicitates suggestions, requirements, and other inputs from the end users of ISA100.11a and WHART through a "Convergence User Requirements Team" or CURT. The user requirements obtained by CURT were then directed to ISA100 and also to NAMUR for implementation of the recommendations. The subcommittee developed a "request for proposals" (RFP), which would be responsible for ultimate convergence of these process control standards. The RFP was passed in September 2010.

20.8 NAMUR PROPOSAL

NAMUR is an international association of process and automation industry end users—mostly European. NAMUR collects inputs from users in the form of drawbacks of existing systems, possible ways out of them, and suggestions with regard to improvements that may be implemented in the future. These inputs are assorted, and the works are published by NAMUR and forwarded to standardization authorities for possible standardization after incorporating any suggestions or alterations as suggested by such authorities.

Three emerging standards in the wireless arena in the process industry sector are IEC/PAS 62591 (WirelessHART), WIA-PA (IEC/PAS 62601), and ISA100.11a. Convergence of these three standards is of prime importance, which would enable the end users to solve the interoperability problem. With this in view, NAMUR prepared the report NAMUR NE 133 (Forbes 2010), which states

> Several wireless standards have appeared in the market place. At least three of these emerging standards specifically address the industrial markets. They are IEC/PAS 62591 (WirelessHART), WIA-PA (IEC/PAS 62601) and ISA100.11a. The recommendation also applies to any future appearing wireless sensor network standard(s), which target process industry use. All of these standards essentially address the same physical space where wireless can be used for industry applications. In the history of wired field networks, supporting multiple network protocols due to competing specifications has increased operational and capital costs. This severely hampered the widespread implementation of wired field networks.

> The user community has therefore strongly recommended that wireless sensor network standards have to converge into one single standard which addresses the long term lifecycle required for an industrial application network installation. This single converged standard has to provide a common network structure that will afford the greatest diversity of wireless instrument applications that can now, or in future, be available to the end users.

Convergence is an issue of utmost importance so that end users are least affected with regard to the standard that is being used by them. It is suggested that a single standard be developed in place of multiple standards that exist at the moment.

20.9 ARCHITECTURE

The detailed architecture of ISA100.11a is shown in Figure 20.2. ISA100.11a is the first wireless protocol from the ISA100 family. It is an evolving standard in the process automation and control field, with reliable and secure wireless communication. Different types of devices have been shown, along with subnets, backbone routers, backbone network, gateway, system manager, security manager, and control network.

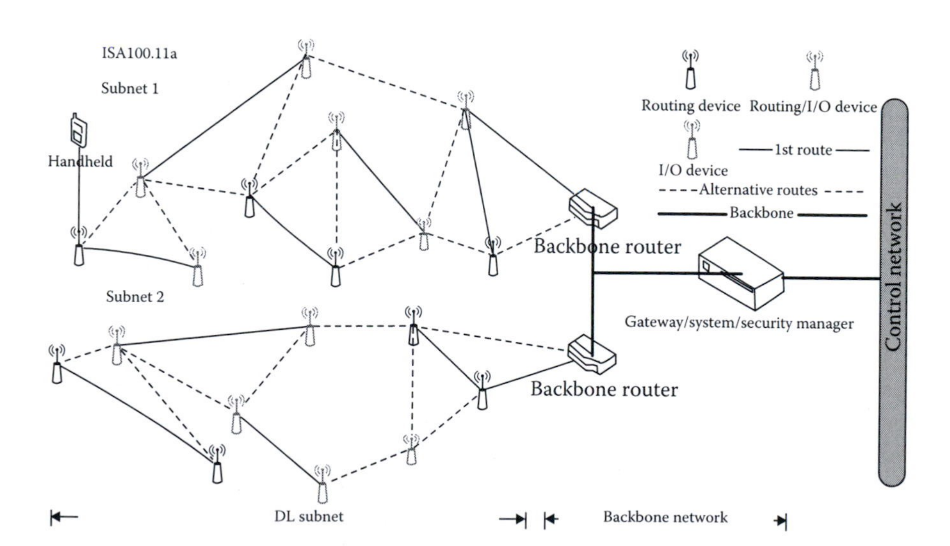

FIGURE 20.2 **(See color insert.)** Detailed architecture of ISA100.11a. (From G. Wang. *Comparison and Evaluation of Industrial Wireless Sensor Network Standards: ISA100.11a and WirelessHART.* Master of Science Thesis, Communication Engineering, 2011.)

Data from a device can reach the router via different paths—depending on the availability and loading pattern of the different paths en route to the router. The security manager is responsible for all the security-related issues for secure communication over the network. It manages the security keys and coordinates with the system manager for proper system operation—both internally and externally. The information related to security between different devices and the security manager is carried out by a proxy security management object (PSMO).

The system manager, which acts as a system administrator, is responsible for system management, device management, resource management, time-related issues, etc. This is implemented by the system manager with the help of a system management application process (SMAP).

The function of the gateway of ISA100.11a is almost identical to the functioning of the gateway of WHART. The gateway directly interfaces the host level applications to the wireless field devices or to wired devices via adapters. The gateway is implemented with a protocol translator and a gateway service access point (GSAP) at the application layer based on the existing protocol suite.

The different devices and their roles and responsibilities are shown in Table 20.3.

TABLE 20.3
Devices and Their Respective Roles/Responsibilities

Device Type	Roles/Responsibilities
Input/output	A source or sink of data
Router	Routes a message through the nodes in the same subnet
Backbone router	Routes data from one subnet to another
System manager	It is central to all operations in the network. It manages all network activities through preconfigured software
Security manager	Enables, controls, and supervises all security issues of all devices present in the network
Gateway	Provides an interface between wireless and plant networks
Provisioning device	Provisions devices with configurations to ensure proper system operation
System time source	Responsible for maintaining the master time source of the network

Source: Instrumentation, Systems and Automation Society (ISA), 2010. The Technology Behind the ISA 100.11a Standard—An Exploration. ISA100 Wireless Compliance Institute. Available at http://www.isa100wci.org/Documents/PDF/The-Technology-Behind-ISA100-11a-v-3_pptx.aspx.

TABLE 20.4
Architecture- and OSI-Based Differences between ISA100.11a and WHART

	ISA100.11a	WHART
Architecture level differences	Mesh, star, or star–mesh network Backbone devices Provisioning device with device provisioning service object (DPSO) QoS with secure communication subnets within the total network The architecture thus becomes scalable. Devices have routing capability	Only mesh network Network access points Handheld used for commissioning Peer-to-peer communication with potential security risks
Digital link layer	Five preprogrammed hopping schemes Configurable length of time slot Active and passive neighbor discovery	One channel hopping scheme Fixed time slot Passive neighbor discovery
Network layer	Three header specifications Based on IPv6 addressing Fragmentation and assembly Compatible with 6LoWPAN	One header specification Based on HART addressing End-to-end security
Transport layer	Connectionless service (UDP), unreliable Compatible with 6LoWPAN End-to-end session security	TCP-like reliable communication
Application layer	Object oriented Support legacy protocol tunneling Standard management objects, dependent, independent objects, ASL services	Command oriented Support HART protocol Predefined data types

Source: G. Wang. *Comparison and Evaluation of Industrial Wireless Sensor Network Standards: ISA100.11a and WirelessHART*. Master of Science Thesis, Communication Engineering, 2011.

20.9.1 Differences with WHART

Table 20.4 shows the architectural and OSI layer based differences between ISA100.11a and WHART.

20.9.2 Routing Ability of Devices

In ISA100.11a, field devices can be implemented with or without router capability. Devices can be assigned flexible roles—i.e., multiple role profiles having different functionalities. A device with router capability can improve the working of network mesh by reducing latency. The router role of the device can be switched off by the system manager to increase the energy savings. Again, in a harsh environment, devices with router capability can improve the mesh networking, ignoring the power-saving aspects.

20.9.3 Subnet

In ISA100.11a, the whole network is divided into smaller ones. Each such network is called a subnet.

An ISA100.11a can have multiple subnets with each subnet having a maximum of 2^{16} devices. Thus, within a specific subnet, a local unique 16-bit address is assigned to every device by the system manager. Data routing takes place both at the local or subnet level using the 16-bit short address and a 128-bit long address at the network level. Thus, scalability in ISA100.11a by having subnets gives a more integrated approach to the overall message transmission in the network.

20.9.4 Provisioning Device

A device intending to join a network has first to be provisioned by a provisioning device. The credentials of the new device are checked by the provisioning device when the new device requests to join via the join process. The join process is carried out by a device provisioning service object (DPSO). Asymmetric key-based over-the-air (OTA) provisioning and open symmetric join key provisioning methods are undertaken for a new device to join the network.

20.9.5 Backbone Routers

A backbone router acts as an interface between the field network and the control network. Backbone routers help in minimizing system latency, increase bandwidth, and improve the quality of service (QoS). They encapsulate network layer data and send the same through the protocol stack of backbone to the destination. Use of backbone routers help in lesser traffic congestion and thus reduces latency. ISA100.11a has already established the configuration between the backbone routers and other devices in the network for proper integration.

20.9.6 Device Management Data Flow

The device management data flow diagram is shown in Figure 20.3. Every device in the network is managed by a device management application process (DMAP), which also oversees the communication aspects. Since the application layer of ISA100.11a is object oriented, DMAP has a series of objects that support device management operations by proper configurations and invoking methods via management service access points at the

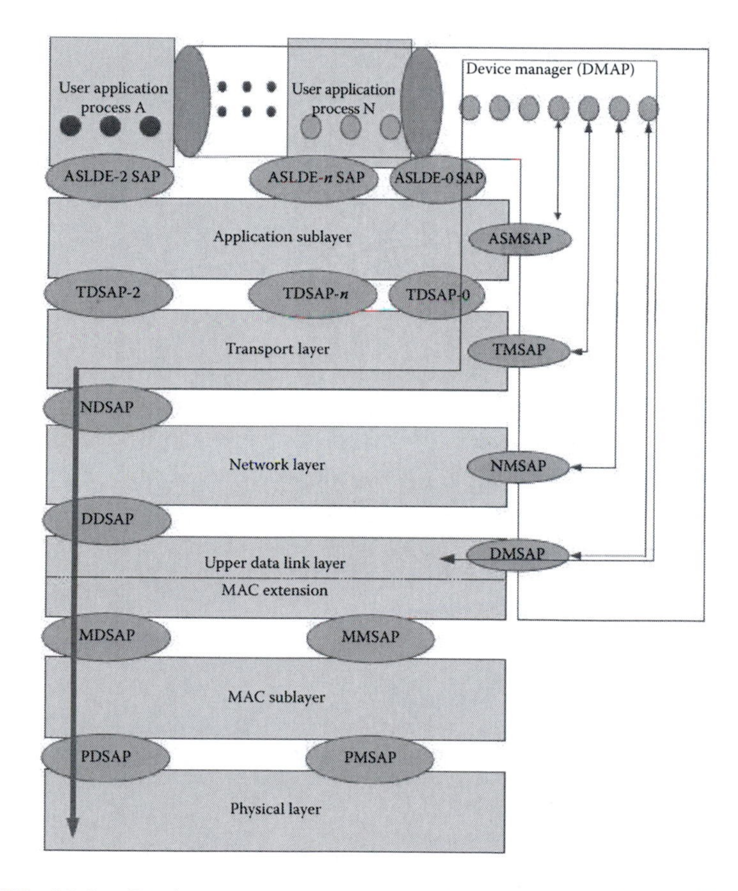

FIGURE 20.3 Device management data flow diagram. (From G. Wang. *Comparison and Evaluation of Industrial Wireless Sensor Network Standards: ISA100.11a and WirelessHART.* Master of Science Thesis, Communication Engineering, 2011.)

respective layers. The network's system manager can also perform device configuration remotely via ASL (application sublayer services).

The management data flows through the protocol layers via the respective data service access points (DSAPs), while, on the other side, DMSAP manages activities up to the level of the upper data link layer.

20.9.7 System Management Architecture

The system management architecture of ISA100.11a is shown in Figure 20.4. Each device belonging to the network can be thought of as comprising two blocks—DMAP and UAP (user application process). DMAP is responsible for the communication aspects and is administered by the system manager via the ASL services. The system manager operates in conjunction with the security manager.

The UAPs are resident in the devices themselves. They are controlled by the host applications via the gateway. DMAP and UAP communicate between themselves for the overall device management.

20.9.8 System Management Application Process

The system manager is in charge of overall system management of the wireless network. The system management application process is shown in Figure 20.5. It acts as the system administrator and is responsible for system management, device management, communication between the devices and the network, timing control, etc.

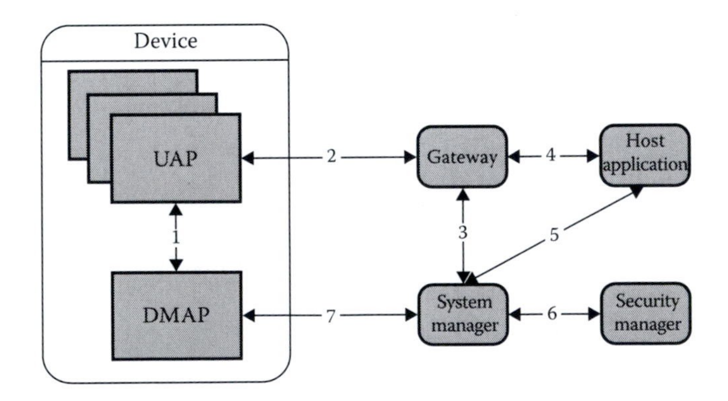

FIGURE 20.4 System management data flow diagram. (From G. Wang. *Comparison and Evaluation of Industrial Wireless Sensor Network Standards: ISA100.11a and WirelessHART*. Master of Science Thesis, Communication Engineering, 2011.)

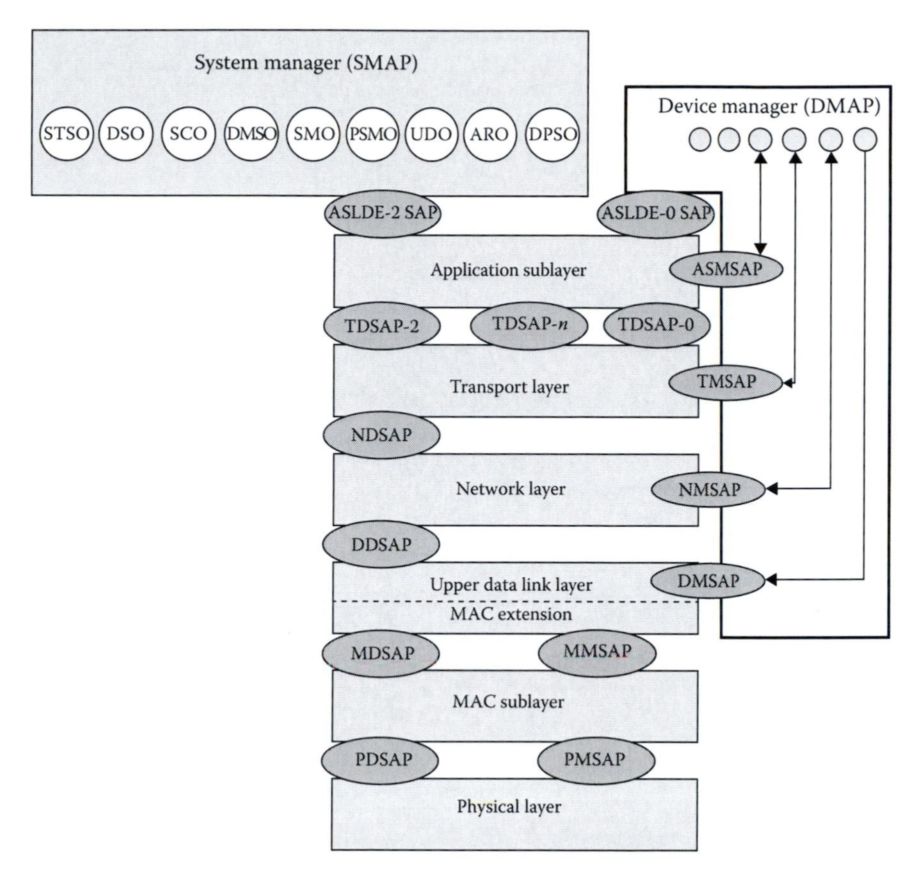

FIGURE 20.5 System management application process. (From G. Wang. *Comparison and Evaluation of Industrial Wireless Sensor Network Standards: ISA100.11a and WirelessHART.* Master of Science Thesis, Communication Engineering, 2011.)

20.10 COMPARISON BETWEEN ISA100.11A AND WHART PROTOCOL STACKS

Figure 20.6 shows the differences in the protocol stack of ISA100.11a and WHART with reference to the seven-layer OSI model. Both ISA100.11a and WHART use the simplified version of the OSI model, and they use neither the presentation layer nor the session layer. The application layer of ISA100.11a is divided into the upper application layer and the application sublayer, while the DLL is divided into the upper DLL, media access control (MAC) extension, and MAC sublayer.

Different standards are adopted at different layers of ISA100.11a. These standards strictly adhere to industry-accepted norms. It is shown in Figure 20.7 along with the OSI layer protocol.

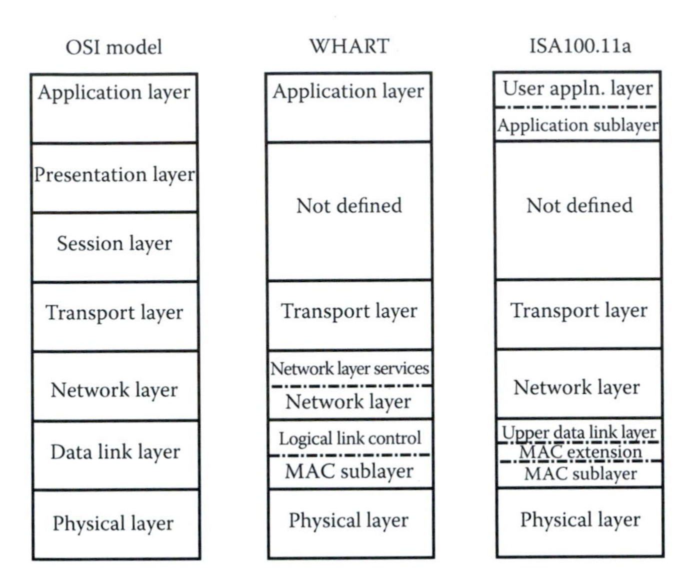

FIGURE 20.6 Comparison of protocol stack of ISA100.11a and WHART.

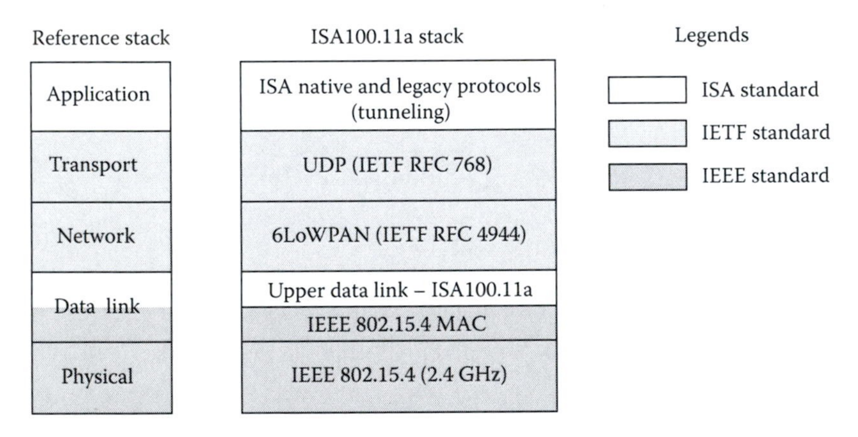

FIGURE 20.7 Different standards in ISA100.11a stack. (From Instrumentation, Systems and Automation Society (ISA), 2010. The Technology Behind the ISA 100.11a Standard—An Exploration. ISA100 Wireless Compliance Institute. Available at http://www.isa100wci.org/Documents/PDF/The-Technology-Behind-ISA100-11a-v-3_pptx.aspx.)

20.11 PHYSICAL LAYER

The physical layer of ISA100.11a is based on IEEE STD 802.15.4. It acts as the interface to the physical medium through which actual message transmission occurs. It transmits and receives raw data packets, performs CCA, provides a control mechanism, and detects RF energy. DSSS along with FHSS techniques are used as modulation techniques. A bit rate of 250 kbits/s with offset quadrature phase-shift keying (O-QPSK) is allowed with a power limit of 10 mW.

It uses channels 11 to 25 with optional channel 26 in the 2.4-GHz ISM (industrial, scientific, and medical) band. Each channel uses a bandwidth of 2 MHz and channel separation of 5 MHz between any two successive channels.

20.12 DATA LINK LAYER

The DLL of ISA100.11a is divided into three sublayers: upper DLL, MAC extension, and MAC sublayer. The DLL's main responsibilities are access and synchronization, handling of acknowledgment frames, and providing a reliable link between two peer-to-peer radio entities.

The MAC sublayer is a subset of IEEE 802.15.4 MAC. Its main responsibility is sending and receiving data frames. The MAC extension includes some additional features that are not supported by IEEE 802.15.4 MAC. This includes changes in the carrier sense multiple access with collision avoidance (CSMA-CA) mechanism by including additional frequency, time, and spatial diversity.

The upper DLL is responsible for routing within the data link (DL) subnet, and takes care of the link and mesh aspects of the subnet level network. The system manager is responsible for DL subnet operations. The DL subnet ends at the backbone router. Networking beyond this is handled by the network layer.

20.12.1 Protocol Data Unit

The protocol data unit of the data link layer (DPDU) is shown in Figure 20.8. It is more complex than WirelessHART DPDU, but offers more flexibility and better performance of the DLL. The DPDU has a subfield called data link layer header (DHR), which takes care of different functions such as routing, congestion control, optional solicitation message, security aspect, mesh, and link. The DHR consists of several subfields: DHDR, DMXHR, DAUX, DROUT, and DADDR. These are now discussed.

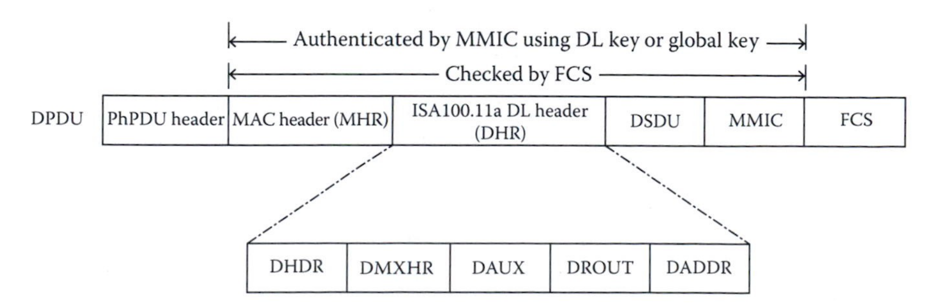

FIGURE 20.8 DPDU of ISA100.11a. (From G. Wang. *Comparison and Evaluation of Industrial Wireless Sensor Network Standards: ISA100.11a and WirelessHART*. Master of Science Thesis, Communication Engineering, 2011.)

DHDR stands for data link layer header subheader. It has the DL version number and DL selections, such as whether acknowledgment (ACK) is needed by the recipient or not. DMXHR is the data link layer media access control extension subheader. It is concerned with security aspects. The 32-bit, 64-bit, or 128-bit keys needed during joining are generated by the DL key. DAUX, the data link layer auxiliary subheader, is used in dedicated advertisements and solicitation messages. Subnet level routing is managed by DROUT, which stands for data link layer routing subheader. It selects the graph ID, source route, or DPDU priority, etc. DADDR is the data link layer address subheader, which contains the network layer source and destination addresses, in addition to explicit congestion control (ECN), last hop (LH), or discard eligible (DE).

20.12.2 Coexistence Issues in DLL

The DLL of ISA100.11a operating in the 2.4-GHz range of the ISM band has to coexist with others such as Bluetooth, ZigBee, or WiMax, of IEEE 802.11 b/g. For proper communication in ISA100.11a, several methods such as channel blacklisting (to avoid channels that are very much in use), TDMA technology (to minimize the possibility of collisions), channel hopping (to reduce multipath fading), data authentication and integrity (to maintain data confidentiality), etc., are followed. ISA100.11a adopts several mechanisms to overcome the coexistence issues so that highly reliable communication takes place in the presence of other communications taking place at the same time in the same ISM band. Coexistence mechanisms utilized for increased data reliability include time diversity (TDMA), collision avoidance technique, frequency diversity and frequency diversity with Automatic Repeat Request (ARQ) (channel hopping), spectrum management by channel blacklisting and adaptive hopping, etc.

20.12.2.1 TDMA

WirelessHART TDMA technology uses a fixed time slot of 10 ms for network-wide time synchronization. ISA100.11a, on the other hand, uses a configurable time slot with duration varying between 10 and 12 ms on a per superframe basis. A variable time slot gives rise to optimized coexistence with shorter time slots using tight synchronized communications, while the longer ones can be used for accommodating serial ACKs from multiple destinations, CSMAs performed at the beginning of a time slot, etc.

20.12.2.1.1 Slotted

The slotted hopping scheme, shown in Figure 20.9, uses different channels in each successive time slot. A DPDU with its ACKs/Negative Acknowledgments (NAKs) is sent in a time slot. Messages, which need

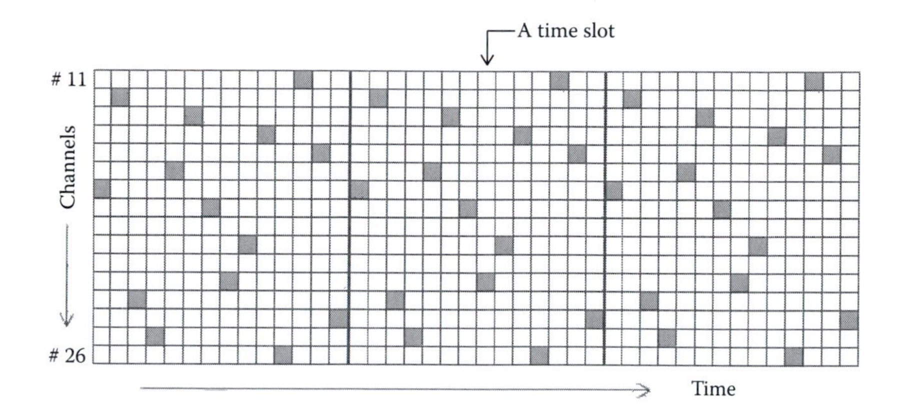

FIGURE 20.9 Slotted hopping scheme. (From G. Wang. *Comparison and Evaluation of Industrial Wireless Sensor Network Standards: ISA100.11a and WirelessHART.* Master of Science Thesis, Communication Engineering, 2011.)

to be sent in a synchronized manner, use slotted hopping. It is also used where energy limitation is of prime importance. All channels, i.e., from 11 to 26, take part in slotted hopping.

In this case, a device is forced to wait until its scheduled turn comes again. This would lead to latency in event-based data transmission.

20.12.2.1.2 Slow

The slow hopping scheme, as the name suggests, hops from one channel to another in a slow manner. That is, in this scheme, a particular channel is used for several consecutive time slots. Thus, a particular channel's usage may range from 100 to 400 ms. The slow hopping scheme is shown in Figure 20.10 and utilizes channel numbers 15, 20, and 25 normally. Channel hopping in this case is slower than slotted hopping. The hopping period is configurable.

Devices having imprecise time settings not requiring tight time synchronization or receivers having considerable energy utilize this scheme for message transmission. This scheme is better suited for event-based traffic, where the occurrence of an event may demand immediate transmission of a data packet. Devices earmarked as receivers are to be kept on for a considerable length of time to listen to incoming traffic, thereby increasing the power consumption of the receivers.

20.12.2.1.3 Hybrid

Hybrid channel hopping, shown in Figure 20.11, is a combination of the slotted and slow hopping schemes. Thus, it enjoys the advantages of both—synchronous and regular message transfer of slotted hopping and less predictable messages, such as alarms, etc., of slow hopping. A combination of slow and slotted hopping is configurable.

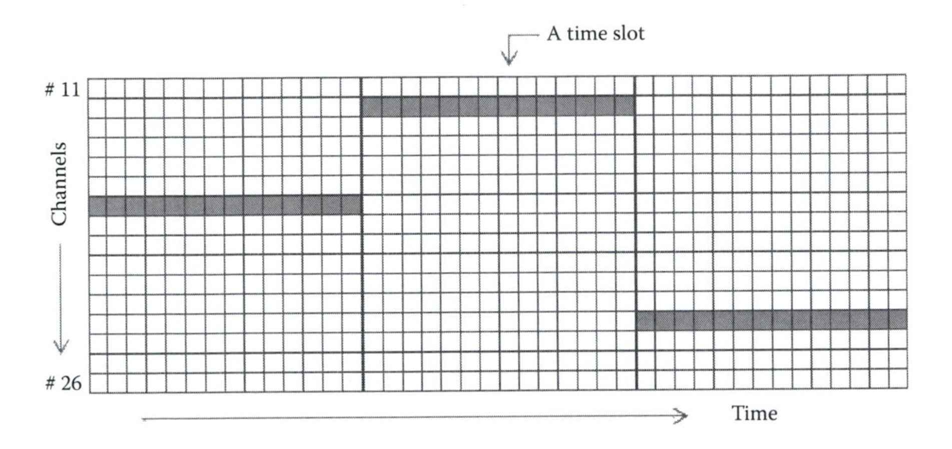

FIGURE 20.10 Slow hopping scheme. (From G. Wang. *Comparison and Evaluation of Industrial Wireless Sensor Network Standards: ISA100.11a and WirelessHART.* Master of Science Thesis, Communication Engineering, 2011.)

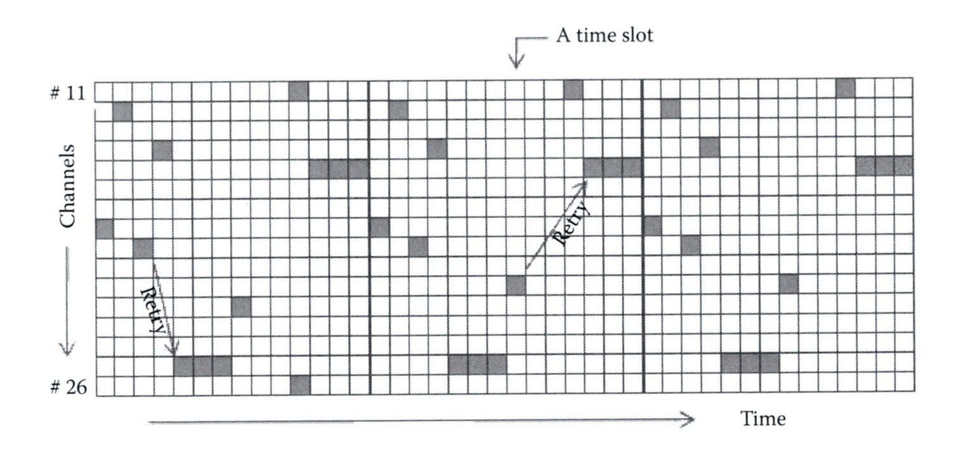

FIGURE 20.11 Hybrid hopping scheme. (From G. Wang. *Comparison and Evaluation of Industrial Wireless Sensor Network Standards: ISA100.11a and WirelessHART.* Master of Science Thesis, Communication Engineering, 2011.)

20.12.2.1.4 Default

ISA100.11a employs five preprogrammed hopping patterns supported by all the devices in the network:

1. Pattern 1: channel numbers 19, 12, 20, 24, 16, 23, 18, 25, 14, 21, 11, 15, 22, 17, 23, and optional 26
2. Pattern 2: pattern 1 in the reverse order
3. Pattern 3: 3, 15, 20, and 25 used for slow hopping channels

4. Pattern 4: pattern 3 in the reverse order
5. Pattern 5: 11, 12, 13, 14, 15, 16, 17, 18, 19, 20, 21, 22, 23, 24, 25, and optional 26 (Petersen and Carlsen 2011)

The system manager configures the network to be used for any one of the above five patterns. Patterns 3 and 4 are meant for slow hopping, while pattern 5 is used for enabling coexistence with WirelessHART.

20.12.2.2 Collision Avoidance

Collision avoidance uses a media contention mechanism using both TDMA and CSMA-CA that considerably increases data transmission reliability. The transmitter employs a CCA mechanism to first ensure that a channel is free for transmission. In case it is otherwise, the transmitter backs off, disallowing transmission. A retransmission is attempted after a random amount of time. Figure 20.12 shows the CSMA-CA-based media contention using CCA.

20.12.2.3 Frequency Diversity

ISA100.11a uses channels 11 to 25 with optional channel 26 for message transmissions in the network. Hopping is undertaken to avoid channels of heavy usage, thus avoiding interference. Therefore, it provides immunity against interference from neighboring channels and also robustness to mitigate multipath interference effects. The hopping sequence determines

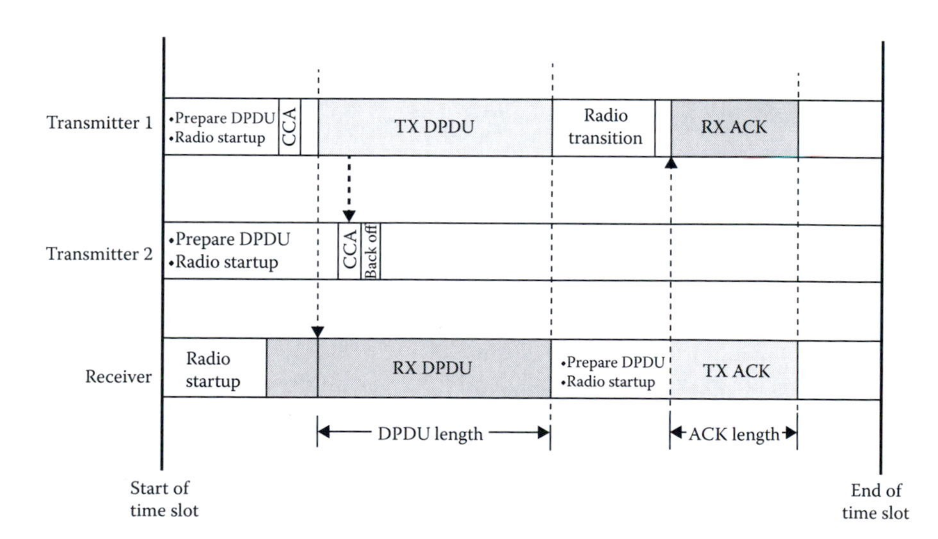

FIGURE 20.12 CSMA-CA-based media contention using CCA. (From Instrumentation, Systems and Automation Society (ISA), 2010. The Technology Behind the ISA 100.11a Standard—An Exploration. ISA100 Wireless Compliance Institute. Available at http://www.isa100wci.org/Documents/PDF/The-Technology-Behind-ISA100-11a-v-3_pptx.aspx.)

the channel that the transmitter is to utilize after completion of the current transmission. ISA100.11a employs several channel-hopping schemes, such as slotted hopping, slow hopping, hybrid hopping, and adaptive hopping, to address different communication needs.

In a subnet, devices use different offsets into the hopping sequence. This results in interleaved hopping. Again, ARQ is enforced to ensure that unacknowledged data packets are retransmitted on a different channel utilizing a different frequency. Thus, frequency hopping, when employed with ARQ, increases reliability of data packet delivery considerably.

20.12.2.4 Spectrum Management

Data delivery in ISA100.11a is vastly improved by properly assessing a channel's availability before data transmission. Each device in ISA100.11a maintains a statistical record of each channel's availability by way of CCA back-off counts, number of attempts made for transmission over a channel, and number of NAKs received. The system manager of the network takes the final decision from the statistical record to blacklist a channel(s) for a particular length of time. Channels that are free from such interferences are used for data transmission purposes. Such spectrum management is very effective when ISA100.11a has to be used in a noisy environment.

Adaptive hopping is a step ahead in the channel blacklisting method where a particular device takes the decision locally by bypassing a channel based on the information collected by the device about the channel. The device in this case can adapt itself so as to alter the hopping sequence, which ultimately increases data transmission reliability.

20.12.3 Routing in DLL

Routing in ISA100.11a is performed at the DL level with the help of graph routing or source routing, which is configured by the system manager. All devices in the network support the two routings at the DL level. The routings are stored in DROUT, which is a field of DHR under DPDU. A message, on reaching a backbone router, will be provided with a network layer header so that the message can move along across the network.

The routing from a device to the backbone router is based on device ID, destination address, or else the default route. Again, a forwarding limit field in the DL header puts the maximum number of times a message can hop. As the message hops from one node to another, this number decreases. If it reaches zero, the corresponding DPDU is cancelled or discarded.

Depending on the neighborhood discovery, the system manager always optimizes the route (the route changes when a device joins/leaves the network) that a message would take to reach the backbone router.

20.12.4 Neighborhood Discovery

Neighborhood discovery helps in joining a new device into a network, and maintains and upgrades the mesh network operation as devices join and leave the network till it is in operation. Its job may primarily be divided into the following:

- Designated advertisement routers advertise the network information. The same is received by the devices desiring to join the network. The new device thus joins the network with the information it receives from the advertising router.
- A device, after it joins the network, sends/receives advertisements to/from neighbors. This helps the system manager to update itself so that network reliability and efficiency is improved. The above is an ongoing process as devices continue to join and leave the network.

For effective neighborhood discovery, two schemes are followed in ISA100.11a: passive listening scheme and active scanning scheme. The passive listening scheme here is the same as in WirelessHART. Active scanning comprises scanning interrogators and scanning hosts. The former seek for advertisements from advertising routers. The interrogators also transmit advertisement solicitations to the routers. The routers, which are called advertisement hosts, advertise the relevant information meant for the interrogators, including subnet information helping the devices join the network. The process of neighborhood discovery is scheduled and configured by the system manager.

Energy consideration is a very important aspect in the neighborhood discovery scheme. The scanning hosts in the active scanning scheme must have enough energy in their receivers for them to remain active during the initial network formation and also during sensitive periods. One advantage of the active scanning scheme is that it reduces the latency involved in passive scanning.

20.12.5 DLL Characteristics

Some of the characteristics associated with DLL are as follows: configurable time slots of 10–12 ms—it thus can accommodate devices with different time requirements; supports unicast, duocast, and broadcast; neighborhood discovery involves active and passive scanning; five predefined hopping patterns; and routing in the network is done via several subnet levels. DL level activities are maintained by the system manager via the DL management object (DLMO).

20.13 NETWORK LAYER

Some of the characteristics associated with the network layer of ISA100.11a are as follows:

- It is influenced by the Internet Engineering Task Force (IETF) 6LoWPAN specification.
- It has a 16-bit short address for DL subnets and a 128-bit long address for backbone routers (IPv6).
- It performs fragmentation and reassembly of data packets.
- It performs mesh-to-mesh routing.
- It has an address translation mechanism.

20.13.1 Functionality

The network layer of ISA100.11a has to take care of addressing and routing at the mesh level and the backbone router level. At the backbone level, addressing and routing are done by IPv6.

The network level is responsible for correct address information—a 16-bit short address for DL subnets and a 128-bit long address for communication at the backbone level. Address conversion from the DL subnet to the backbone is done by the backbone router. This address conversion is done by an *address translation table* maintained at the network level. Each device in the network maintains and updates their own address translation table with the help of the system manager. Each entry in this table has a DL subnet address and its corresponding backbone address. Fragmentation and assembling of data packets are also carried out at this level, with packet lengths more than that allowed at the DLL level. Fragmentation with proper size is done where backbone level routing switches to DL level routing.

20.13.2 Header Formats

ISA100.11a supports both subnet level routing (mesh level) and backbone level routing. Communication can take place at the intra-subnet level or between two subnets taking the help of backbone routers in the process. For the former, a 16-bit address is all that is needed for data transfer, requiring much less energy and bandwidth in the process. A 128-bit address needed for data transfer across two subnets require extra power and bandwidth.

Three different header formats are employed to address such diverse addressing needs. Depending on the level of service and the routings required, designs can be optimized for both low-power consumption and bandwidth.

20.13.2.1 Basic

The basic header format of the network layer is shown in Figure 20.13. The size of the octet is 1 only. It is used when the UDP header is needed to be fully compressed.

20.13.2.2 Contract Enabled

The contract-enabled header format is shown in Figure 20.14. The size of the octet varies from 1 to 6. It is used when the UDP header is either fully compressed or else a full header format is not realized.

octets	bits							
	7	6	5	4	3	2	1	0
1	Dispatch							
(variable)	Network payload							

FIGURE 20.13 Basic header format. (From G. Wang. *Comparison and Evaluation of Industrial Wireless Sensor Network Standards: ISA100.11a and WirelessHART*. Master of Science Thesis, Communication Engineering, 2011.)

octets	bits							
	7	6	5	4	3	2	1	0
1	LOWPAN_IPHC dispatch			LOWPAN_IPHC encoding (bits 8-12)				
2	LOWPAN_IPHC encoding (bits 0-7)							
3 (opt)	Octet alignment				FlowLabel (bits16-19)			
4 (opt)	Flow Label (bits 8-15)							
5 (opt)	Flow Label (bits 0-7)							
6 (opt)	HopLimit							
(variable)	Network payload							

FIGURE 20.14 Contract-enabled header format. (From G. Wang. *Comparison and Evaluation of Industrial Wireless Sensor Network Standards: ISA100.11a and WirelessHART*. Master of Science Thesis, Communication Engineering, 2011.)

octets	bits							
	7	6	5	4	3	2	1	0
1	Version				TrafficClass (bits 7-4)			
2	TrafficClass (bits 3-0)				FlowLabel (bits 19-16)			
3	FlowLabel (bits 15-8)							
4	FlowLabel (bits 7-0)							
5	PayloadLength (bits 15-8)							
6	PayloadLength (bits 7-0)							
7	NextHeader							
8	HopLimit							
9 -24	Destination address							
25-40	Source address							
(variable)	Network payload							

FIGURE 20.15 Full IPv6 header format. (From G. Wang. *Comparison and Evaluation of Industrial Wireless Sensor Network Standards: ISA100.11a and WirelessHART*. Master of Science Thesis, Communication Engineering, 2011.)

20.13.2.3 Full IPv6

The full IPv6 header format of the network layer is shown in Figure 20.15. It has an octet size of 40. It is used when data is to be transported from one subnet to another, via the backbone.

20.13.3 Summary of Header Differences

The three header formats used depend on the type and complexity of the data communication involved. The size of octet varies from 1 (for basic header), to 1–6 (for contract-enabled header), to 40 (for full IPv6 header). The basic header and the contract-enabled header belong to the DLL, while the full IPv6 header belong primarily to the network layer.

The basic header is used when the UDP header is fully compressed, while the contract-enabled header is used when the UDP header is either fully compressed or else a full header format is not realized. The third or the full IPv6 is used when data or message needs to be transferred from one subnet to another via the backbone.

20.13.4 6LoWPAN

6LoWPAN stands for "IPv6 over low-power wireless area networks." The transport and network layers of ISA100.11a are both 6LoWPAN compatible. 6LoWPAN aims at encapsulating and header compression techniques so that IPv6 data packets can be transported over IEEE STD 802.15.4 low-power wireless area networks. IPv6 compatibility would enable direct communication between Internet hosts and the wireless sensor nodes, allowing very fast data communication.

As regards its size, an IPv6 has a maximum of 1280 octets—termed maximum transmission unit (MTU). Since an IEEE STD 802.15.4 frame cannot accommodate such huge data, the same must be fragmented before it is pushed into the lower layers. The payload of IEEE STD 802.15.4 frame is 81 octets for upper layer data and the IPv6 header size is 40 octets in size. Thus, a maximum of 41 octets only is made available as the combined payload for transport and application layer data. Again, for an 8-octet header size for the transport layer, only 33 octets are left to be utilized by the application layer.

Header compression and fragmentation are done in 6LoWPAN by inserting an additional layer—called the *adaption layer*. This new layer is placed between the digital link layer and the transport layer, which would help in sending large IPv6 packets over the IEEE STD 802.15.4 network.

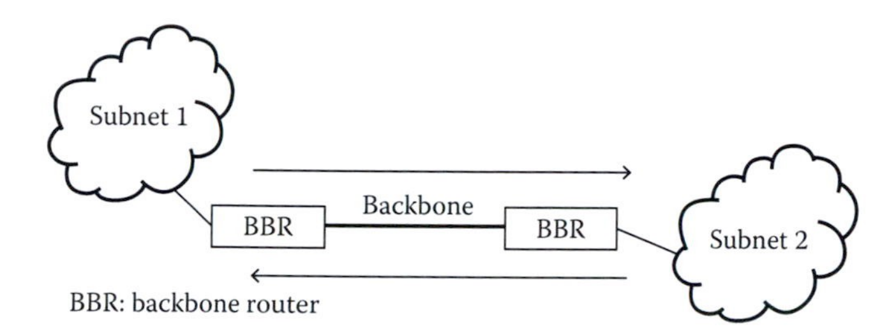

FIGURE 20.16 Data flow between two subnets in a network. (From G. Wang. *Comparison and Evaluation of Industrial Wireless Sensor Network Standards: ISA100.11a and WirelessHART.* Master of Science Thesis, Communication Engineering, 2011.)

20.13.5 Data Flow between Two Subnets

An ISA100.11a network has a number of subnets—the number is dependent on the physical placement of the devices and also on the total number of devices in the network. Data flow between two devices belonging to the same subnet takes place via the 16-bit DL address. Whenever data has to move from one subnet to another within the same overall network, the 16-bit short DL address of the device in the subnet is converted into a 128-bit long address by the backbone router for communication via the backbone. This is shown in Figure 20.16. The receiving end router converts the 128-bit long address with the 16-bit DL address of the device. Fragmentation and reassembling are done at the point where backbone level routing switches over to DL level routing and vice versa, respectively.

20.14 TRANSPORT LAYER

The transport layer transfers data between end systems or host to host, but not on the routers. It is responsible for end-to-end error recovery. It uses the User Datagram Protocol (UDP) for connectionless service over IPv6 with optional compression as defined in IETF 6LoWPAN specifications. It offers better data integrity, encryption, and data authentication. Some of the jobs performed by the transport layer are as follows: reliable but unacknowledged service, flow control, segmentation and reassembly, and basic and enhanced security service, which is optional using session key.

|← Authenticated by TMIC using session key →|

Uncompressed UDP header	Security header	AL payload encrypted using session key	TMIC

FIGURE 20.17 TPDU of ISA100.11a. (From G. Wang. *Comparison and Evaluation of Industrial Wireless Sensor Network Standards: ISA100.11a and WirelessHART.* Master of Science Thesis, Communication Engineering, 2011.)

20.14.1 Protocol Data Unit

The transport layer protocol data unit (TPDU) is shown in Figure 20.17 and consists of four fields. The UDP at this layer is normally not compressed but optionally can be compressed when the TPDU is passed on to the network level. It should be noted that the uncompressed UDP header requires a mandatory checksum in the form of IPv6.

The security header contains several security aspects in the form of different security levels and key index, mentioning the current session key and the timing information. The third field is the application layer payload that is encrypted by using the session key. The fourth field corresponds to the transport layer message integrity code (TMIC), which is generated using the session key used for data authentication.

20.14.2 Security

AES128_CCM in conjunction with UDP provide security in the transport layer for various types of communication services and for various levels of security requirements. Various options are enforced for the above:

- TMIC is computed with the help of the session key for end-to-end data integrity protection; however, the UDP checksum is not enforced for full header compression as required in 6LoWPAN.
- During the join process, the UDP checksum is incorporated for end-to-end data integrity; however, TMIC cannot be enforced owing to lack of availability of the session key.

Data encryption, data authentication, and the level of security to be enforced are all dependent on the security policy that is already in place during a contract.

The transport layer provides the required security against any replay attack by using a time stamp in the nonce, which indicates the time of formation of the data packet. If the data packet was created earlier than the configurable time limit, the receiving device simply rejects it.

20.14.3 Session and Contract

Communication between two devices is executed with the help of a security manager, which is assigned a contract for the same. Several contracts may be in place and exist in the system manager to support different communication needs. Similarly, there must be a session between a source and the destination (end-to-end devices). The session key is granted permission by the system manager for data flow to take place. The master key in each device is used for any new session key or updating the same.

20.15 APPLICATION LAYER

The application layer provides the necessary communication facilities for object-to-object communication between applications residing in different devices. It supports user-defined application processes. It also supports object-oriented software and predefined data structures for effective, robust, and manageable communication between devices.

An efficient messaging system is the prime consideration of the application layer. It would result in faster configuration uploads and would reduce traffic congestion, which leads to a very effective and energy-efficient communication resulting in enhanced battery life. Enhancement of battery life in wireless communication for sensors/devices is in itself a huge advantage.

20.15.1 Structure

The application layer of ISA100.11a is subdivided into the upper application layer (UAL) and the application sublayer (ASL). The UAL contains user application processes (UAPs) that address and perform user-specific problems. It also supports computational functions, protocol tunneling, input/output hardware needs, etc. Object-oriented communication between and routing between peer objects belonging to the same or different UAPs across the network is carried out by ASL.

20.15.2 Protocol Data Unit

The application layer PDU (APDU) is shown in Figure 20.18. In ISA100.11a, the application layer takes the responsibility of addressing a particular object belonging to a particular application process. The header of the APDU contains the object identifier information for addressing the object, identification of service type for specific service needed to be provided, etc. The object identifier addressing scheme becomes very efficient by

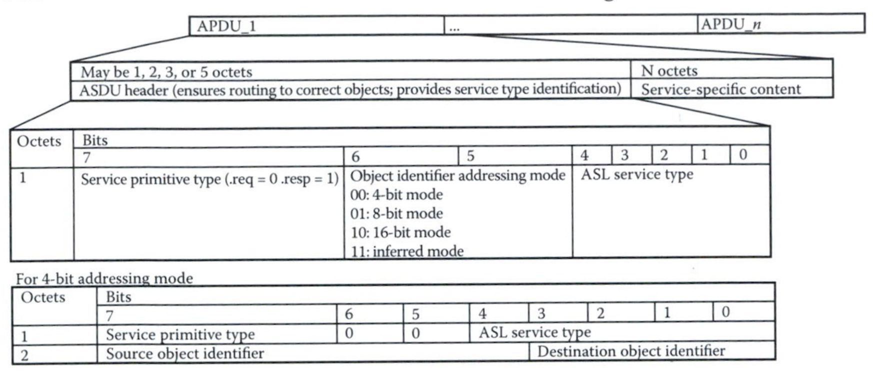

FIGURE 20.18 APDU of ISA100.11a. (From G. Wang. *Comparison and Evaluation of Industrial Wireless Sensor Network Standards: ISA100.11a and WirelessHART*. Master of Science Thesis, Communication Engineering, 2011.)

previously saving the coding bits of the identifiers of both the source and the destinations.

When APDUs need to be concatenated, an inferred addressing mode is used. In this, a reference is made between the source and destination object identifiers of APDUs with the most recently transmitted source and destination object identifiers of the APDUs in the same concatenation. This thus saves the overhead bits of the APDUs headers.

20.15.3 Communication Model

The communication model of ISA100.11a has three different communication formats: unidirectional buffered communication, queued unidirectional communication, and queued bidirectional communication.

Unidirectional buffered communication is used by the publish–subscribe service for unconfirmed scheduled periodical and aperiodical data publishing.

Queued unidirectional communication is used by the source–sink service for unconfirmed, unscheduled alert reporting, but without any flow or rate control. However, an alert acknowledgment is sent back to the alert source. If no acknowledgment is received, alert reporting is repeated.

Queued bidirectional communication is used by the client–server service for bidirectional communication in cases of on-demand one-to-one aperiodic communication having retries and flow control facilities.

20.15.4 Objects, Their Addressing, and Merits

The application layer of ISA100.11a defines standard objects, standard attributes, and interoperability issues. The industry-specific and vendor-specific standard objects can also be included in the same layer. These objects are contained in UAPs. A user application process management object (UAPMO) is a must in a UAP. The UAPMO contains the number of objects, UAP status, etc. The UAP would also contain an upload/download object (UDO) if the UAP has a self-upgrade feature.

Because there are so many objects in the application layer, addressing the individual objects is a must for properly accessing and identifying an object. For this, each object has a 16-bit object identifier ID. By default, UAPMO is assigned a reserved ID of 1 in every UAP and an ID of 2 is reserved when UDO is implemented.

Objects residing in the UAPs at the application layer have manifold advantages, as follows: interoperability with different field devices and also legacy systems, modular approach enables add-ons for either industry- or vendor-specific objects within the same UAP, and better network management.

The user application layer consisting of several UAPs is shown in Figure 20.19. Each UAP consists of several objects, with each object having its own attributes, methods, and alerts. The application layer of ISA100.11a is modular, object oriented, extensible, and can accommodate a variety of applications. Multiple applications are supported through UDP port addressing.

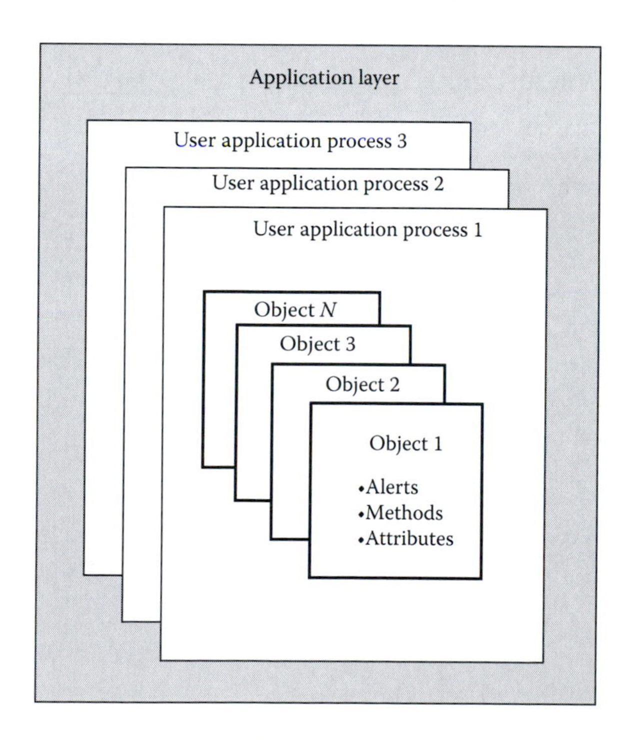

FIGURE 20.19 Application layer. (From Instrumentation, Systems and Automation Society (ISA), 2010. The Technology Behind the ISA 100.11a Standard—An Exploration. ISA100 Wireless Compliance Institute. Available at http://www.isa100wci.org/Documents/PDF/The-Technology-Behind-ISA100-11a-v-3_pptx.aspx.)

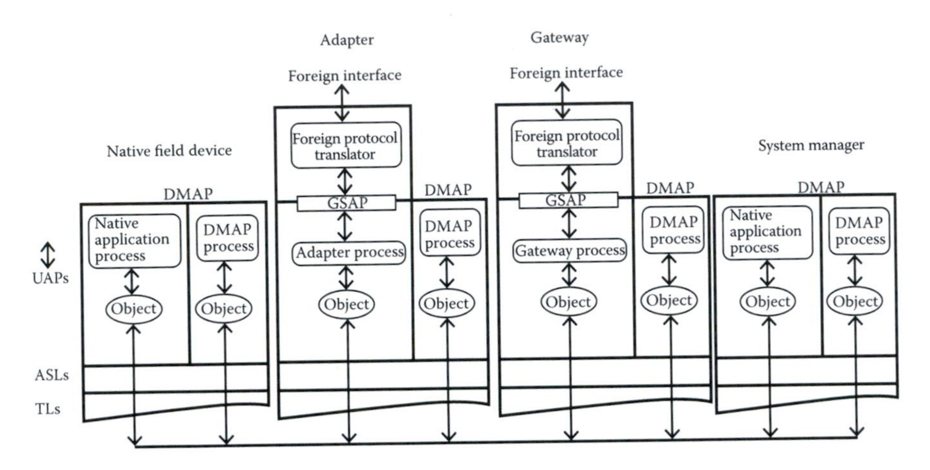

FIGURE 20.20 Gateway service access point. (From G. Wang. *Comparison and Evaluation of Industrial Wireless Sensor Network Standards: ISA100.11a and WirelessHART.* Master of Science Thesis, Communication Engineering, 2011.)

20.15.5 Gateway

To support gateway functionalities, a gateway has a specific UAP in ISA100.11a. A gateway is required to interface host level applications to the wireless field devices or else to the wired field devices via an adapter, which acts as a protocol translator. The adapter is used for converting from a wireless protocol to a wired protocol and vice versa.

20.15.5.1 Gateway Service Access Point

A gateway service access point (GSAP), also called a gateway high side service interface, is a mandatory requirement that provides the requisite functionalities for connecting the field devices to host level applications. A GSAP is shown in Figure 20.20.

The foreign protocol translator, residing on the top of GSAP, helps in converting the foreign host level protocols, such as tunneling to communicate with the wireless field devices. Object-to-object communication is provided by the ASL services with the help of GSAP and a gateway application process.

20.16 KEYS IN ISA100.11A

ISA100.11a supports both symmetric and asymmetric key-based join processes. The join process is governed by the administrator of the network, which helps in providing the new devices with enough information to join and participate in the network. Once joining is over, the new devices gain access to network communication resources and get the relevant bandwidth that ensures normal operations in the network activity. There are different types of keys: the data link key is used for message authentication, the master key is used between security manager and device, while the session key is used for normal data transmissions.

20.16.1 Joining by Symmetric Key—A Comparison between ISA100.11a and WHART

Both ISA100.11a and WHART support symmetric key-based joining. Table 20.5 shows a comparison in the symmetric key-based join process between the two wireless protocols.

TABLE 20.5
Symmetric Key-Based Joining—A Comparison between ISA100.11a and WHART

ISA100.11a	WHART
The device desirous to join the network must be provisioned with necessary information, such as join key and network information.	Identical to ISA100.11a
The device acquires advertisements from existing routers in the network. The new device develops the join request from the joining advertisement that contains the configuration specifications.	Identical to ISA100.11a
The join request from the new device is forwarded by the advertising router to the *system manager*. It (join request) contains both security and nonsecurity information.	The join request from the new device is forwarded by the advertising router to the *network manager*. It (join request) contains both security and nonsecurity information.
The system manager along with the security manager processes the join request and checks whether the new device has enough security features to join the network.	The network manager along with the security manager processes the join request and checks whether the new device has enough security features to join the network.
After the join request is approved, the system manager responds with a join request to the new device. It is then formally inducted into the network.	After the join request is approved, the network manager responds with a join request to the new device. It is then formally inducted into the network.

(*continued*)

TABLE 20.5 (Continued)
Symmetric Key-Based Joining—A Comparison between ISA100.11a and WHART

ISA100.11a	WHART
For hop-to-hop security, a global key is used for data integrity at the DL level before obtaining the DL key from system manager. This DL key is utilized for subsequent secure communications along with the new keys from system manager.	For hop-to-hop security, a well-known key is used for data integrity at the DL level before obtaining the network key from network manager. This network key is utilized for subsequent secure communications along with the new keys from the network manager.
Join messages separate the security and nonsecurity information in the different services. The security manager handles the security information and the system manager handles the nonsecurity information.	Not so in WHART.
Session level security is carried out with UDP checksum (integrity) at the transport level.	Join messages are protected by join key. This is temporarily used as session key to ensure end-to-end security (encryption and authentication) required at the transport level.
During the joining process, a challenge–response is exchanged between the new device and the system/security manager to validate mutual authentication. This ensures that the two parties are alive and agree on the same shared key.	Not available in WHART.
It supports asymmetric key-based joining.	Does not have this facility.

20.16.1.1 Protection of Join Messages

At the transport level, a UDP checksum replaces the session level security. In ISA100.11a, there is no session key during the time of joining a device; the security sublayer does not process any outgoing join request from a device and any outgoing join response from the advertising router. The security header size thus becomes zero. The join key in ISA100.11a provides security cover to the application layer data and other key distribution schemes.

In the key-based join process, the device initializes a *Security_Sym_Join.Request()* for sending security information to the security manager. The join key is used to compute the MIC of the data from *Security_Sym_*

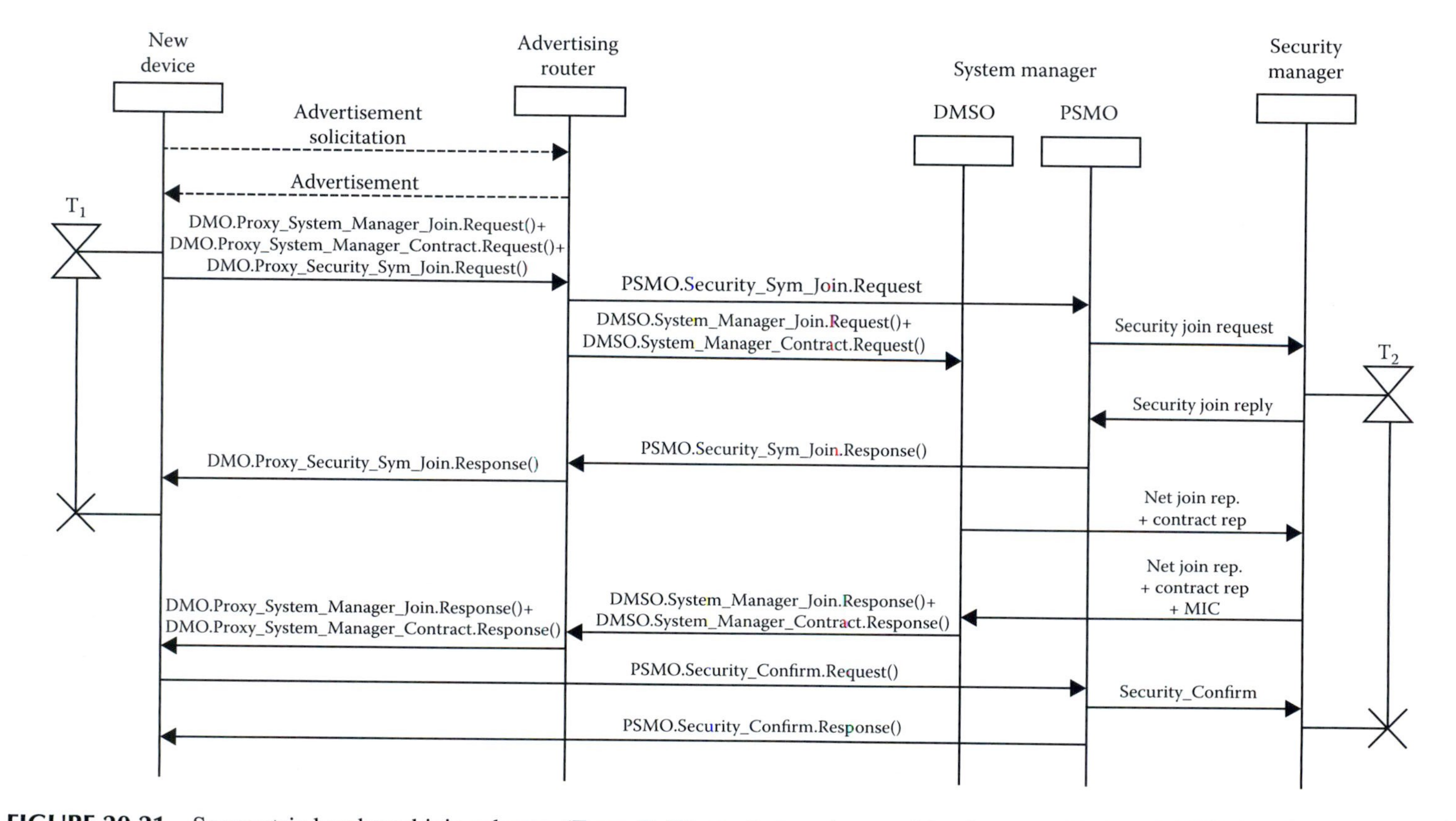

FIGURE 20.21 Symmetric key-based join scheme. (From G. Wang. *Comparison and Evaluation of Industrial Wireless Sensor Network Standards: ISA100.11a and WirelessHART*. Master of Science Thesis, Communication Engineering, 2011.)

Join.Request() for its authenticity checking. Figure 20.21 shows the manner in which the symmetric key-based join process is carried out.

20.16.1.2 Key Agreement and Distribution

ISA100.11a generates a Secret Key Generation (SKG), which is used as a master key between a device and the system manager. The master key is then used to derive a DL key and a session key in the encrypted message from system/security manager via *Security_Sym_Join.Response()*.

Figure 20.22 shows the sequential steps needed for key distribution-*cum*-agreement scheme required in a symmetric join process. This is realized by SKG methods involving a join key: master key = $SKG_{joinkey}$ [A||B||device EUI-64||security manager EUI-64]. The device first sends its EUI-64 to the system/security manager via *Security_Sym_Join.Request()* and the 128-bit AES. The SKG generated by the security manager is sent back to the device via *Security_Sym_Join.Response()*. The security manager also sends its own 128-bit AES. After the join key authenticates the MIC of other related data, the master key encrypts the session and DL keys. The device shares the same master key as that of the security manager. This is confirmed by

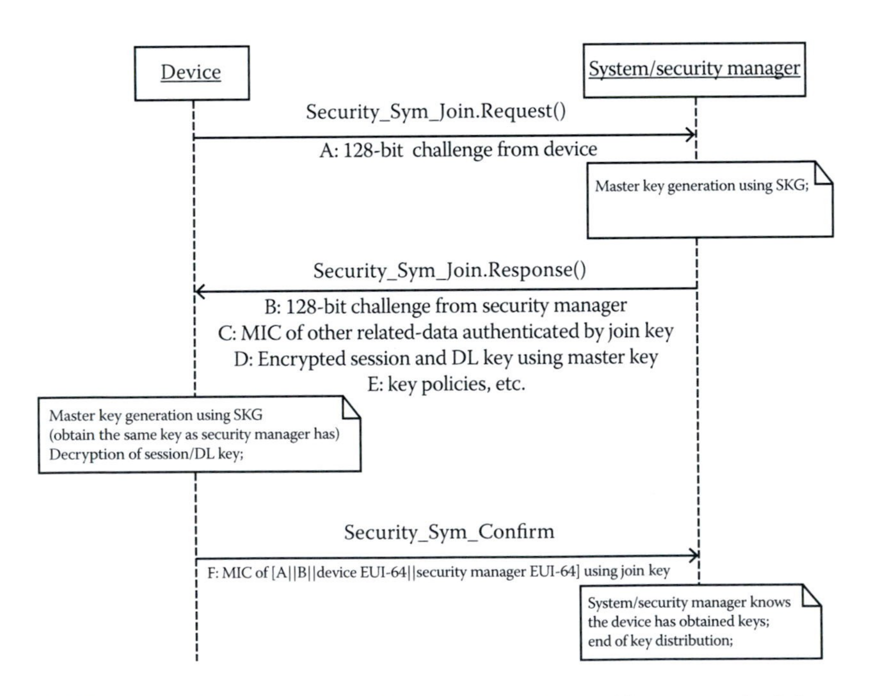

FIGURE 20.22 Key agreement/distribution scheme in symmetric join process. (From G. Wang. *Comparison and Evaluation of Industrial Wireless Sensor Network Standards: ISA100.11a and WirelessHART*. Master of Science Thesis, Communication Engineering, 2011.)

the device via *Security_Sym_Confirm* with both the device and the security manager sharing the same join key.

20.16.2 Asymmetric Keys

Unlike WHART, ISA100.11a supports asymmetric keys, although it is optional in nature. Different keys are used to encrypt and decrypt a message. Each device has two asymmetric keys: a public key and a private key. The public key is openly circulated, while the private key, as the name suggests, is kept a secret. A message is encrypted with the public key and decrypted by the secret key. There are two asymmetric keys in ISA100.11a: CA_root and Cert-A. The former is a public key certified by some authority authorized to certify the device's asymmetric key. Cert-A is the asymmetric key certificate for device A and is used to authenticate the asymmetric key establishment protocol.

20.16.2.1 Asymmetric Key-Based Join Process

An asymmetric key-based join process is initiated after the new device is certified by the certification authority with regard to its credentials, which include device identity EUI-64. The new device is able to integrate itself into the wireless network after the join process is over with the support from the system/security manager. The device then undertakes normal activity in the network, such as data publishing, network maintenance, and messaging services.

20.16.2.2 Key Agreement and Distribution

The key distribution and agreement scheme in an asymmetric join process is shown in Figure 20.23. The key distribution scheme in this case is identical to that in the symmetric case, although the key agreement scheme differs in the two join methods.

The device first sends the *Security_Pub_join.Request()* to the system/security manager along with its temporary ECC point and the device's EUI-64. The security manager responds with *Security_Pub_join.Response()* with the temporary ECC points and its own implicit certificate. A temporary key pair is generated by either device for key agreement. The shared secret master key is generated by either. The device now sends the *Security_Pub_join.Request()* along with the MIC over previously communicated information using the master key. The security manager in its turn sends back the *Security_Pub_join.Response()* along with the MIC over previously communicated information using the master key. The new device is now in a position to derive the DL key within the same subnet and the session key so that it can have a session with the system manager.

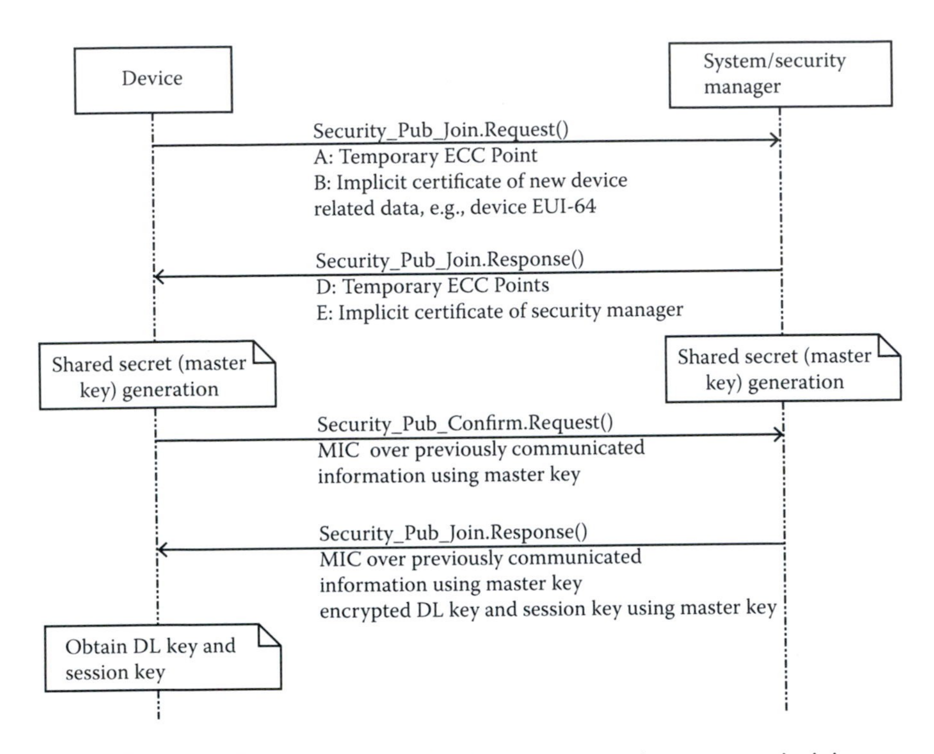

FIGURE 20.23 Key agreement/distribution scheme in asymmetric join process. (From G. Wang. *Comparison and Evaluation of Industrial Wireless Sensor Network Standards: ISA100.11a and WirelessHART.* Master of Science Thesis, Communication Engineering, 2011.)

20.16.2.3 Security Policy

ISA100.11a, which is based on IEEE 802.15.4, can have different levels of security for data authentication and encryption at its data link and network layers. The security manager controls policies related to cryptographic material that it generates. Table 20.6 shows the encryption based on AES-128 block ciphers.

TABLE 20.6
Encryption Based on AES-128 Block Ciphers

Security Policy	Authentication Message Integrity Code (MIC) Length	Encryption
MIC-32	4 bytes	Off
MIC-64	8 bytes	Off
MIC-128	16 bytes	Off
ENC-MIC-32	4 bytes	On
ENC-MIC-64	8 bytes	On

20.17 PROVISIONING OVERVIEW

A new device, while joining the network, must first be provisioned before its formal induction into the network. This device is called the *device to be provisioned* (DBP). The wireless sensor network that the device intends to join is called the *target network*. A device that is able to provision a new device into the network so that the latter can technically become part of the network is called the *provisioning device* (PD). The system/security manager and a handheld device can act as a provisioning device. The PD would implement the device provisioning service object (DPSO), which would provide information to the device provisioning object (DPO) in the DBPs.

The network between the PD and the DBP is a type A field medium for provisioning a device OTA. It can be a temporary, isolated small network or else a separate logical network that works along with the target network consisting of system manager or a security manager. The logical network can have different forms depending on priority and security levels.

The provisioning procedures undertaken at the site may instead be done at the factory premises if the device manufacturer is delegated with such provisioning powers. In that case, the device can straightway join the network with the preinstalled symmetric key K_Join/PKI (Public Key Infrastructure) certificate and other network-related information. It should be noted that the system/security manager is synchronized with the same secret keys. The EUI-64 keys of the device, K_Global and K_Join (Join Key), are thus preinstalled at the factory premises so as to skip the same procedure at the site.

20.17.1 Different Keys

A secret AES symmetric key is used for data confidentiality, which can either be preinstalled at the factory or can be provisioned during network initialization. Only the system/security manager of the network can have knowledge of this key.

K_Global is an asymmetric key-based OTA provisioning used for authentication of device credentials and reading of device identity and configuration settings. This key is used for data integrity during the provisioning process. In case the target network does not have asymmetric cryptographic support, the DBP joins the target network with the help of PKI certificates and the K_Join key.

K_Join is an open join key and is set to a nonsecret value by default. This is used for joining the network provisioning in an unsecured manner. The joining by this open symmetric key is rejected by both the PD and the system manager.

20.17.2 Configuration Bits

Four configuration bits, A1 to A4, are all preset to 1 by default and have their own functionality. They are as follows:

- A1: It allows OOB (out of band) provisioning.
- A2: It allows asymmetric key-based provisioning.
- A3: It allows default joining (joining a default network with default join key).
- A4: It allows reset to factory defaults.

The default settings of a device can be changed during the pre-provisioning stage.

20.17.3 Provisioning Data Flow between PD and DBP

Figure 20.24 shows the provisioning data flow in the protocol suite between the provisioning device (PD) and the device to be provisioned (DBP). During provisioning, the attributes and settings required to join the DPO of DBP of the target network is either read or written by the DPSO of PD, in terms of target network ID, EUI-64 of target security manager, superframes, hop patterns, etc.

20.17.4 Requirement for Joining

For a device to be able to join a network, both network-related and trust-related information must be provisioned into the device. The network-related information required are the network ID, 128-bit address of system manager, subset of frequencies within which the device is able to listen and respond to correct frequencies, join time, and DL configuration settings. Again, the trust-related information required are the EUI-64 of the security manager, specific Key_Join to join a specific network, and types of network join methods supported. The process of joining and leaving a network is shown in Figure 20.25.

Several steps are necessary before a device can formally join a network: The application-specific codes and defaults are first assigned, followed by installations of security and network information. Then, the device is installed and it joins the network.

20.17.5 Differences in Provisioning between ISA100.11a and WHART

Provisioning depends on wired connections for WHART, while it is OTA for ISA100.11a, both for symmetric and asymmetric key-based provisioning. Again, OOB provisioning and preinstallation at the manufacturer's premises

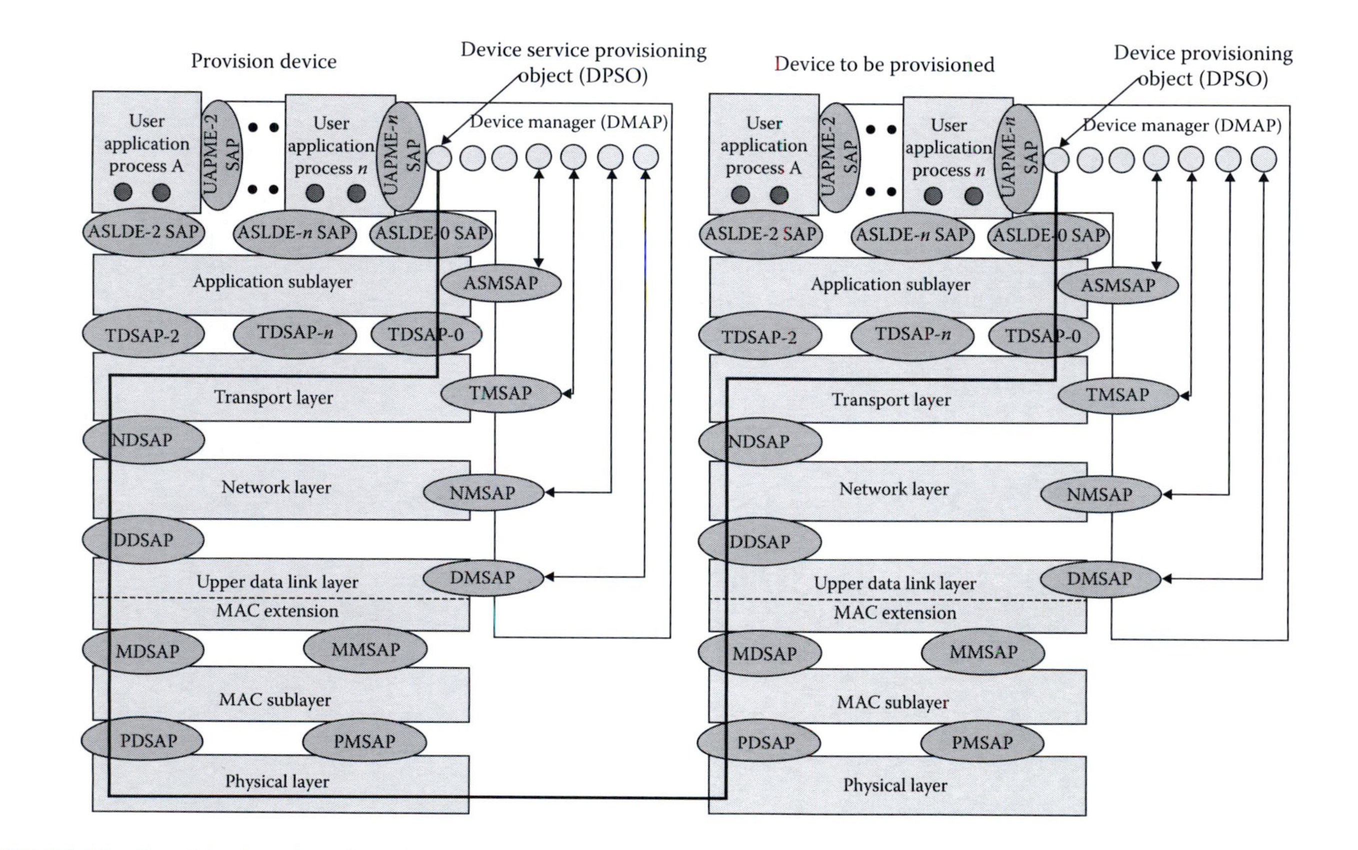

FIGURE 20.24 Provisioning data flow between PD and DBP. (From G. Wang. *Comparison and Evaluation of Industrial Wireless Sensor Network Standards: ISA100.11a and WirelessHART*. Master of Science Thesis, Communication Engineering, 2011.)

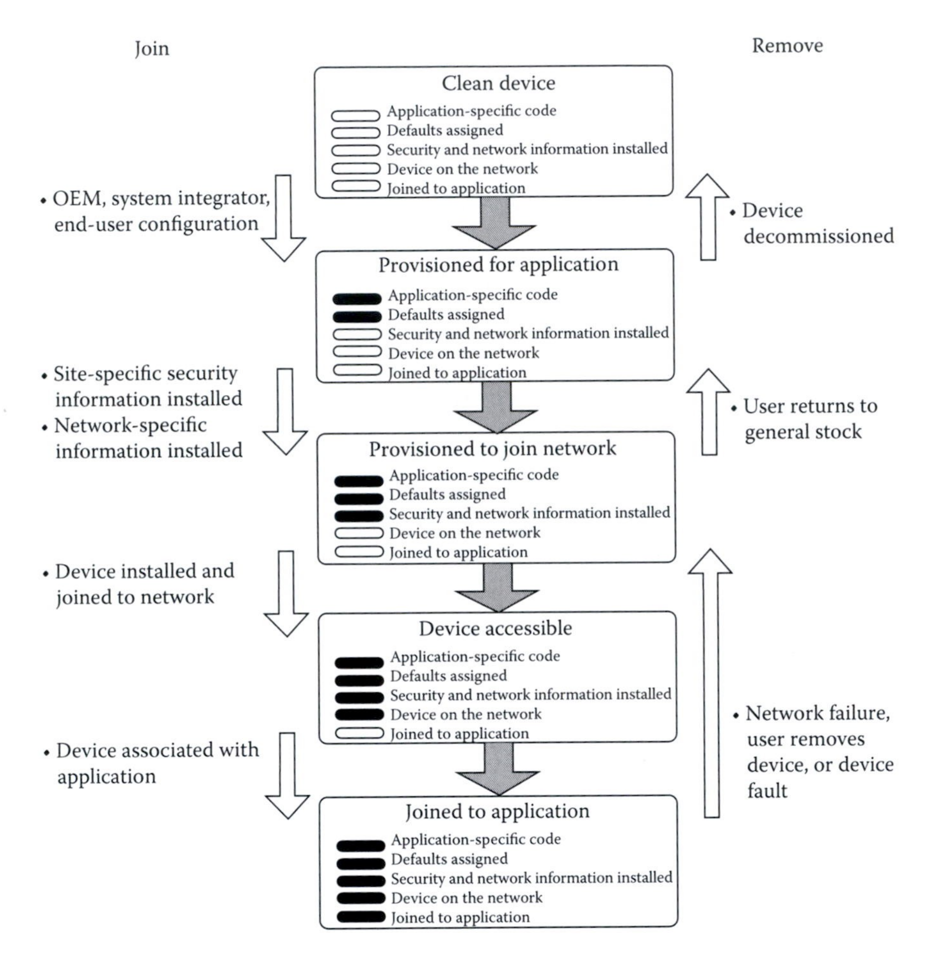

FIGURE 20.25 Process of joining and leaving the network. (From W. W. Manges. *ISA100.11a Principles of Operation Overview*, ISA Expo 2007, Oak Ridge National Laboratory, September 11, 2007.)

are some of the other facilities available with ISA100.11a. WHART relies on a handheld device for distributing provisioning information to the device intending to join the network. The handheld terminal with built-in storage of join keys, or having the software for key generation, is placed very close to the DBP. The handheld will then be in a position to start key distribution. In ISA100.11a, the system/security manager can set up the provisioning of new devices in enciphered text OTA. Thus, in this case, provisioning can be carried out from the office with a HMI (human–machine interface) facility and without any need to go to the plant site where the DBPs are placed. Another advantage with ISA100.11a is that after provisioning but before joining the network, the application level objects of the DBPs may be preconfigured.

However, coexistence issues from nearby RF channels along with security threats need to be taken into consideration during provisioning with ISA100.11a for secure and reliable operation. During OTA provisioning, it has to coexist with other standards such as WHART, Bluetooth, and ZigBee operating in the 2.4-GHz ISM band. Unintentional collisions with others can be avoided by CCA at the start of a time slot. However, if it is intentional, CCA would not work because of the high level of noise and interfering signals, which would cause jamming. This would lead to a blocked channel, interruption in scheduled communication, etc. In the worst-case scenario, all the channels can be flooded with intentional interference, such as "join request flooding." Thus, not only will the join requests be impeded but it will also hamper normal data communication because the system/security manager will be forced to be preoccupied with such multifake join requests. Again, the system may be vulnerable to spoofing attacks wherein a malicious attacker may pose to be a PD and would thus be able to advertise in DBPs. In such a scenario, the attacker may get a chance to join the network, thereby launching even more damaging attacks on the network. The spoofing attack may be negated by a challenge–response mechanism (i.e., mutual authentication of the DBP and the system/security manager). In the end, however, there is no sure-shot way to get rid of such attacks. Strict and enhanced network monitoring along with better access control are the only ways to combat such attacks from degrading the network.

The handheld terminal in WHART is equivalent to the provisioning device in ISA100.11a. When a handheld device is connected to the WHART network, the handheld itself needs to join the network first, like a new device. However, once the joining process of the handheld is over, it can provision as many devices as desired. In the case of ISA100.11a, the PD plays the same part as the handheld in WHART. It actually acts as the system/security manager of the DBP in the implementation of DPSO.

In WHART, the join key is downloaded in plaintext to the device to be provisioned. This poses a grave security risk since the same can easily be eavesdropped by a malicious intruder. In the case of ISA100.11a, on the other hand, the symmetric join key—either the K_Global or K_Join—is used for provisioning purposes in OTA. K_Global = ISA100 is a well-known value; it can offer data integrity for devices that do not have a secret join key. Again, with K_Global (Join Provisioning Network and Establish a Contract), an updated encrypted join key can be obtained with the help of asymmetric cryptographic modules of the DBP. This new enciphered join key offers enhanced security for data protection and confidentiality to ward off malicious attacks and eavesdropping. However, since enciphering by a cryptographic method involves both time and energy consumption, provisioning by plaintext (as in WHART) or cipher text (as in

ISA100.11a) is all decided by the amount of security requirement and the hostile environment in which the system is to operate.

20.18 DATA DELIVERY RELIABILITY

Path diversity and duocast are implemented to increase data delivery reliability. Path redundancy in the form of redundant routes is continuously followed to adapt to existing spectrum conditions. It addresses coexistence issues without user intervention. Also, the devices make adaptive routing to enhance reliability of data delivery.

ISA100.11a supports mesh topology based on graph routing. Graphs are preconfigured data routes set by the system manager, and data packets follow inbound and outbound routes to reach the final node, i.e., destination.

In duocasting, a data packet sent by a device is received by multiple devices within the same time slot. It is normally implemented for devices having direct connection with routers but can also be applied for devices that operate deeper in a subnet. Successful data delivery increases exponentially when duocast (or n-cast) is employed.

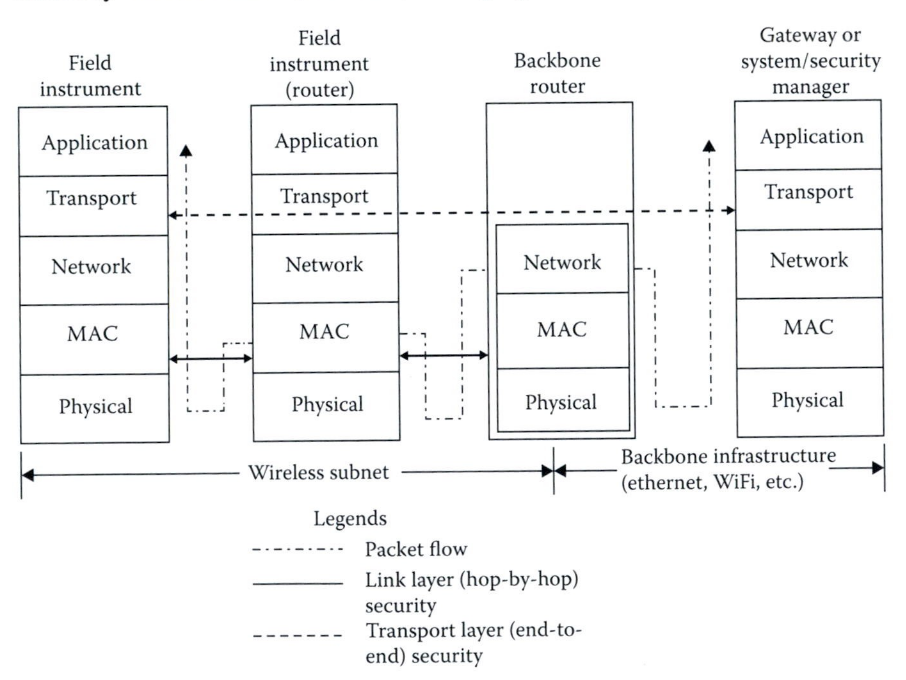

FIGURE 20.26 Two-layer security. (From Instrumentation, Systems and Automation Society (ISA), 2010. The Technology Behind the ISA 100.11a Standard—An Exploration. ISA100 Wireless Compliance Institute. Available at http://www.isa100wci.org/Documents/PDF/The-Technology-Behind-ISA100-11a-v-3_pptx.aspx.)

20.19 TWO-LAYER SECURITY

A two-layer security is enforced in ISA100.11a for transmission of data message—for hop-to-hop authentication and encryption, DLL security is enforced while for end-to-end authentication and encryption, transport layer security is enforced. For hop-to-hop transportation of a data message from one node to another in the same subnet, a MAC (DLL)-to-MAC transfer takes place. Along the transmission path, the routers authenticate and encrypt/decrypt data packets as they progress to the destination. This is shown in Figure 20.26.

For end-to-end security of data messages, authentication and encryption/decryption of the same is done by the originating device at the transport layer. Only the destination device has the capability of authenticating and decrypting the received data message at its transport layer.

20.20 COMMUNICATIONS IN ISA100.11A

Communication in ISA100.11a can take place in various ways: between two nodes belonging to the same subnet, between two nodes across two subnets, from a node to the control system, and from a legacy node to the control system.

The simplest communication process in ISA100.11a is when communication takes place between two nodes belonging to the same subnet. This is shown in Figure 20.27. The message does not have to flow through either the backbone router or the gateway to reach the final destination.

The process of information flow starts at the application layer of the first node and travels through different layers before reaching the router. It travels through the physical and data link layers of the router before reaching the application layer of the second node after entering it through the physical layer.

Figure 20.28 shows the example when the message from the application node travels all the way to the control system via the field router and the ISA100.11a gateway. At the gateway, there is a translator that converts from the ISA100.11a application to the control system application and vice versa.

Again, Figure 20.29 shows a data communication scheme that involves data transfer from a legacy device to the control system. It involves the use of two translators—one for translating legacy application to ISA100.11a application and the second for converting ISA100.11a application to control system application. The whole communication process involves three networks: legacy network, ISA100.11a network, and plant network.

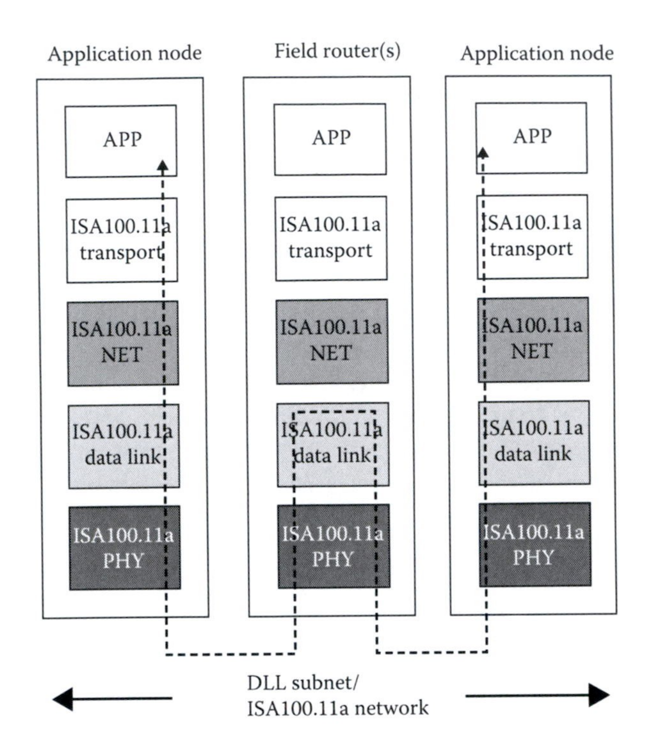

FIGURE 20.27 Communication in the same subnet. (From W. W. Manges. *ISA100.11a Principles of Operation Overview*, ISA Expo 2007, Oak Ridge National Laboratory, September 11, 2007.)

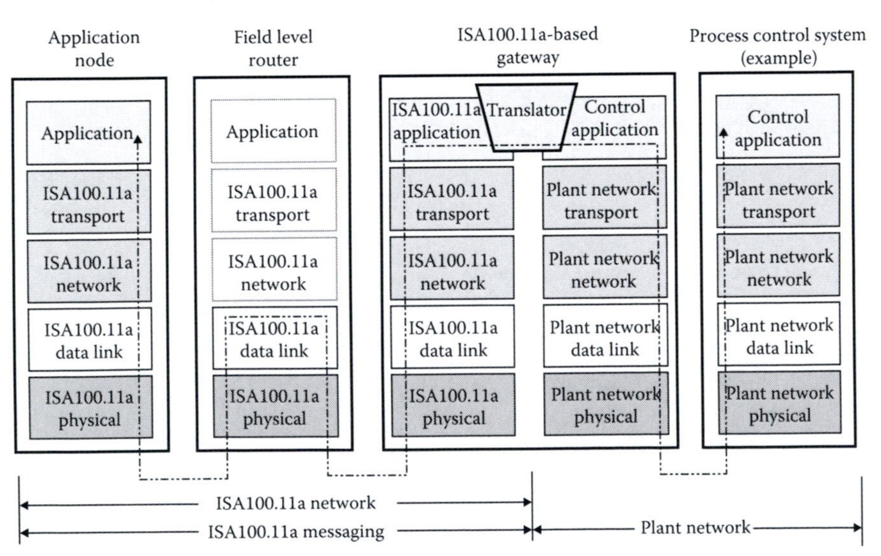

FIGURE 20.28 Communication between a node and the gateway. (From W. W. Manges. *ISA100.11a Principles of Operation Overview*, ISA Expo 2007, Oak Ridge National Laboratory, September 11, 2007.)

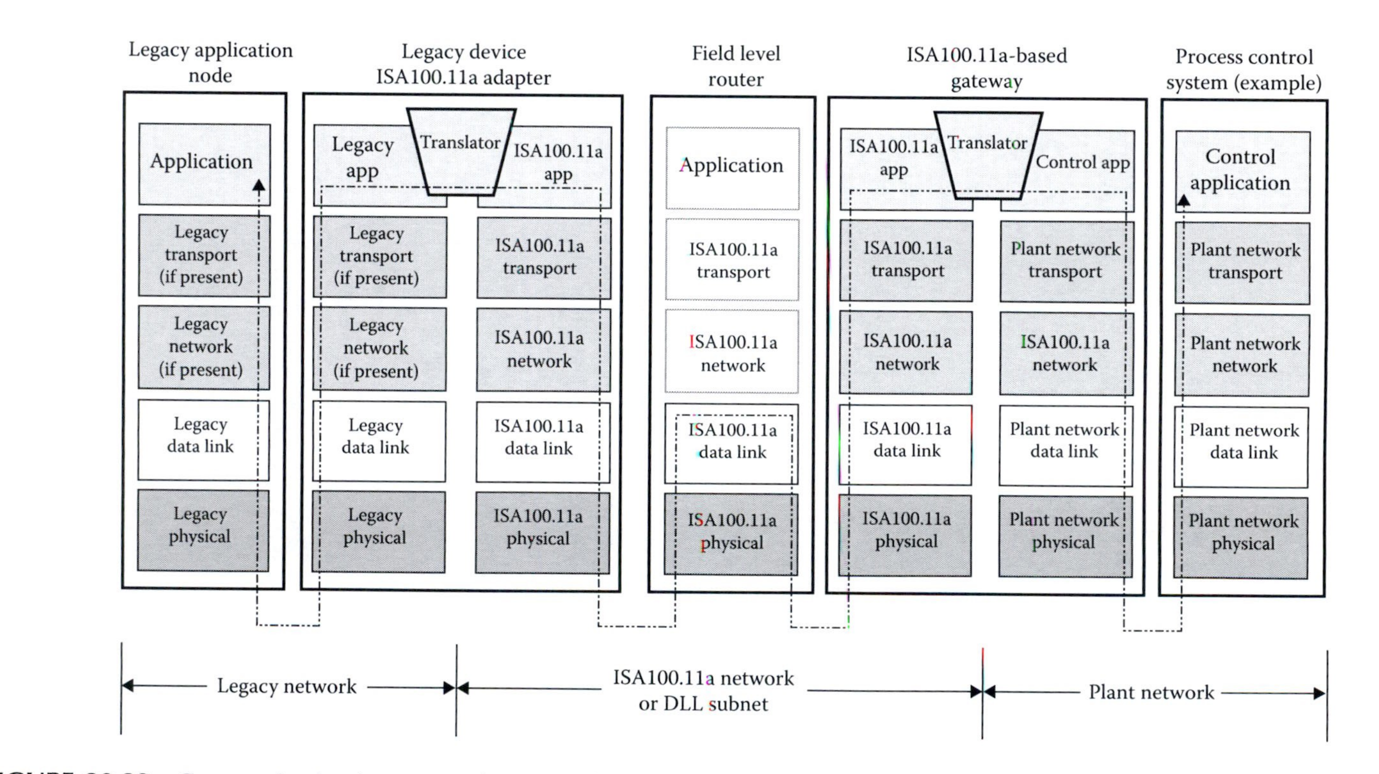

FIGURE 20.29 Communication between a legacy node and the gateway. (From W. W. Manges. *ISA100.11a Principles of Operation Overview*, ISA Expo 2007, Oak Ridge National Laboratory, September 11, 2007.)

TABLE 20.7
Comparison between WHART and ISA100.11a

ISA100.11a	WHART
A mesh, star, or mesh–star network	A mesh network
Field devices can act as end nodes with no routing capability and/or as router nodes with routing capability	Field devices and adapters can act as routers capable of forwarding packets to and from other devices
Wireless provisioning	Wired provisioning
Not all devices in the network are necessarily capable of provisioning other devices to join the network	All devices are capable of provisioning other devices to join the network
Application layer subdivided into upper application layer and application sublayer	Application layer not subdivided
Network layer not subdivided	Network layer subdivided into network layer services and network layer
Data link layer subdivided into upper data link layer, MAC extension, and MAC sublayer	Data link layer subdivided into logical link control and MAC sublayer
Operation defined only in 2.4-GHz ISM band. Uses channels 11–25, with channel 26 as optional	Operation defined only in 2.4-GHz ISM band. Uses channels 11–25
Each channel bandwidth is 2 MHz; channels are spaced 5 MHz apart	Each channel bandwidth is 2 MHz; channels are spaced 5 MHz apart
A combination of DSSS along with FHSS is used as modulation technique	A combination of DSSS along with FHSS is used as modulation technique
Data rate: 250 kbits/s; maximum transmitted power: 10 mW	Data rate: 250 kbits/s; maximum transmitted power: 10 mW
TDMA with frequency hopping for channel access	TDMA with frequency hopping for channel access
Only one superframe is operational at a time. The same can be added/deleted while the network is operational	Only one superframe is operational at a time. The same can be added/deleted while the network is operational
Time slot value is configurable and set to a specific value by the system manager when a device joins the network	A time slot has a fixed value of 10 ms
It defines five preprogrammed hop patterns	Does not explicitly define the frequency hop pattern

(continued)

TABLE 20.7 (Continued)
Comparison between WHART and ISA100.11a

ISA100.11a	WHART
Supports counter with cipher block chaining message authentication code (CCM mode) in conjunction with Advanced Encryption Standard (AES-128) block cipher using symmetric keys for message authentication and encryption	Supports counter with cipher block chaining message authentication code (CCM mode) in conjunction with Advanced Encryption Standard (AES-128) block cipher using symmetric keys for message authentication and encryption
Security features are optional	Security features are mandatory
Generation and management of security keys and authentication of new devices are managed by the security manager	Generation and management of security keys and authentication of new devices are managed by the security manager
Joining by symmetric and asymmetric methods	Joining by symmetric method only

20.21 ISA100.11A AND WHART—A COMPARISON

Table 20.7 shows a comprehensive comparison between WHART and ISA100.11a.

20.22 CONCLUSION

Compared with WHART, ISA100.11a offers a wider coverage for process automation needs. Scalability of the network is derived by dividing the whole network into smaller ones—called subnets, which are joined by backbone routers. Subnets minimize latency in the network. Compatibility of ISA100.11a with IPv6 ensures that various Internet technologies can better be embraced. Network management of the system is made easier by object orientation, and helps improve better interoperability of legacy systems. Better control of field devices is made possible because of the contract between devices and the system manager. Variable time slots help manage different communication needs.

As of the present moment, WHART is better positioned compared with ISA100.11a because of the huge installed base of WHART-compatible devices. It is possible to integrate the two wireless standards by having a dual protocol stack over the MAC layers, which would understand messages from different sources and are in compliance to either of the two standards.

References

Axiomatic Global Electronic Solutions. *Q&A—What is CAN?* Application Note, 2006. Available at www.axiomatic.com/whatisCAN.pdf.

Beck A. and Hennecke A. *Intrinsically Safe Fieldbus in Hazardous Areas*, Technical White Paper, EDM TDOCT-1548_ENG, Pepperl+Fuchs GmBH. Mannheim, Germany, 2008. Available at files.pepperl-fuchs.com/selector_files/navi/productInfo/doct/tdoct1548a_eng.pdf.

Bentje H. et al. "Wireless in Automation" Working Group. *Coexistence of Wireless Systems in Automation Technology: Explanations on Reliable Parallel Operation of Wireless Radio Solutions*, 1st Edition. ZVEI-German Electrical Manufacturers' Association, Automation Division, Frankfurt, Germany, April 2009.

Berge J. *Fieldbuses for Process Control: Engineering, Operation and Maintenance*. ISA, 2002.

Berge J. *Fieldbuses for Process Control: Engineering, Operation and Maintenance*. ISA, Research Triangle Park, NC, 2004.

Bernecker + Rainer Industrie Elektronik GmbH. Available at www.br-automation.com/cps/.

Cisco. Available at www.learningnetwork.cisco.com.

CISCO Systems Inc. *Integrating an Industrial Wireless Sensor Network with Your Plant's Switched Ethernet and IP Network*, White Paper. New York, 2009, pp. 1–18. Available at www.cisco.com/web/strategy/docs//manufacturing/swpIIWS_Emerson_wp.pdf.

Department of Informatics and Automation. An Introduction to the ControlNet Network, System Overview, Release 1.5, 1997, Rockwell Automation—Allen-Bradley.

Department of Informatics and Automation. An Introduction to the ControlNet Network. Available at www.dia.uniroma3.it/autom/Reti_e_Sistemi_Automazione/PDF/ControlNetDetails.pdf.

EazyNotes. Available at www.eazynotes.com/notes/computer-networks/slides/osi-model.pdf.

Emerson Electric Co. Available at www2.emersonprocess.com/siteadmincenter/PM%20Central%20Web%20Documents/EMR_WirelessHART_SysEngGuide.pdf.

Finneran M. F. *WiMax Versus Wi-Fi: A Comparison of Technologies, Markets and Business Plans*. Copyright dBrn Associates, Howlett Neck, NY, 2004.

Forbes H. *ISA 100 and Wireless Standards Convergence, ARC Brief*. ARC Advisory Group, Dadham, MA, October 1, 2010.

Forouzan B. A. *Data Communications and Networking*, 4th Edition, Special Indian Edition. Tata McGraw Hill Companies Inc., New Delhi, India, 2006.

Glanzer D. A. *Technical Overview, Foundation Fieldbus*, FD-043, Rev 3.0, 1996 (Rev. 1998, 2003). Fieldbus Foundation, Austin, TX. Available at www.fieldbus.org/images/stories/technology/developmentresources/development_resources/documents/techoverview.pdf.

HMS Industrial Networks. Available at www.hms.se/technologies/devicenet.shtml.

HART Communication Foundation. Control with WirelessHART. Austin, TX 78759.

HART Communication Foundation. *Application Guide HCF LIT 34. HART Field Communications Protocol.* Austin, TX, 1999. Available at www.pacontrol.com/download/hart-protocol.pdf.

HART Communication Foundation. *System Redundancy with WirelessHART*. Austin, TX.

HART Communication Foundation. WirelessHART Device Types—Gateways, HCF_LIT-119 Rev. 1.0. Gerrit Lohmann—Pepperl+Fuchs. Austin, TX, June 23, 2010.

Hill J. L. System architecture for wireless sensor networks. PhD Dissertation, Department of Computer Science, University of California, Berkeley, CA, Spring, 2003.

IBM Redbooks. Available at www.redbooks.ibm.com/redbooks/pdfs/gg243376.pdf.

Ikram W. and Thornhill N. F. *Wireless Communication in Process Automation: A Survey of Opportunities, Requirements, Concerns and Challenges*. Control 2010, Coventry, UK, 2010.

Industrial Text & Video Company. I/O Bus Networks-Including Device Net, 1999. Available at www.idc-online.com/technical_references/pdfs/instrumentation/I_O_BusNetworks.pdf.

Instrumentation, Systems and Automation Society (ISA), 2010. The Technology Behind the ISA 100.11a Standard—An Exploration. ISA100 Wireless Compliance Institute. Available at http://www.isa100wci.org/Documents/PDF/The-Technology-Behind-ISA100-11a-v-3_pptx.aspx.

International Society of Automation. Available at www.isa.org/Template.cfm?Section=Books1&template=Ecommerce/FileDisplay.cfm&ProductID=6959&file=Chapter1_Profibus.pdf.

Krammer L. *Wireless in Automation: WirelessHART, ISA 100*. Institute of Automation, Automation Systems Group, Vienna, Austria, 2012.

Langmann R. *INTERBUS Basics*. Process Control Laboratory, University of Applied Sciences, Fachhochschule Dusseldorf, pp. 1–76.

Liptak B. G. *Instrument Engineers' Handbook—Process Software and Digital Networks*, 3rd Edition. CRC Press, Boca Raton, FL, 2002.

Mackay S., Wright E., Park J. and Reynders D. *Practical Industrial Data Networks, Design, Installation and Troubleshooting*. Newnes, An imprint of Elsevier, UK, 2004.

Manges W. W. ISA 100.11a *Principles of Operation Overview, ISA Expo 2007.* Oak Ridge National Laboratory, Oak Ridge, TN, September 11, 2007.

Mathivanan N. *PC Based Instrumentation: Concepts and Practice*. New Delhi: Prentice-Hall of India, 2007.

Maxim Integrated. Available at www.maximintegrated.com/app-notes/index.mvp/id/1890.

Mindteck. Available at www.mindteck.com/images/Wireless%20HART%20%20Overview.pdf.

Modbus. Available at www.modbus.org/.

Modbus. Modbus Application Protocol Specification, V1.1b, December 28, 2006. Available at www.modbus.org/docs/Modbus_Application_Protocol_V1_1b.pdf.

National Instruments. Available at www.natinst.com.

ODVA. Available at www.odva.org.

Park J., Mackay S. and Wright E. *Practical Data Communications for Instrumentation and Control*. Newnes, An imprint of Elsevier, UK, 2003.

Petersen S. and Carlsen S. WirelessHART Versus ISA100.11a: The format war hits the factory floor. *IEEE Industrial Electronics Magazine*, pp. 23–24, December 2011. Digital Object Identifier 10.1109, MIE.2011.943023.

Profibus. Available at www.profibus.com.

PROFIBUS. PROFIBUS Technology and Application: System Description. PROFIBUS International Support Center, Karlsruhe, Germany, Copyright by PNO 10/02, October 2002. Available at www.pacontrol.com/download/profibusoverview.pdf.

PROFIBUS. PROFIBUS PA System Description: Open Solutions for the World of Automation. Karlsruhe, Germany, August 2007. Available at www.profibus.jp/tech/document/PROFIBUS-PA-system-descr_e_Aug07.pdf.

Raza S. *Secure Communication in WirelessHART and its Integration with Legacy HART*. Swedish Institute of Computer Science, Technical Report T2010:01, ISSN: 1100-3154, SICS Networked Embedded Systems (NES) Laboratory, Kista Sweden, January 13, 2010.

Real Time Automation, Inc. Available at www.rtaautomation.com/devicenet/.

Relcom Inc. *Fieldbus Wiring Guide*, 4th Edition. Austin, TX, Doc. No. 501–123, Rev.: E.0. Available at www.relcominc.com/pdf/501-123FieldbusWiringGuide.pdf.

Renesas Electronics America Inc. Introduction to CAN, REJ05B0804-0100/Rev. 1.00. Renesas Electronics America Inc., USA, 2006.

Roberts R. ABC's of Spread Spectrum—A Tutorial. Spread Spectrum Scene Magazine. Available at www.sss-mag.com/ss.html. Revised August 8, 2012.

Rodriguez J. Multiplexing and demultiplexing. Available at comsci.liu.edu/~jrodriguez/cs154fl08/Slides/Lecture5.pdf.

Rosemount Inc. *The Basics of Fieldbus*, Technical Data Sheet. Chanhassen, MN, 1998. Available at www.rosemount.com.

Router Alley. Available at www.routeralley.com.

SMAR. *Fieldbus Tutorial—A Foundation Fieldbus Technology Overview*. USA, pp. 1–29. Available at www.smar.com/PDFs/catalogues/FBTUTCE.pdf.

Samson AG. Available at www.samson.de.

Samson A. G. PROFIBUS PA Technical Information, Part 4 Communications, L453EN.

Singhal R. CISCO Systems and E. Rotvold, Emerson Process Management. *Coexistence of Wireless Technologies in an Open Standards based Architecture for in-Plant Applications*. September 6, 2007.

Square D. SERIPLEX: Design, Installation and Troubleshooting Manual, Bulletin no. 30298-035-01A, Raleigh, NC, January, 1999. Available at www.guilleviniag.com/downloads/Products/Schneider/Seriplex.pdf.

Stallings W. *Data & Computer Communications*, 6th Edition. Pearson Inc., New Delhi, India, 2000.

The Technology Behind the ISA 100.11a Standard—An Exploration. ISA 100 Wireless Compliance Institute, ISA—Instrumentation, Systems and Automation Society, 2010. Available at http://www.isa100wci.org/Documents/PDF/The-Technology-Behind-ISA100-11a-v-3_pptx.aspx

Texas A&M University-Kingsville. Available at www.engineer.tamuk.edu/cleung/EEEN4329/FDM.pdf.

Tomasi W. *Advanced Electronic Communications Systems*, 6th Edition. Pearson Inc., p. 263, New Delhi, 2004.

Verwer A. A. Introduction to PROFIBUS. Manchester Metropolitan University. Department of Engineering, Automation Systems Centre, PROFIBUS Competence Centre, August, 2005.

Wang G. Comparison and evaluation of industrial wireless sensor network standards: ISA100.11a and WirelessHART. Master of Science Thesis, Communication Engineering, Department of Signals and Systems, Chalmers University of Technology, Gothenburg, Sweden, Report No. EX036, 2011. Available at www.publications.lib.chalmers.

Yamamoto S., Emori T. and Takai K. Field Wireless Solution based on ISA100.11a to innovate Instrumentation. *Yokogawa Technical Report, English Edition*, Vol. 53, No. 2, pp. 69–74, 2010.

Yokogawa Electric Corporation. *Fieldbus Book—A Tutorial*, Technical Information TI 38K02A01-01E, pp. 1–33, 2000.

ZVEI German Electrical Manufacturers' Association. Coexistence of Wireless Systems in Automation Technology: Explanations on Reliable Parallel Operation of Wireless Radio Solutions, 1st Edition. Automation Division, Frankfurt, Germany, April 2009.

Index

Page numbers followed by f and t indicate figures and tables, respectively.

B

C

D

E

F

G

H

I

M

O

U

Z